MW00763435

Handbook of Modern Biophysics

Series Editor
Thomas Jue
University of California Davis
Davis, California

For other titles published in this series, go to
www.springer.com/series/7845

Thomas Jue
Editor

Biomedical Applications of Biophysics

Volume 3

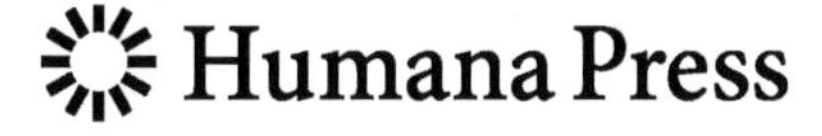

Editor
Thomas Jue, Ph.D.
Department of Biochemistry and Molecular Medicine
University of California Davis
One Shields Avenue
Davis, CA 95616, USA
tjue@ucdavis.com

ISBN 978-1-60327-232-2 e-ISBN 978-1-60327-233-9
DOI 10.1007/978-1-60327-233-9
Springer New York Dordrecht Heidelberg London

Library of Congress Control Number: 2010932766

© Springer Science+Business Media, LLC 2010
All rights reserved. This work may not be translated or copied in whole or in part without the written permission of the publisher (Humana Press, c/o Springer Science+Business Media, LLC, 233 Spring Street, New York, NY 10013, USA), except for brief excerpts in connection with reviews or scholarly analysis. Use in connection with any form of information storage and retrieval, electronic adaptation, computer software, or by similar or dissimilar methodology now known or hereafter developed is forbidden.
The use in this publication of trade names, trademarks, service marks, and similar terms, even if they are not identified as such, is not to be taken as an expression of opinion as to whether or not they are subject to proprietary rights.
While the advice and information in this book are believed to be true and accurate at the date of going to press, neither the authors nor the editors nor the publisher can accept any legal responsibility for any errors or omissions that may be made. The publisher makes no warranty, express or implied, with respect to the material contained herein.

Printed on acid-free paper

Humana Press is part of Springer Science+Business Media (www.springer.com)

PREFACE

In the *Biomedical Applications of Biophysics*, Volume 3 of the *Handbook of Modern Biophysics,* the authors have added to the topics introduced in Volume 1, *Fundamental Concepts in Biophysics*. These additional topics help trace the broad field of biophysics.

Patrice Koehl starts the book with an introduction to protein structure prediction based on energetics, homology modeling, and ab-initio calculations. Dickey and Faller follow with a "how-to approach" to model biomembranes. Brynda and Ames present the principles of magnetic resonance techniques, which researchers often use to solve protein and biomembrane structure. Brynda encapsulates the theoretical and methodological concepts of electron paramagnetic resonance spectroscopy. James Ames covers correspondingly the theory and application of biomolecular NMR spectroscopy. Because optical techniques can also reveal biomolecular structure, Jie Zheng discusses the commonly used fluorescence resonance energy transfer (FRET) in determining molecular distance. Chu and Lebrilla then turn their attention to a pivotal method of analytical chemistry, mass spectrometry. Green and Cheng show how transmission electron microscopy and computer-aided image processing can help visualize macromolecules in three dimensions. Finally, Weeks and Huser introduce to the reader the use of inelastic scattering of Raman spectroscopy to investigate the living cell.

Each chapter presents the fundamental physics concepts, describes the instrumentation or technique, and illustrates the application in studying current biomedical problems. With the addition of problem sets, further study, and references, the interested reader can use the chapters to launch an independent exploration of the ideas presented.

Thomas Jue
Biochemistry & Molecular Medicine
University of California Davis

CONTENTS

1 Protein Structure Prediction

Patrice Koehl

2 Molecular Modeling of Biomembranes: A How-to Approach

Allison N. Dickey and Roland Faller

3 Introduction to Electron Paramagnetic Resonance Spectroscopy

Marcin Brynda

4 Theory and Applications of Biomolecular NMR Spectroscopy

James B. Ames

5 FRET and Its Biological Application as a Molecular Ruler

Jie Zheng

7 Transmission Electron Microscopy and Computer-Aided Image Processing for 3D Structural Analysis of Macromolecules

Dominik J. Green and R. Holland Cheng

8 Raman Spectroscopy of Living Cells

Tyler Weeks and Thomas Huser

Contributors

James B. Ames (chap. 4)
Department of Chemistry
University of California Davis
One Shields Avenue
University of California Davis
Davis, CA 95616-8635, USA
ames@chem.ucdavis.edu

Marcin Brynda (chap. 3)
Department of Chemistry
University of California Davis
One Shields Avenue
Davis, CA 95616-8635, USA
mabrynda@ucdavis.edu

R. Holland Cheng (chap. 7)
Department of Molecular
& Cellular Biology
University of California Davis
007 Briggs, CBS
Davis, CA 95616-8536, USA
rhch@ucdavis.edu

Caroline Chu (chap. 6)
Department of Chemistry
University of California Davis
One Shields Avenue
Davis, CA 95616, USA
cschu@ucdavis.edu

Allison N. Dickey (chap. 2)
Department of Chemical Engineering
& Materials Science
University of California Davis
Bainer Hall, One Shields Avenue
Davis, CA 95616, USA
Current address:
Chemical & Biological Engineering
Northwestern University
Evanston, IL 60208, USA

Roland Faller (chap. 2)
Department of Chemical Engineering
& Materials Science
University of California Davis
Bainer Hall, One Shields Avenue
Davis, CA 95616, USA
rfaller@ucdavis.edu

Dominik J. Green (chap. 7)
Department of Molecular
& Cellular Biology
University of California Davis
007 Briggs, CBS
Davis, CA 95616-8536, USA
djgreen@ucdavis.edu

Thomas Huser (chap. 8)
NSF Center for Biophotonics
University of California, Davis
2700 Stockton Boulevard, Suite 1400
Sacramento, CA 95817, USA
trhuser@ucdavis.edu

Patrice Koehl (chap. 1)
Department of Computer Science
and Genome Center
University of California Davis
Bainer Hall, One Shields Avenue
Davis, CA 95616, USA
koehl@cs.ucdavis.edu

Carlito B. Lebrilla (chap. 6)
Department of Chemistry
University of California Davis
One Shields Avenue
Davis, CA 95616, USA
cblebrilla@ucdavis.edu

Tyler Weeks (chap. 8)
NSF Center for Biophotonics
University of California, Davis
2700 Stockton Boulevard, Suite 1400
Sacramento, CA 95817, USA

Jie Zheng (chap. 5)
Department of Physiology
& Membrane Biology
School of Medicine
University of California Davis
One Shields Avenue
Davis, CA 95616, USA
jzheng@ucdavis.edu

1

PROTEIN STRUCTURE PREDICTION

Patrice Koehl
Department of Computer Science and Genome Center,
University of California Davis

1.1. INTRODUCTION

The molecular basis of life rests on the activity of large biomolecules, mostly nucleic acids (DNA and RNA), carbohydrates, lipids, and proteins. While each of these molecules has its role, there is something special about proteins, as they are the lead performers of cellular functions. This was dramatized by Jacques Monod, who stated that "*C'est à ce niveau d'organisation chimique que gît, s'il y en a un, le secret de la vie*," i.e., that it is at this level of organization that lies the secret of life, if there is one [1]. To understand how these molecules function we first need to know their shapes; consequently, structural molecular biology has emerged as a new line of experimental research focused on revealing the structure of these biomolecules. This branch of biology has recently experienced a major uplift through the development of high-throughput structural studies, the structural genomics projects, aimed at developing a comprehensive view of the protein structure universe. All these initiatives are expected to help us unravel the connections between the sequence, structure, and function of a protein. Experimental data at a molecular level are scarce, however; this has led to the development of many modeling initiatives to shed light on these connections. Probably the most famous is the study of the protein-folding problem — the "holy grail" for the structural biology community. Its elusive goal is to predict the detailed three-dimensional structure of a protein from its sequence as well as to decipher the sequence of events the protein goes through to reach its folded state. This chapter is dedicated to the first part of this task, namely the protein structure prediction problem. We first emphasize the need to study structures and then describe how efforts to solve the protein structure prediction problem benefit from two different approaches to science, which differ in the importance they give to experimental data.

Address correspondence to Patrice Koehl, Department of Computer Science and Genome Center, University of California, Davis, One Shields Avenue, Davis, 95616, USA, <<koehl@cs.ucdavis.edu>>, <<http://koehllab.genomecenter.ucdavis.edu>>.

T. Jue (ed.), *Biomedical Applications of Biophysics*,
Handbook of Modern Biophysics 3, DOI 10.1007/978-1-60327-233-9_1,
© Springer Science+Business Media, LLC 2010

1.1.1. The Importance of Shape

A finding that has crystallized over the last few decades is that geometric reasoning plays a major role in our attempt to understand the activities of biomolecules in general. Molecular structure (i.e., its shape) and chemical reactivity are highly correlated as the latter depends on the positions of the nuclei and electrons within the molecule. Indeed, chemists have long used three-dimensional models to understand the many subtle effects of structure on reactivity and have invested in experimentally determining the structure of important molecules. The same applies to biochemistry, where structural genomics projects are based on the premise that the structure of biomolecules implies their function. Here we focus on the shape and function of proteins.

The sequence of amino acids of a protein defines its primary structure. This linear chain of amino acids becomes functional only when it adopts a three-dimensional shape, its tertiary structure. This is by no means different from the macroscopic world: proteins serve as tools in the cell and as such must have a defined shape to function, much like the tools we use are designed according to the shape they need to perform. At the molecular level, the premise that shape and function are related is supported by a number of specific and quantifiable correlations.

Enzymes fold into unique structures and the three-dimensional arrangement of their sidechains determines their catalytic activity;

- There is theoretical evidence that the mechanisms underlying protein complex formation depend on the shapes of the biomolecules involved [2];
- The folding rate of many small proteins correlates with a gross topological parameter that quantifies the difference between distance in space and along the mainchain [3–6].
- There is also evidence that the geometry of a protein plays a major role in defining its tolerance to mutation [7].

It is noteworthy that the parallel between macroscopic tools and microscopic tools such as proteins extends also to the concept of flexibility. Most macroscopic tools are specialized, and their shape is usually fixed to satisfy a specific function. "Universal" tools exist, however, that adapt to a task; the hand is probably the best example of this, as we can adapt its shape, mimicking a pincher if we need to grab something or a hook if we need to carry something. While most proteins do not have the flexibility of a hand, they are not rigid objects and they can adapt to their partners upon interaction. Understanding the dynamics of protein structure is therefore key to understanding its function (this will be discussed below).

1.1.2. The Data Revolution

Science has long been hypothesis driven, as scientists would formulate models and data were subsequently sought for either to confirm or infirm these models. This traditional view of science is probably best illustrated by the famous quote of Lord Rutherford: "Science is either physics or stamp collecting." The amassing of large amounts of data in the multiple genomics projects has led, however, a number of scientists to envision a future in which data-mining techniques, or "data-driven discovery," will eventually replace this traditional hypothesis-driven discovery in the biological sciences. The field of protein structure prediction is an excellent example, however, which illustrates that these two approaches to science are not competitive but complementary and synergetic. Others have addressed the complementary nature of these two approaches in detail, and I refer the reader to their papers for a discussion on how large-scale data and informatics change the current state of biological sciences [8,9].

Early research from the physics and chemistry communities has provided significant insight into the nature of atoms and their arrangements in small chemical systems; most if not all of this research was and is hypothesis driven. The current focus of structural biology is to understand the structure and function of biomolecules; these usually large molecules serve as storage for genetic information (the nucleic acids), and as key actors of cellular functions (the proteins). Direct applications of the ideas that have been used for modeling small chemical systems have not yet been successful on these much larger molecules; as a consequence, structural biology currently relies mostly on experiments. Large-scale experimental projects are being performed as collaborative efforts involving many laboratories in many countries to provide maps of the genetic information of different organisms (the *genome projects*), to derive as much structural information as possible on the products of the corresponding genes (the *structural genomics projects*), and to relate these genes to the function of their products, usually deduced from their structure (the *functional genomics projects*). As of November 2008, more than 880 whole genomes have been fully sequenced and published, corresponding to a database of over four million gene sequences [10] and more than 3000 other genomes are currently being sequenced (http://www.genomesonline.org/). In parallel, the repository of biomolecular structures [11,12] contains more than 54,000 entries of proteins and nucleic acids as of November 2008. These data-driven approaches, however, are not successful on their own: it is clear from the data given above that the number of protein structures known experimentally represents a small fraction of the number of protein sequences available from the genome projects. If modeling and experimental approaches cannot solve the structure of all proteins on their own, the hope today is that the solution will come from a combination of these two techniques.

Outline. Given the introduction to the importance of protein structure and an update on the progress of research on proteins in structural biology, we can now examine protein structure prediction, the topic of this chapter. The next section describes proteins and surveys their different levels of organization, from their primary sequence to their quaternary structure in cells. The following section covers the physics-based empirical force fields for protein structure prediction. It starts with a definition of the stability of a protein, which is followed by descriptions of the internal energy of a protein and its interactions with its environment. The two following sections survey the two main approaches to protein structure prediction, namely homology modeling and ab-initio protein structure prediction. Finally, the tutorial concludes with a discussion of the future of protein structure modeling. The bibliography is by no means exhaustive, but an attempt was made to quote most recent relevant papers and reviews in the field of protein structure prediction.

1.2. BASIC PRINCIPLES OF PROTEIN STRUCTURE

While all biomolecules play an important role in life, there is something special about proteins, which are the products of the information contained in the genes. In this section, the basic principles that govern the shapes of protein structures are briefly reviewed. More information on protein structures can be found in protein biochemistry textbooks, such as those of Schulz and Schirmer [13], Cantor and Schimmel [14], Branden and Tooze [15], and Creighton [16]. We also refer the reader to the excellent review by Taylor and collaborators [17].

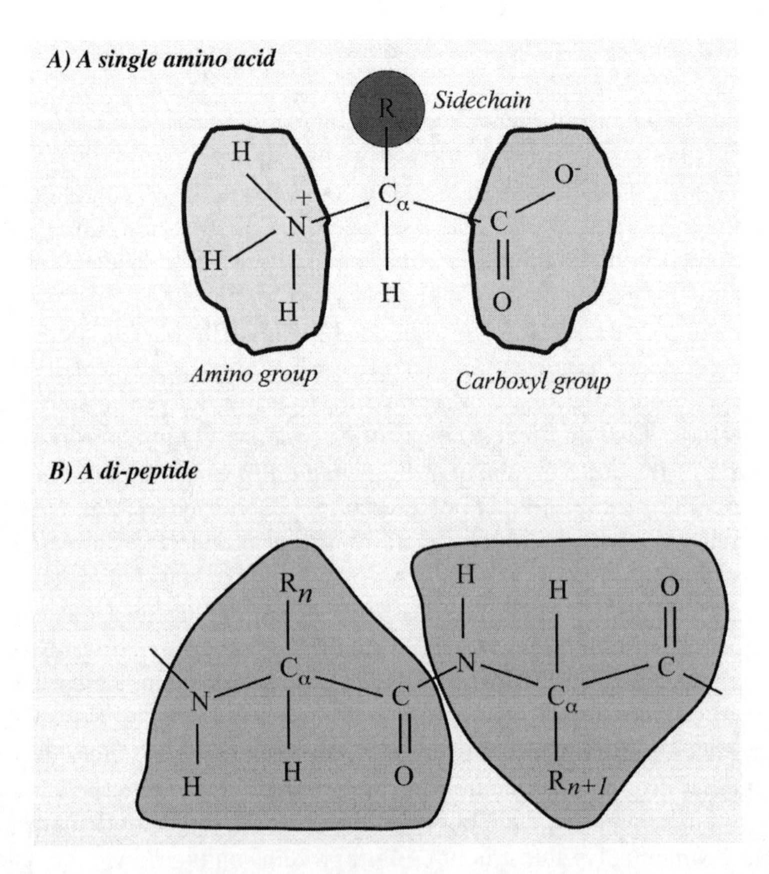

Figure 1.1. Amino acids: the building blocks of proteins. (A) Each amino acid has a mainchain (N, C_α, C, and O) on which is attached a sidechain schematically represented as R. The mainchain can itself be partitioned into three groups: the amino group, the central C_α group, and the carboxyl group. Note that even though the amino group and the carboxyl group are charged at neutral pH, the amino acid is neutral: we say that it is a *zwitterion.* (B) Amino acids in proteins are attached through planar peptide bonds, connecting atom C of the current residue to atom N of the following residue. Please visit http://extras.springer.com/ to view a high-resolution full-color version of this illustration.

1.2.1. Protein Building Blocks

Proteins are heteropolymer chains of amino acids, often referred to as *residues*. There are twenty naturally occurring amino acids that make up proteins. With the exception of proline, amino acids have a common structure, shown in Figure 1A. Substituents on the alpha carbon, called *sidechains*, range in size from a single hydrogen atom to large aromatic rings. Those substituents can be charged or they may include only non-polar saturated hydrocarbons [18] (see Table 1.1). Non-polar amino acids do not have concentration of electric charges and are usually not soluble in water. Polar amino acids carry local concentration of charges, and are globally neutral, negatively charged (acidic), or positively charged (basic). Acidic and basic amino acids are classically referred to as electron acceptors and electron donors, respectively, and can associate to form salt bridges in proteins.

Table 1.1, Classification of the 20 Amino Acids [18]

Classification	Amino acid[a]
Non-polar	Glycine (G), alanine (A), valine (V), leucine (L), isoleucine (I), proline (P), methionine (M), phenylalanine (F), tryptophan (W)
Polar	Serine (S), threonine (T), asparagine (N), glutamine (Q), cysteine (C), tyrosine (Y)
Acidic (polar)	Aspartic acid (D), glutamic acid (E)
Basic (polar)	Lysine (K), arginine (R), histidine (H)

[a] The one-letter code for each amino acid is given in parentheses.

1.2.2. Protein Structure Hierarchy

Condensation between the $-NH_3^+$ and the $-COO^-$ groups of two amino acids generates a peptide bond and results in the formation of a dipeptide (Fig, 1.1B). Protein chains correspond to an extension of this chemistry, resulting in long chains of many amino acids bonded together. The order in which amino acids appear defines the *sequence* or *primary structure* of the protein. In its native environment, the polypeptide chain adopts a unique three-dimensional shape, referred to as the *tertiary* or *native structure* of the protein. The amino acid backbones are connected in sequence, forming the protein *mainchain*, which frequently adopts canonical local shapes or *secondary structures*, mostly α-helices and β-strands (see Fig. 1.2). The α-helices form a right-handed helix with 3.6 amino acids per turn, while the β-strands form an approximately planar layout of the backbone. Helices often pack together to form a hydrophobic core, while β-strands pair together to form parallel, or anti-parallel, β-sheets (Fig. 1.2).

1.2.3. Three Main Types of Proteins

Protein structures come in a large range of sizes and shapes. They can be divided into three major groups: *fibrous* proteins, *membrane* proteins, and *globular* proteins (Fig. 1.3).

Fibrous proteins are elongated molecules in which the secondary structure is the dominant structure. Because they are insoluble in water they play a structural or supportive role in the body, and are also involved in movement (such as in muscle and ciliary proteins). Fibrous proteins often (but not always) have regular repeating structures. Keratin, for example, which is found in hair and nails, is a helix of helices, and has a seven-residue repeating structure. Silk, on the other hand, is composed only of β-sheets, with alternating layers of glycines, and alanines and serines. In collagen, the major protein component of connective tissue, every third residue is a glycine, and many of the others are prolines.

Membrane proteins are restricted to the phospholipid bilayer membrane that surrounds the cell and many of its organelles. These proteins cover a large range of sizes and shapes, from globular proteins anchored in the membrane by means of a tail, to proteins that are fully embedded in the membrane. Their function is usually to ensure transport of ions and small molecules like nutrients through the membrane. The structures of fully embedded membrane proteins can be placed into two major categories: the all-helical structures, such as bacteriorhodopsin, and the all-beta structures, such as porins.

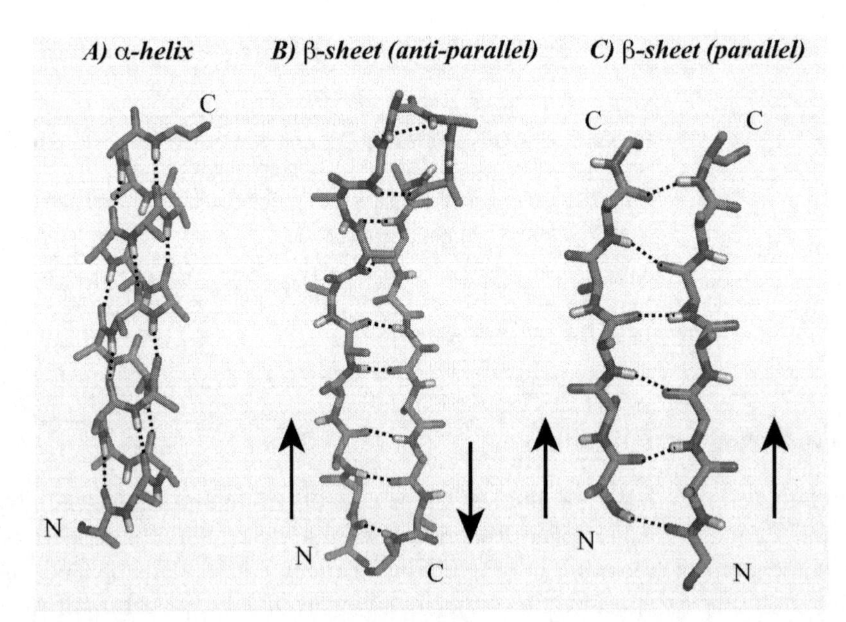

Figure 1.2. The three most common arrangements of secondary structure elements (SSE) found in proteins. (A) The regular α-helix is a right-handed helix, in which all residues adopt similar conformations. The α-helix is characterized by hydrogen bonds between the oxygen O of residue i, and the polar backbone hydrogen HN (bound to N) of residue $i + 4$. Note that all C=O and N–HN bonds are parallel to the main axis of the helix. (B) An anti-parallel β-sheet. Two strands (stretches of extended backbone segments) are running in an anti-parallel geometry. The atoms HN and O of residue i in the first strand hydrogen bond with the atoms O and HN of residue j in the opposite strand, respectively, while residues $i + 1$ and $j + 1$ face outward. (C) A parallel β-sheet. The two strands are parallel, and the atoms HN and O of residue i in the first strand hydrogen bond with the O of residue j and the HN of residue $j + 2$, respectively. The same alternating pattern of residues involved in hydrogen bonds with the opposite strand, and facing outward is observed in parallel and anti-parallel β-sheets. A strand can therefore be involved in two different sheets. For simplicity, sidechains and non-polar hydrogens are ignored. Figure drawn using Pymol (http://www.pymol.org). Please visit http://extras.springer.com/ to view a high-resolution full-color version of this illustration.

Globular proteins have a non-repetitive sequence. They range in size from one hundred to several hundred residues, and adopt a unique compact structure. In globular proteins, non-polar amino acid sidechains have a tendency to cluster together to form the interior, hydrophobic core of the proteins, while the hydrophilic polar amino acid sidechains remain accessible to the solvent. In the tertiary structure, β-strands are usually paired in parallel or anti-parallel arrangements, to form β-sheets. On average, a globular protein mainchain consists of about 25% of residues in α-helix formation, 25% of residues in β-strands, with the rest of the residues adopting less-regular structural arrangements [19].

1.2.4. Geometry of Globular Proteins

Our knowledge of protein structure comes from years of experimental studies, primarily using X-ray crystallography or NMR spectroscopy. The first protein structures to be solved were those of myoglobin and hemoglobin [20,21]. As of November 2008, there are 50,000 protein structures in the PDB database [11,12] (see http://www.rcsb.org). It is to be noted that this number overestimates the actual number of different structures available because the PDB is redundant, i.e., it contains several copies of the same proteins, with minor mutations in the sequence and no changes in the structure.

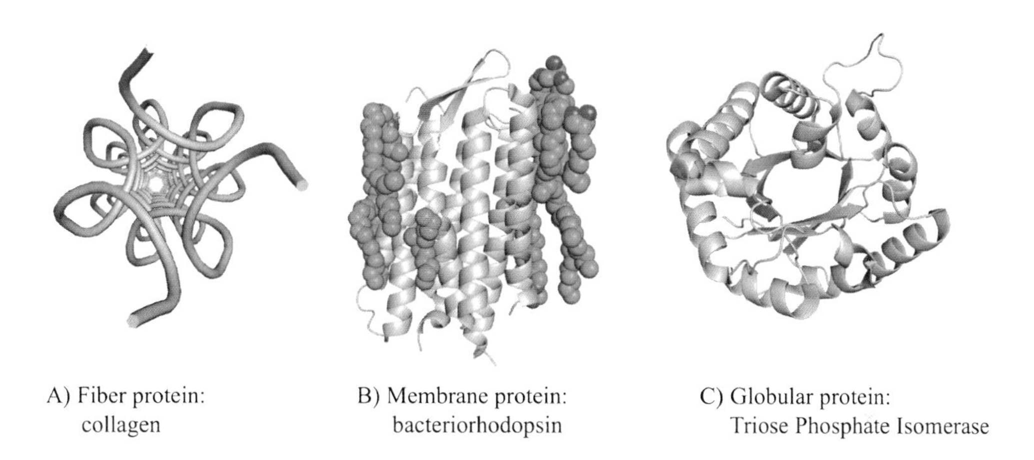

A) Fiber protein: collagen   B) Membrane protein: bacteriorhodopsin   C) Globular protein: Triose Phosphate Isomerase

Figure 1.3. The three main types of proteins. (A) Collagen is the main protein of connective tissues in animals and the most abundant protein in mammals, making up close to 30% of their body protein content. It is a fiber protein, with each fiber made up of three polypeptide strands possessing the conformation of left-handed helices. These three left-handed helices are twisted together into a right-handed coiled coil, a cooperative quaternary structure stabilized by numerous hydrogen bonds. (B) Bacteriorhodopsin is a mainly α-protein, containing seven helices, that crosses the membrane of a cell (a few lipids of the membrane are shown as a space-filling diagram in green). It serves as an ion pump, and is found in bacteria that can survive in high salt concentrations. (C) TIM is a globular protein that belongs to the α–β class. The protein chain alternates between β and α secondary structure type, giving rise to a barrel β-sheet in the center surrounded by a large ring of α-helix on the outside. This structure, first seen in the triose phosphate isomerase of chicken, has been observed in many unrelated proteins since then. Figure drawn using Pymol (http://www.pymol.org). Please visit http://extras.springer.com/ to view a high-resolution full-color version of this illustration.

Because only two types of secondary structures (α and β) exist, proteins can be divided into three main structural classes [22]: mainly α proteins [23], mainly β proteins [24–26], and mixed α–β proteins [27]. There has been significant work on predicting a protein folding class based on its sequence, the details of which can be found in [28–36].

The mainly α class, the smallest of the three major classes, is dominated by small proteins, many of which form a simple bundle of α-helices packed together to form a hydrophobic core. A common motif in the mainly α class is the four-helix bundle structure. The most extensively studied α structure is the globin fold, which has been found in a large group of related proteins, including myoglobin and hemoglobin. This structure includes eight helices that wrap around the core to form a pocket where a heme group is bound [20].

The mainly β class contains the parallel and anti-parallel β structures. The β-strands are usually arranged in two β-sheets that pack against each other and form a distorted barrel structure. Three major types of β-barrels exist: up-and-down barrels, Greek key barrels [37], and jelly roll barrels. Most of the known anti-parallel β structures, including the immunoglobulins, have barrels that include at least one Greek key motif. The two other motifs are observed in proteins of diverse function, where functional diversity is obtained by differences in the loop regions connecting the β-strands. The β structures are often characterized by the number of β-sheets in the structure and the number and direction of the strands in the sheets.

The α–β protein class is the largest of the three classes. It is subdivided into proteins having an alternating arrangement of α-helices and β-strands along the sequence, and those with more segregated secondary structures. The former subclass is itself divided into two groups: one with a central core of parallel β-strands arranged as a barrel surrounded by α-helices, and a second group consisting of an open, twisted parallel or mixed β-sheet, with α-helices on both sides.

1.3. THE ENERGETICS OF PROTEIN STRUCTURE

The stability of a protein is usually measured as the free energy difference between its native structure and an unfolded conformation. We start this section by defining the free energy of a protein. The rest of the section discusses various methods of computing the free energy.

1.3.1. Stability of Protein Structures

The stability of a protein sequence P is a thermodynamics quantity. It is measured as the difference $\Delta G(P)$ in free energy between its native state, N, and an unfolded state, U:

$$\Delta G_{U\to N}(P) = G_N(P) - G_U(P)\,. \tag{1.1}$$

Note that P here refers to the "solvated" protein, i.e., accounts for the protein and its surrounding solvent and ionic atmosphere. G refers to the Gibbs free energy of the system; it is defined as the sum of three terms: the internal energy U of the protein, its entropy S, where the entropy is a measure of disorder, and the product pV, where p and V are the pressure and volume of the system, respectively:

$$G = U - TS + pV\,, \tag{1.2}$$

where T is the temperature of the system. For a protein in solution, the pressure p and the volume V can be considered constant; Eq, (1.1) can therefore be rewritten as

$$\Delta G_{U\to N}(P) = \Delta U_{U\to N}(P) - T\Delta S_{U\to N}(P)\,. \tag{1.3}$$

A native-like, "stable" structure has a minimum (negative) $\Delta G(P)$: ideally, this is reached when $\Delta U(P)$ is minimum and $\Delta S(P)$ is maximum. However, $\Delta U(P)$ is minimal when the native state has many stabilizing contacts, which requires an organized structure; whereas the entropy is maximal when the structure has a low level of organization. Stability therefore is reached through a compromise between these two effects. To estimate the free energy of a protein, we need to compute its internal energy U, and sample its conformational space to measure the entropy S. Note again that the system we need to consider includes the protein of interest, the water molecules that surround it, the ions in the solvent, and any ligands of the protein.

1.3.2. Internal Energy

A typical semi-empirical energy function used in computational structural biology to compute the internal energy of a biomolecule has the form:

$$U = \frac{1}{2}\sum_b K_b (b-b_0)^2 + \frac{1}{2}\sum_\theta K_\theta (\theta-\theta_0)^2 + \sum\sum_n K_n \left(1+\cos(n\phi-\delta_n)\right)$$
$$+\sum_{i<j}\left(\frac{A_{ij}}{r_{ij}^{12}} - \frac{B_{ij}}{r_{ij}^{12}} + \frac{q_i q_j}{\varepsilon r_{ij}}\right). \tag{1.4}$$

This energy form implies all interactions are additive. The first three sums on the right side of Eq. (1.4) account for bonded interactions (covalent bonds, valence angles, and torsions around bonds), while the last term represents non-bonded interactions.

1.3.2.1. Bonded Interactions

Bond stretching. When a bond of length b is stretched or compressed, its energy goes up. For small deviations from equilibrium, Hooke's law approximates the functional form for this energy, i.e., it is a quadratic function of the change in bond length. The reference bond length b_0 is often called the *equilibrium* bond length.

Bond angle bending. In the ground state, the angle θ between two bonds attached to the same central atom assumes an equilibrium value, θ_0. While this energy increases when the angle deviates from the ground state, there is no consensus on its functional form: some force fields use a direct harmonic term as in Eq. (1.4) above, while others use a cosine harmonic potential. The difference between the functional forms may be significant if the bond angles vary significantly; it is expected to have negligible effects in structure prediction problems.

Torsional stress. Intramolecular rotations, i.e., rotations about torsion angles, require energy. This energy is different from the energies observed for bond stretching and bond angle bending. First, the energy barriers are low, and changes in torsion angles can be large; second, the energy is periodic with a period of 360°. Torsional energies are usually expressed as a Fourier series:

$$E(\phi) = \sum_n K_n \left(1+\cos(n\phi-\delta_n)\right), \tag{1.5}$$

where K_n is the torsional rotation constant; the phase angles δ_n are usually chosen so that terms with positive K_n have minima at $\phi = 180°$ (i.e., $\delta_n = 0°$ for n odd and 180° for n even), and n is the multiplicity, which reflects the type of symmetry in the torsion angle. A CH_3–CH_3 bond, for example, repeats its torsional energy every 120°.

1.3.2.2. Non-Bonded Interactions

By definition, the non-bonded interactions act between atoms that are not linked by a chemical bond. In practice, all pairs of atoms connected by two bonds (1–3 interactions) are also discarded and only accounted for by bonded interactions (see above). In addition, interactions between atoms connected by three bonds (1–4 interactions) are often scaled down. There are two types of non-bonded interactions that make significant contributions: electrostatics and van der Waals (vdW).

Electrostatics interactions. As a first "classical" approximation, the interaction between two atoms i and j carrying charges q_i and q_j, respectively, is governed by Coulomb's law:

$$U_{\text{elec}} = \frac{q_i q_j}{4\pi\varepsilon_0 \varepsilon_r r_{ij}}, \tag{1.6}$$

where ε_0 is the dielectric permittivity of free space, ε_r is the relative dielectric permittivity of the medium in which the two atoms are placed (usually water, in which case $\varepsilon_r \approx 80$), and r_{ij} is the separation between the two atoms. When more than two charges interact, the total electrostatic energy of the system is derived as a sum of its pairwise Coulomb interactions.

vdW interactions: The van der Waals term covers all interactions that are not accounted for by bonded and electrostatic interactions. It includes induced dipole–dipole interactions, London dispersion forces, as well as repulsion forces that prevent electrons from non-bonded atoms to occupy the same space. The attractive part of vdW interactions has a $1/r^6$ dependence, where r is the interatomic distance and is dominated by dipole–dipole interactions. The repulsive part of vdW interactions is usually modeled with a $1/r^{12}$ dependence, although there are no physical reasons that the exponent should be 12. Together, they lead to the classical Lennard-Jones potential for vdW interactions:

$$U_{\mathrm{vdW}} = \frac{A_{ij}}{r_{ij}^{12}} - \frac{B_{ij}}{r_{ij}^{6}}, \qquad (1.7)$$

where A_{ij} and B_{ij} are parameters that depend on the nature of atoms i and j, and r_{ij} is the distance between i and j.

1.3.3. The Entropy of a Protein Structure

The total entropy of a molecule in solution can be separated into two parts: solvent entropy associated with solvent motions, and a solute, or configurational, entropy associated with protein motions. Computing these entropy terms requires in principle complete sampling of the conformational space accessible by the protein and the solvent molecules. This is usually achieved through molecular dynamics simulation [38].

1.3.4. The Denatured State

Equation (1.3) requires computing the free energy of a protein in its "unfolded" state, which is usually taken to be the fully extended structure. The denatured "state" of a protein is known to be a distribution of different molecular conformations (for review, see [39]). Though distant residue contacts do exist in the denatured states, their free energies are dominated by local interactions within the protein and by interactions with the solvent. Therefore, the free energy of a denatured protein depends mostly on its sequence rather than on its structure itself.

Protein structure prediction problems usually resort to comparing the free energies of two or more conformations for the protein of interest; the denatured state being the same for all these conformations, it is usually enough to compare their internal energies.

1.4. HOMOLOGY MODELING

Homology modeling, also known as comparative modeling, predicts the structure of a protein (the target or query) by inference from a homologous protein whose structure is known (the template). Its success rests on (i) the existence of a homologue with known structure, (ii) our ability to detect this homologue, and (iii) the quality of the model-building process once the homologue is detected. Steps (i) and (ii) have greatly benefited from the different genomics projects presented in the introduction to this chapter: with the additional structures from structural genomics, sampling of the protein structure space is becoming finer, improving the chance that

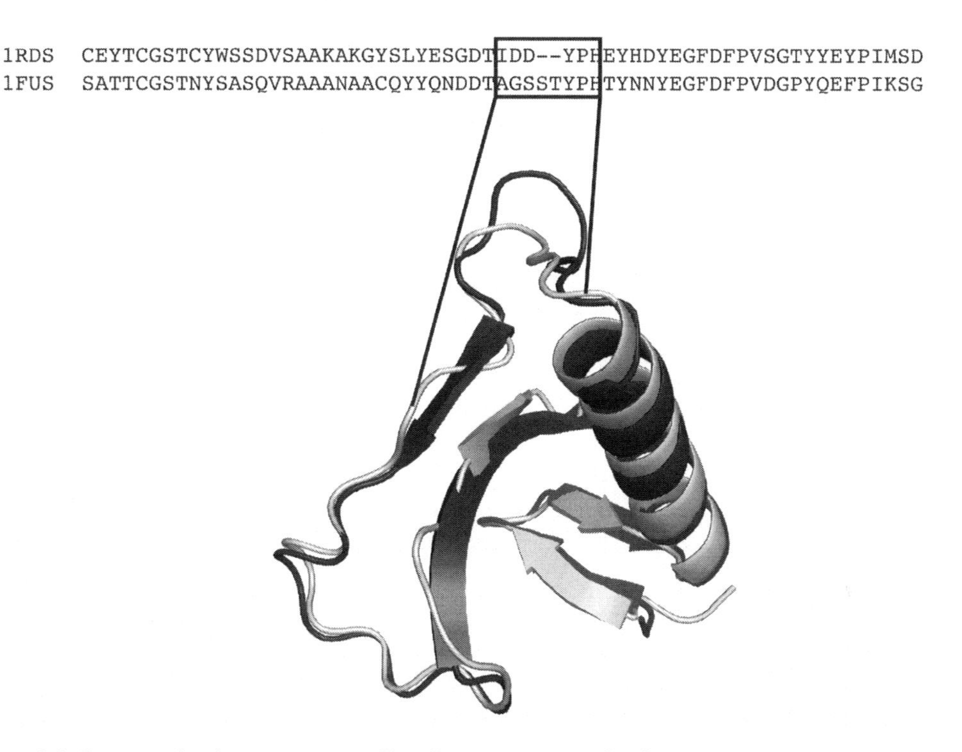

Figure 1.4. Conservation in sequence usually relates to conservation in structure. The sequence alignment between FUS and RDS (see Fig. 1.5) is compared to the structural alignment between their corresponding structures, 1FUS and 1RDS (shown as gray and black, respectively). The structural alignment was computed using STRUCTAL [171]. For the sake of clarity, only the region between residues 1 and 60 of FUS is shown. Conserved regions in the sequence alignment correspond to conserved regions in the structural alignment (coordinate root mean square [cRMS] = 0.8 Å over 102 residues). The two structures differ in the loop region between two secondary structures (highlighted in bold in the sequence alignment). The loop in FUS is longer by two residues. This figure was generated using PYMOL.

a structural homologue exists for any given protein sequence. With the parallel increase in the number of sequences available in genomic databases and the development of meta-servers to analyze and query these sequence databases, there has been significant improvement in the detection of homology [40]. There is hope that in the near future homology modeling will reach its ultimate goal: the generation of model protein structures as accurate as those determined by high-resolution experimental studies.

The technique of homology modeling is based on the fundamental assumption that two proteins with similar sequences fold into similar structures. Chothia and Lesk [41] were the first to report that the extent of the structural changes observed between two proteins is directly related to the extent of sequence changes. Figure 1.4 illustrates this point on a simple example comparing two related small proteins of a hundred amino acids. Recent studies have confirmed Chothia and Lesk's findings using large sets of proteins [42–44].

Almost all homology-modeling techniques proceed in four steps: (i) identification of one or more known protein structures likely to resemble the structure of the query sequence, (ii) production of an alignment that maps residues in the query sequence to residues in the template sequence, (iii) generation of a structural model for the query protein using this alignment, and

(iv) evaluation of the final model. In this section of this chapter we describe the techniques used for these four steps.

1.4.1. Finding a template

The first step in homology modeling is to detect some or all proteins whose structures are known that are expected to be similar to the target protein. As the structure of the target protein is unknown, "similarity" here refers to sequence similarity; a naïve approach to solve this first step is then simply to compare the sequence of the target protein with each of the sequences of the proteins whose structure is available in the protein structure database, using a pairwise sequence comparison program. If this search identifies one or more proteins with a high level of similarity, these proteins can serve as templates for homology modeling. FASTA [45] and BLAST [46] are two programs that are frequently used to perform this search, with BLAST slowly becoming the method of choice, as it is much faster. Both programs have been extensively tested for their ability to detect similarity at the sequence level that reflects at the structure level [47–49].

The correspondence between sequence similarity and structure similarity is well established only when sequence similarity is high (above 40% identity for long sequences) [41,50]; it is known, however, to become unreliable at a lower level of similarity, such as in the twilight zone (20 to 35% sequence identity), and marginal at even lower sequence identity: more than 90% of all protein pairs with a sequence identity larger than 30% are found to be structurally similar, while in the midnight zone (less than 10% sequence identity) this number falls below 10%. In his study of the twilight zone for sequence comparison, Rost underlined the problem of the explosion of false negatives, i.e., the existence of many pairs of proteins that are structurally similar but whose sequences are not significantly similar based on an internal threshold of the sequence alignment technique [50]. This clearly indicated that there was a need for more sensitive methods to detect sequence similarity by changing the scoring of an alignment and/or by introducing more information than the two sequences to be compared. It is the latter that has resulted in the biggest improvement for template selection for homology modeling through the introduction of the evolution information contained in multiple sequence alignments [51]; it is in fact at the root of all profile methods used in current database searching programs such as PSIBLAST [52] and HMMER [53]. The basic idea behind these methods is simple: two sequences may be marginally similar, but if we were to compare instead the families of sequences to which they belong, we may have a better chance to detect similarity. PSIBLAST, for example, implements this idea as follows. For a given sequence, PSIBLAST collects an initial set of homologues in the database; these sequences are aligned together with the target sequence using a weighted multiple sequence alignment program, and the alignment is converted into a position-specific scoring matrix. The matrix is subsequently used to search the database for new homologues. The procedure is iterated until no new homologues are found. PSIBLAST usually finds twice as many homologues compared to BLAST [51].

In addition to simple sequence comparison (using, e.g., BLAST), or sequence comparison using multiple sequence alignments (such as PSIBLAST), threading methods have emerged as a third approach to the template or fold recognition problem (for reviews, see [54–57]. These methods rely on a direct comparison of a sequence with a protein structure, using scoring functions that measure the compatibility of an amino acid with a given environment in the structure.

The target sequence is compared with a library of folds, and the best matching fold is identified as a possible template. The success of these methods is strongly tied to the belief that the size of the protein structure space is much smaller than the size of the protein sequence space: it is commonly assumed that there are only a thousand different protein folds [58]. The fact that the structure space is finite has given rise to the hope that it is possible to build a library that contains representatives of all possible structures; this library would then be ideal for threading methods.

The different methods described above select, when successful, a list of protein structures expected to be similar to the structure of the target protein. It is then necessary to choose among these structures those that can serve as appropriate templates for the given modeling problem. Usually, the template with the higher overall sequence similarity to the target sequence is chosen; this may not always be the best choice, and several other factors should be taken into account:

- *Quality of experimental structure*: the resolution and R-factor for X-ray structures, the number of constraints per residue for an NMR structure, the experimental conditions, and the number of residues absent from the structure are all important factors indicative of the structure accuracy.
- *Choice of sequence subfamily*: the template and target should belong, if possible, to the same branch of the phylogenetic tree of the protein family they belong to.

It should be noted that it is not necessary to select only one template: modeling methods such as MODELLER [59] were designed to handle multiple templates, as using those was found to increase accuracy compared to using a single template [60].

1.4.2. Aligning the Target and Template Sequences

There has been significant progress in fold recognition, as the corresponding techniques have greatly benefited from the different genomics projects. Finding the correct alignment between the target sequence and the parent homologue has now become the bottleneck to homology modeling. Errors during this step are crucial to the quality of the final model: a shift of one residue in the alignment results in a distortion of 3.8 Å of the backbone of the final model. Most fold-recognition techniques produce an alignment between the target sequence and the templates they identify. This alignment is unfortunately not always reliable: there are documented cases of two homologues being detected although part or sometimes the entire alignment was wrong [61]. As a consequence, special care must be given to the alignment, and specialized methods have been designed for this task, most of them relying on dynamic programming. For closely related sequences (i.e., with identity greater than 40%), the problem is easily solved and any dynamic programming methods usually produce the correct alignment. The situation is, however, not the same in the twilight zone: regions of low sequence similarity become common, and the alignment usually contains a large number of gaps and is more prone to errors. In these difficult cases, it is often beneficial to rely on multiple sequence and structure alignment [62].

Generating a good alignment between the target and template sequences is hindered by the fact that it is difficult to evaluate its correctness. As it is usually easier to evaluate a structural model, current strategies for homology modeling proceed by generating 3D models for all alternate alignments available, evaluate these models (see below), and pick the best model based on these evaluations rather than on a measure of alignment accuracy.

1.4.3. Generating a Model

Once a template structure has been identified and an alignment between the target and template sequences has been generated, model building usually proceeds in three steps. First, a framework is derived using the regions of the template structure that match regions outside gaps in the alignment. Second, the missing regions in the framework are generated using a loop-building procedure. Finally, the sidechains are added, using a sidechain-modeling procedure. The whole procedure is illustrated in Figure 1.5.

1.4.3.1. Loop Building

Many loop-building procedures have been proposed. These methods can be divided into ab-initio methods, and database search techniques. The ab-initio loop-building methods are based on a conformational search, often guided by a scoring function. For short loops, the search can be deterministic, following some simple geometric rules [63–65]. For long loops, ab-initio methods rely on heuristics, and sampling becomes a crucial issue. Sampling is usually performed in torsion angle space [66–68]. The initial set of loops derived from this sampling are further optimized using either Molecular Dynamics simulations [69], Monte Carlo searches [70], genetic algorithms [71], bond scaling and relaxation [72], multicopy searches [73], or one of many robotics techniques [74–76].

An alternative approach to ab-initio methods is to search a database of protein structures for loop candidates, based on geometric fitness criteria. Figure 1.6 illustrates this approach for the small toy problem described in Figure 1.5. This method was originally introduced by Jones and Thirup [77]. It has the advantage of rapidly finding fragments that have protein-like conformations. In 1994, however, John Moult and colleagues concluded that the use of fragments from the PDB to generate loops was useful for loops up to a length of 4 residues, as the completeness of the database degrades rapidly with increasing length [78]. This limit of 4 residues can now be safely extended to 9 residues, as a result of the large increase in the size of the protein structure database [79]. For longer loops, Kolodny and colleagues developed a new technique that proceeds by concatenating small fragments of protein chosen from a small library of representative fragments to form a chain that closes the gap [79]. This approach has the advantages of ab-initio methods as it is based on a systematic enumeration of all candidate loops in the discrete approximation of the conformational space accessible to the loop, as well as the advantage of database search since the use of fragments of known protein structures guarantees that the backbone conformations are protein-like. It is guaranteed to find among all loops generated conformations that are close to the native structure, where the meaning of "close" is related to the coarseness of the library of fragments.

Generating high-quality conformations is, however, only the first part of a solution to the loop-closure problem. Once many conformations are available, we need to select those that are native-like. In essence, this is a special case of the problem of picking native-like conformations among decoys, a common task in protein structure modeling. The problem is not simpler for loop selection, as some loops can be long enough that their prediction becomes a small structure

```
1RDS  CEYTCGSTCYWSSDVSAAKAKGYSLYESGDTIDD--YPHEYHDYEGFDFPVSGTYYEYPIMSD
1FUS  SATTCGSTNYSASQVRAAANAACQYYQNDDTAGSSTYPHTYNNYEGFDFPVDGPYQEFPIKSG
```

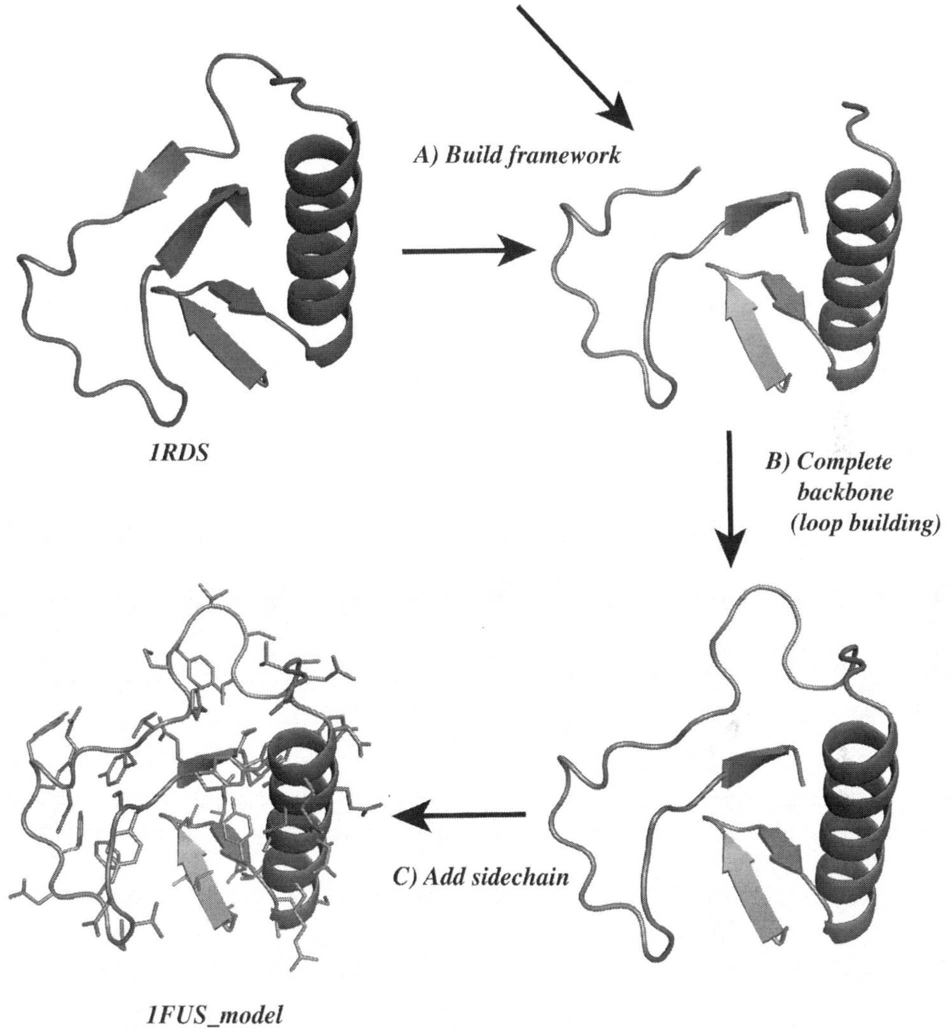

A) Build framework

1RDS

B) Complete backbone (loop building)

C) Add sidechain

1FUS_model

Figure 1.5. **Homology modeling of the Ribonuclease F1 from *Fusarium monoliforme* (FUS)**. RF1 is a small protein of 106 residues. Its sequence is found to be similar to the sequence of Ribonuclease MS from molsin (RDS) (Fasta *E*-value: 1e-22; 57,8% sequence identity in 102 amino acid overlap); the structure of RDS is known (PDB code 1RDS). A toy experiment designed to predict the structure of RF1 proceeds via a number of steps. (A) *Building the framework*. A framework for FUS is built by using the backbone regions of RDS corresponding to the regions in which the sequence is conserved. The framework has a gap of nine residues. (B) *Completing the backbone*. Fragments of nine residues with proper end-to-end geometry are either selected from a database of protein segments, or built using small libraries of protein building blocks. (C) **Completing the model**. Once a protein segment is selected to fill the gap, the sidechains are added to yield a full atom model for the target protein FUS. This figure was generated using PYMOL.

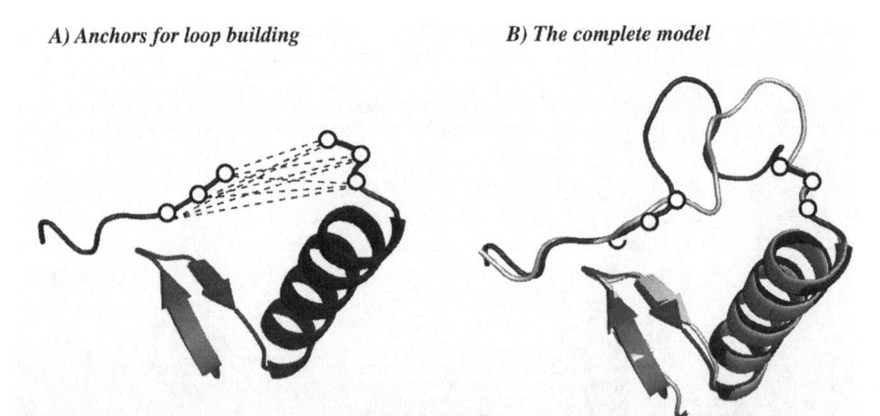

Figure 1.6. A database approach to loop modeling. In the homology-modeling problem described in the legend of Figure 1.5, the framework for the target protein has a gap of nine residues. (A) The position of the three residues preceding and following the gap are marked with balls centered at their Cα. These residues define the "stems" of the loop to be modeled. The geometry of this step is defined based on the 15 inter-distances between the centers of the balls. (B) Fragments of nine residues with proper end-to-end geometry are selected from a database of protein segments of length nine. The best fragment selected is shown in black. For reference, the actual structure for FUS is shown in white; the best selection is 5.6 Å away from the correct conformation for the loop; this discrepancy is not unusual for large loops. This figure was generated using PYMOL.

prediction problem by itself. Solutions to this problem are related to the choice of energy functions and to refinement techniques, which we discuss in the section on ab-initio protein structure prediction.

1.4.3.2 Sidechain Modeling

Predicting protein sidechain conformation has been an active field of research in computational biology for at least 15 years. It usually assumes a fixed backbone, a pairwise energy function (similar to the one given in Eq. (1.4)), and a discrete set of possible conformations for each sidechain, called rotamers [80,81]. The goal is clearly stated: choose a rotamer for each residue in the protein, so that the total energy of the molecule is minimized. This discrete approximation of the problem remains difficult because of its combinatorial nature: it was shown theoretically that it is not only NP-complete [82] but also "inapproximable" [83], i.e., it is unlikely that there exists a polynomial time method that guarantees a good solution for all instances of the problem. These theoretical hardness results for sidechain positioning may not always hold in practice and have not deterred scientists from searching for ways to solve this problem.

This is clearly demonstrated by the considerable progress made in the development of both exhaustive and heuristic techniques for this problem. Successful methods in this field include application of the dead-end elimination theorem [84–88], simulated annealing [89–91], Monte Carlo search [92], and graph theoretical approaches [93–95]. SCWRL3 [94], developed by Dunbrack and colleagues, has become a method of choice for predictors at CASP (see §1.6). It uses a detailed backbone-dependent rotamer library and a graph theoretical approach to select the best rotamer for each sidechain. Its implementation is fast and it can easily be incorporated into other modeling packages. I have developed my own approach, CONFMAT, which is based on self-consistent mean-field theory [96,97] and whose principle is depicted in Figure 1.7. CONFMAT2000, the most recent version of the software, only differs from the

A) Multicopy sampling

P1

P2

P3

P1+P2+P3=1

B) The chimera molecule

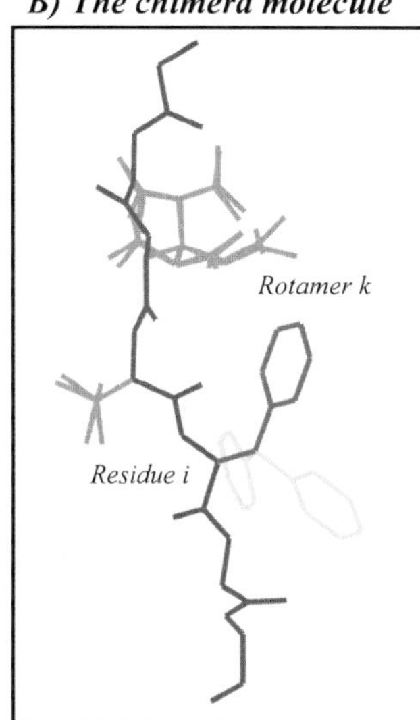

C) Mean field energy

$$E(i,k) = U(i,k) + U(i,k,Backbone) + \sum_{j=1,j\neq i}^{N} \sum_{l=1}^{Nrot(j)} P(j,l)U(i,k,j,l)$$

D) Updating the probabilities

$$P_{new}(i,k) = \frac{\exp\left[-\frac{E(i,k)}{kT}\right]}{\sum_{l=1}^{Nrot(i)} \exp\left[-\frac{E(i,l)}{kT}\right]}$$

Figure 1.7. A self-consistent mean field (SCMF) approach to the problem of predicting sidechain conformation. (A) *The multicopy approach.* Let us assume that residue *i* in the protein of interest is a phenylalanine, and that this phenylalanine can adopt three possible conformations. A systematic enumeration of all possible sidechain conformations in the protein would require that all three conformations of phenylalanine *i* be considered. If the protein contains 100 residues, each with three possible conformations, the size of the corresponding conformational space is 3^{100}, a number out of reach of modern computers. As an alternative, we construct a chimera molecule, where sidechains are represented as an ensemble of discrete conformation: phenylalanine *i* is now represented with 3 conformations, each with a weight *P*(*i*,*j*), such that the sum of the weights is 1. (B,C) *The mean field.* The chimera molecule considered contains all conformations of all sidechains in the proteins. The energy of conformation *k* for residue *i* includes the internal energy for conformation *k*, the energy of interaction of conformation *k* for *i* with the backbone, and all interactions with all conformations of the remaining sidechains of the protein, each weighted with their probabilities. (D) *Updating the probabilities.* The initial probabilities are chosen to be uniform. Using the equations given in (C) we get the energies of all conformations of all residues in the chimera protein. These energies are then used to update the probabilities of these conformations. We have shown that updating the probabilities using a Boltzmann law is equivalent to minimizing the total free energy of the chimera molecule [97]. The new probabilities are then used to compute new energies; this procedure is repeated until we reach convergence ("self-consistency"), i.e., when the probabilities and energies do not change anymore. For each residue, we choose the conformation with the highest converged probability as its predicted conformation. Please visit http://extras.springer.com/ to view a high-resolution full-color version of this illustration.

original CONFMAT in its energy function: a Coulomb energy function for electrostatics has been added to the original Lennard-Jones potential. SCWRL3 and CONFMAT2000 were compared over a range of 1400 non-redundant high-resolution protein structures. No differences were found in the overall quality of the prediction of these two methods. SCWRL3 and CONFMAT2000 predicted 75.9 and 75.4% of the $\chi 1$ angle of all 183640 sidechains in the test set, respectively.

Note that the self-consistent mean-field approach included in CONFMAT can be adapted to the problem of building loops in protein for homology modeling [98].

1.4.3.3. Model Refinement

Even if the best possible structural template was identified, an error-free alignment between the sequence of the target protein and this template was obtained, and the model building was successful, these steps are not expected to generate crystal-quality structures of proteins. There is a definitive need for refinement of the model, and this need increases as the similarity between the template and the target protein decreases. To reach high accuracy in model building depends on two things. First, one needs an energy or discrimination function that has a deep minimum at or very near the native structure. Second, one needs a method that is able to change the coordinates of any model sufficiently near the native structure so as to bring it significantly closer to that structure. Such improvement of model structures has been long known as "energy refinement" [99], and the difficulty of achieving it is termed the "refinement problem." Standard methods for model refinement usually proceed by minimization of an energy function such as the one given in Eq. (1.4), using standard minimization techniques or molecular dynamics simulation to avoid local minima. Sadly for the computational biologists, these techniques are currently not successful, and it remains difficult to generate a model closer to the native structure than the template used to build it [100,101]; much remains to be done on this aspect of model building.

1.4.4. Evaluation of Models

The quality of the structure derived by homology modeling defines its usefulness; it is therefore of great importance to assess it correctly. There are many factors that need to be taken into account when estimating this accuracy. Here we describe a few of these factors:

- The model should have good stereochemistry. This term regroups the normality of bond lengths and bond angles, the planarity of the peptide bonds and aromatic rings, and the chirality of the amino acids. In addition, the model should be free of atomic "clashes," i.e., of severe overlaps between atoms. There are many programs that evaluate the stereochemistry of proteins, such as PROCHECK [102] and WHATCHECK [103].
- The model should have common properties found in proteins. This includes the presence of a hydrophobic core, the correct positioning of polar residues either in proximity of other polar residues or accessible to solvent, hydrogen bonding, local geometry of the backbone, as well as the conservation of atomic volumes, among others. The idea is to check if the model is compatible with the sequence of the target protein. Many of the criteria listed here have been implemented in statistical

methods for testing the quality of proteins, such as VERIFY3D [104–106] and PROSA [107–109].
- The model should be consistent with biological data available for the target protein, if any. This mainly concerns the position of active residues that are expected to be accessible, either at the surface or in a pocket of the protein. It also relates to the putative binding of ions: if the template corresponds to an ion-free protein, while the target protein is known to bind a specific ion (such as calcium or zinc), the model is likely to be incorrect.

This list is by no means exhaustive, and we refer the reader to the excellent review by Sali and colleagues [62] for a thorough analysis of the errors that can be seen in homology models, and how to assess them.

1.4.5. Applications of Homology Modeling

The shape of a protein defines its function; unfortunately, experimental determination of protein structure remains a lengthy and costly process that is not always successful (i.e., some proteins do not crystallize or are not soluble at a level high enough for NMR studies). Homology modeling is a viable option for these situations, and its success has led to the development of many applications for structural and functional studies.

One of the most promising applications of homology modeling in structural biology is related to molecular replacement for X-ray crystallography studies [110–112]. Diffraction data on the crystal of a protein of interest are scalar values providing the intensities of the diffracted waves. To reconstruct an electron density map, however, we need the phases of these waves. Several methods exist to solve the *phase problem* [113], among which molecular replacement plays an important role. This approach is based on the availability of at least one close structural homologue to the protein of interest. A Patterson map is computed from the structural homologue and compared to the Patterson map of the protein under study: from this comparison it is possible to derive the phases and therefore to build an electron density map. As the Protein Data Bank (PDB) is rapidly growing, the probability that such a structural homologue exists increases accordingly. When a structural homologue is not available, however, in the PDB, homology models have been proposed as viable alternatives [110,114–117]. In a similar spirit, homology models have proved useful for electron-microscopy [118] and small angle X-ray scattering (SAXS) studies [119]. Note that in these two cases the interest is mutual: the experimental techniques benefit from the homology models and, conversely, the accuracy of the homology models can be greatly improved if some experimental data are available.

The prime function of homology modeling remains to provide the biologists with useful structural information on the proteins they are working with. Multiple applications of homology models have been published over the years, including identification of functionally important residues in proteins [120], analysis and prediction of protein–protein interactions [121], as well as multiple applications for drug design [122–125].

Homology modeling is expected to help derive a full picture of the protein structure space and as such to play an essential role in the structure genomics initiative. The aim of the latter is to determine the 3D structure of all the proteins encoded in the various genomes whose genes have been identified and sequenced. This aim cannot be achieved by experimental techniques such as NMR and X-ray crystallography alone: the current success rate of the latter, for example, is around 5% (i.e., only 5% of the genes proposed for structural analysis have had their structures determined by X-ray crystallography and deposited in the PDB database; see, e.g., http://targetdb.pdb.org/statistics/TargetStatistics.html). The hopes of structural genomics rests on the fact that proteins are not random: their sequences cluster into families and their structures can be regrouped into a small number of folds, estimated to be on the order of 1000 [58]. In addition, it is well recognized that two proteins with similar sequences adopt similar structures (though there are documented exceptions to this rule, such as the recent discovery of two CRO proteins whose sequences have 40% sequence identity but whose structures are significantly different [126]). It has been estimated that approximately one third of all sequences are recognizably related to one or more known protein structures (see, e.g., [62]). As the November 2008 release of Uniprot/TrEMBL (the database of protein sequences) contains more than six million sequences, this would mean that homology modeling could be used to predict the structure of more than two million proteins by the end of 2008; this number should be compared to the 50,000 experimental protein structures available in the Protein DataBank as of the same date.

1.5. AB-INITIO PROTEIN STRUCTURE PREDICTION

Ab-initio structure prediction and homology modeling have the same goal: predict the structure of a protein given its amino acid sequence. The latter proceeds by induction, using the known structure of a homologue to build a model for the target protein. The former, on the other hand, refers to all problems in which it was not possible to detect such a homologue; the prediction is then "ab initio," i.e., it relies solely on basic and established properties of proteins and not on specific knowledge. Note that these properties can be derived from fundamental laws of physics or chemistry, or inferred using information from the database of known protein structures. Ab-initio structure prediction is more difficult than homology modeling, and while there are signs that this technique is making progress [127], there is still much room for improvement, especially for large proteins (i.e., those with more than 150 residues). Ab-initio structure prediction has long been a purely academic and prestigious problem that has fascinated the scientific community at large, from computer science to biology, with contributions from mathematics, physics, and chemistry: it is usually referred to as the "holy grail" of computational biology. I will review here a few representative approaches to ab-initio protein structure prediction in order to highlight the challenges and future possibilities. This overview is by no means exhaustive, and I refer the reader to some excellent reviews for a more complete representation of the field [128–133].

1.5.1. Simple Models: Lattice Studies

Lattice models provide a simplified representation of the geometry of a protein chain that allows for simple and fast computation of its stability (for review, see [134]). In such models, each amino acid is represented as a bead and the connecting bonds are represented by lines. The background lattice serves to divide space into sites that may be either empty or filled by one bead. Bond angles have only a few discrete values, dictated by the structure of the lattice. Many different types of lattices are possible, in both two and three dimensions; while the cubic lattice is usually preferred because of its simplicity, face-centered cubic lattices provide better approximation to the geometry of a protein [135,136]. Figure 1.8 shows the representation of a 27-residue protein on a 3D cubic lattice. The lattice models of protein structures consider several

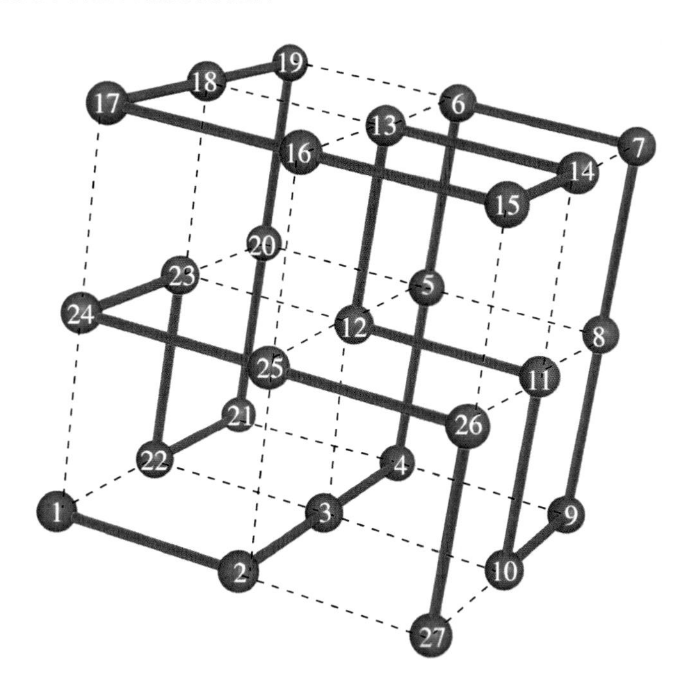

Figure 1.8. Lattice model of a protein structure. The figure depicts an example of a compact self-avoiding structure of a protein chain of 27 "residues" on a regular cubic lattice. This structure contains 28 contacts between non-sequential residues (shown as dashed line). The total energy of this conformation is the sum of the energies over these contacts. Please visit http://extras.springer.com/ to view a high-resolution full-color version of this illustration.

types of sequences. The HP model [137] is based on a binary alphabet with H for hydrophobic and P for polar amino acids; in this model, HH contacts are favorable, while HP and PP are neutral. In the perturbed homopolymer model [138], all monomers are strongly attracted to each other, and the effects of sequence are treated as relatively small perturbations to this large net attraction. Other lattice models consider all 20 types of amino acids that interact either with a Go potential (a specific potential derived from the native conformation of the protein studied, [139]) or based on statistical potentials such as the Miyazawa-Jernigan potential [140].

The advantages of lattice models derive from their simplicity. It is possible to enumerate all possible conformations and sequences of small enough protein chains on lattices; this has been used for protein structure prediction and for protein-folding studiesm as well as for protein sequence design (for an exhaustive review of all possible applications, see [134]). Lattice simulations, for example, predict that the collapse of polymer chains helps drive the formation of secondary structures, both helices and sheets [141,142]. The disadvantages of lattice models, however, are clear. Resolution is lost. The details of protein structures and energetics are not accurately represented. As computers become more powerful, simulations of proteins based on off-lattice realistic representations become more accessible and gradually replace lattice simulations.

1.5.2. Protein Structure Prediction Based on First Principles

Maybe the most intellectually pleasing approach to solving the protein structure prediction problem is to use purely physics-based methods, without knowledge derived from databases (such as statistical energy functions). These methods only assume that the native conformation of a protein corresponds to the minimum of its free energy (Eq. (1.3)). The most natural approach to reach this state is standard molecular dynamics (MD) simulations [143], which numerically integrate Newton's equation of motion for the protein. MD simulations explore the energy landscape accessible to the protein, favoring regions with low internal energy. These methods are currently limited by inaccuracies in the force fields and by huge computational requirements. Nevertheless, there have been some notable successes in the past decade. The first milestone was a supercomputer simulation by Duan and Kollman [144] in 1998 of the 36-residue villin headpiece in explicit solvent. Starting from an unfolded conformation and following its dynamics over nearly a microsecond, the simulation reached a collapsed state 4.5 Å from the NMR structure. A few years later, the IBM Blue Gene group of Pitera and Swope [145] folded the 20-residue Trp-cage peptide in implicit solvent to within 1 Å. More recently, Duan and coworkers pushed the limits of their simulations of the villin headpiece and reached accuracies below 0.5 Å [146]. Several groups are addressing the huge computational requirements for protein folding by considering distributed grid computing. The idea is to combine the power of computers from throughout the world to solve a large-scale problem that cannot be solved on a single computer. Folding@home (http://folding.stanford.edu) was the first of such distributed grid computing systems dedicated to the protein structure prediction problem. Using this system, Pande et al. folded villin to an RMSD distance of 1.7 Å [147]. It is worth noting that folding@home became in 2007 the "world's largest supercomputer" according to the Guinness book of records. Following the success of folding@Home, several other distributed computing initiatives have emerged for solving the structure prediction problem, including rosetta@home (http://boinc.bakerlab.org/rosetta/) and predictor@home (http://predictor.chem.lsa.umich.edu/).

1.5.3. Bioinformatics Approaches to ab-initio Structure Prediction

We use the term "bioinformatics" here to refer to techniques that mine the existing biological databases to assist in the process of predicting protein structure, in opposition to the "physics-only" approach described above. Techniques that fall within this category usually proceed in three steps: (i) prediction of the secondary structure of the protein, (ii) generation of a large collection of putative conformations for the protein (the decoys), and (iii), selection of hopefully native-like conformations among all the decoys that were generated. We describe briefly these three steps.

1.5.3.1. Secondary Structure Prediction

Protein structures are hierarchic: their backbones organize locally into the so-called secondary structures (helices and strands; see Fig. 1.2), which are subsequently packed to form the tertiary structures. While proteins probably do not fold following this hierarchy, i.e., folding usually start with a hydrophobic collapse rather than with the formation of the secondary structures, knowledge of the position and nature of the secondary structures of a protein would greatly help its 3D structure prediction.

Early methods for secondary structure prediction use single sequences only. They are based on analyses of the relative frequencies of each amino acid to be in α-helices, β-sheets, and turns based on known protein structures in the PDB. From these frequencies *propensities* are derived that quantify the appearance of each amino acid in each secondary structure type. The propensity of finding an amino acid of type i in the secondary structure of type S is the ratio of the probability of finding i in S divided by the probability of finding amino acid type i anywhere in a protein; this propensity is greater than 1 if i has a preference for the structure S, lower than 1 if i tends to avoid S, and close to 1 if there are no marked preferences. Table 1.2 lists the propensities of all 20 amino acids for the three types of secondary structures; namely helices, strands, and turns. Chou and Fasman proposed a method that uses these propensities to predict the probability that a given sequence of amino acids would form a helix, a β-strand, or a turn in a protein [148]. Their rules to predict the helical content of a protein proceed as follows:

- *Search for seed regions*: any six consecutive residues, with at least four of these six having a propensity to be in a helix, $P(\alpha)$, greater than 1.
- *Extension of seed regions*: extend each seed helical region in both directions until a set of 4 contiguous residues has an average $P(\alpha)$ smaller than 1 (breaker)
- *Prediction*: For each extended seed region, if the average $P(\alpha)$ over the whole region is greater than 1, it is predicted to be helical.

The rules for strands are the same, with the exception that the seed region contains five consecutive residues, with at least three having a propensity to be in a strand, $P(\beta)$, greater than 1. In cases of ambiguities, i.e., if the same region is predicted to form both a helix and a strand, the prediction with the highest average propensity is retained.

Table 1.2. Amino Acid Propensities to Be in Helices, Strands, or Turns

Amino acid	Helix	Strand	Turn
Ala	1.29	0.90	0.78
Cys	1.11	0.74	0.80
Leu	1.30	1.02	0.59
Met	1.47	0.97	0.39
Glu	1.44	0.75	1.00
Gln	1.27	0.80	0.97
His	1.22	1.08	0.69
Lys	1.23	0.77	0.96
Val	0.91	1.49	0.47
Ile	0.97	1.45	0.51
Phe	1.07	1.32	0.58
Tyr	0.72	1.25	1.05
Trp	0.99	1.14	0.75
Thr	0.82	1.21	1.03
Gly	0.56	0.92	1.64
Ser	0.82	0.95	1.33
Asp	1.04	0.72	1.41
Asn	0.90	0.76	1.23
Pro	0.52	0.64	1.91
Arg*	0.96	0.99	0.88

Amino acids shown in bold, italic, and regular fonts have higher propensities to be in a helix, a strand, or a turn, respectively. *Arg, a special case, shows no preference.

Garnier and coworkers later improved upon the Chou and Fasman method by introducing local correlations when computing the propensities; their method — named the GOR method — was shown to yield better predictions [149]. Both methods have since gone through several cycles of improvement. They are still available for education purposes (see, for example, http://fasta.bioch.virginia.edu/fasta_www/chofas.htm) but have been superseded by multiple sequence methods for research (for review, see [150–152]). Indeed, the availability of sequences for large families of homologous sequences has revolutionized secondary structure prediction. The Chou and Fasman method as well as the GOR method proved much more accurate at identifying core secondary structure elements when applied to a family of proteins rather than a single sequence. The combination of sequence data with sophisticated computing techniques such as neural networks and hidden Markov models has led to accuracies well in excess of 70% [153]; a recent extension that includes structural data gives us hope that the success of these methods will reach levels above 80% accuracy [154]. Note that all secondary structure prediction methods only account for local interactions; as such, there is likely to be an upper limit on their accuracy.

1.5.3.2. Conformational Sampling for ab-initio Structure Prediction

Protein structures are deceptively simple, with most of their chains involved in secondary structures that are tightly packed to form their globular shapes. Many methods for generating decoy structures make use of this description: starting from the predicted secondary structures of the protein, they test different packing options that lead to compact structures. The packing is generated using either preferred arrangements of secondary structures observed in experimental structures [155,156], energetically favorable interactions between the secondary structures [157], or heuristic sampling using Monte Carlo or simulated annealing methods. Packing of the secondary structures can be constrained by imposing contacts predicted from correlated mutation analysis of multiple sequence analysis. Such analyses are not accurate [158]; they have been shown, however, to improve decoy generation [159].

The methods described above rely on secondary structure prediction for a protein of interest. They have consequently two inherent limitations: they include no constraints on loop regions and their success is tightly related to the quality of secondary structure prediction. Fragment-based approaches emerged as a way to alleviate these limitations, with Rosetta being certainly the most successful of these [160]. The Rosetta procedure makes the assumption that the distribution of configurations sampled by a peptide segment are reasonably well approximated by the distribution of configurations observed in the protein structure database for segments with the same sequence. Tertiary structures are generated using a Monte Carlo search of the possible combinations of likely local structures, minimizing a scoring function that accounts for nonlocal interactions such as compactness, hydrophobic burial, specific pair interactions (disulfides and electrostatics), and strand pairing. Using Rosetta, David Baker and colleagues have shown that high-resolution structure prediction (<1.5 angstroms) can be achieved for small protein domains (<85 residues) [127]. Rosetta has now become a complete modeling platform, with applications in most areas of macromolecular modeling [161].

1.5.3.3. Scoring Protein Decoys

Once a large collection of decoys has been generated, it is essential to filter them in order to retain only those whose structure is native-like: this is the role given to scoring functions. Designing and testing scoring functions are major concerns in computational biology; here I only provide an overview of the two main classes of scoring functions for proteins; for a more complete description, see the excellent reviews of Lazaridis and Karplus [162], Huang et al. [163], Bonneau and Baker [128], and Ngan et al. [164].

The first class of scoring functions is based on the physics of the molecular interactions, using energy computed with equations similar to Eq. (1.4). While the different terms involved in computing the internal energy of the protein are usually well understood (see §1.3), the situation is more complex when it comes to treating properly the interactions of the protein with its environment, i.e., both solvent and ions or cofactors. Water is easily incorporated into a molecular dynamics simulation by explicitly adding solvent molecules around the protein; this approach, however, is not appropriate for scoring a large quantity of protein decoy structures, as it would be much too costly in computing time. To alleviate this problem, most scoring functions rely on implicit solvent models in which water is treated implicitly using potential of mean forces (for review, see [165,166]).

The second class of scoring functions is knowledge based. Each of these functions try to capture some properties of protein structure (such as the propensities of some residues to form contact with one another or with the solvent) and translate them into a score that can then be used to evaluate whether a model satisfies these properties. These knowledge-based functions are usually compiled based on statistics derived from the database of known protein structures; their formulation is either grounded on statistical physics, in which case they are referred to as potential of mean forces [167,168], or directly on statistical models, in which case they are referred to as statistical potentials [169]. The two formulations lead to the same master equation:

$$E(x,S) = -k \ln\left(\frac{P(x \mid S)}{P(x)}\right). \tag{1.8}$$

where the score $E(x,S)$ of observing that a specific subset S of a protein takes the value x for a property P is derived as the log of the ratio of the probability $P(x|S)$ that subsets S in known proteins take the value x, normalized with the reference probability $P(x)$ that the property P has value x, independent of the subset considered. The property P refers to any feature of protein structures that can be parameterized, including and not limited to solvent exposure, pairwise interatomic distances, geometric arrangement of two, three, or four atoms, pseudo-dihedral angles, ... For a more complete list of properties implemented in knowledge-based potentials, see the review of Ngan et al. [164].

Note that almost all current protein structure prediction techniques include both physics- and knowledge-based potentials.

1.6. THE CASP EXPERIMENT

Testing the accuracy of a protein structure prediction method is not as easy as one would think: as this method most likely relies on some experimental data (usually structural data coming from the PDB), one must carefully remove proteins that are homologous to the proteins included in the test set from all databases used by the method in question. Any oversights could lead to overestimates of success. For that reason, the Critical Assessment of Structure Prediction (CASP) experiment, a biannual, community-wide blind test of prediction methods, was conceived and implemented by John Moult [170]. It was first held in 1994, and the eighth edition was undertaken in the summer of 2008.

CASP works as follows. Protein crystallographers and NMR spectroscopists are solicited for proteins whose structures are likely to be completed before the next CASP meeting. The sequences of these target proteins are then made available on a web server. Researchers interested in taking part in CASP then submit up to five predictions for each target before a given deadline. Assessors chosen by the organizers critically analyze these predictions. The results are then presented at a conference (originally held in Asilomar, CA, USA, and now held in alernately in Asilomar and at a site in Europe) and papers by the assessors and the successful groups chosen to speak are published in a special issue of *Proteins: Structure, Function and Genetics*. The first meeting (CASP1) took place in 1994, assessing 135 predictions made by 35 groups for 33 different protein targets. The success of CASP can be ascertained from the fact that it has been growing steadily ever since, with increased participation of both experimentalists providing test structures and predictors. The most recent CASP was held in the summer of 2008, assessing more than 55,000 predictions made by 161 groups for 121 different protein targets. The CASP results are published in special issues of the scientific journal *Proteins*, all of which are accessible through the CASP website (http://predictioncenter.org). These reports provide an extensive and extremely valuable picture of the current state of the art in protein structure prediction.

1.7. CONCLUSIONS

Proteins are key molecules in all cellular functions. Nature has extensively explored their sequences and structures in order to build the library of functions needed for the diversity of life, taking into account all external constraints and the corresponding adaptation. The wealth of information encoded in the protein sequences and structures therefore provides clues needed to unravel the mysteries of life and its evolution and adaptation over time. The success of the genomics project has led to an exponential explosion in the number of known protein sequences; we are only at the beginning of extracting the information they contain. Structural genomics projects, on the other hand, have not yet been as successful: protein structure databases cover only a small fraction of the known proteins. In response, modeling techniques are attracting a lot of attention in the hope they become viable alternatives to experimental methods for studying protein structures. In this chapter we have reviewed the main techniques for protein structure prediction. We have shown that ab-initio techniques have enjoyed significant progress in recent years, especially for small proteins [127]; homology-modeling techniques, however, remain the most promising methods for structure prediction, and there is hope that in the near future they will generate models with crystal structure quality.

Recent success in homology modeling, and more specifically in its initial step of detecting homologous templates, are often seen as a consequence of the exponential increase in protein sequence data and continuing accumulation of experimental three-dimensional structures of proteins [101]. The success related to the use of these large datasets on sequence and structure and the knowledge that many more datasets are available as a consequence of the different "-omics" projects are both reason for optimism in this field. Sheer accumulation of data, however, may not prove as useful as expected, because of inherent difficulties in integrating the information they contain. All these data are derived using different experimental procedures. As a result, there exists now a vast array of heterogeneous data resources distributed over different Internet sites that cover genomic, cellular, structure, phenotype, and other types of biologically relevant information. The complexity of the molecular biology domain makes the modeling, handling, and exchange of these data very difficult. In order to alleviate these problems, controlled vocabularies and ontologies have become essential tools in modern molecular biology. They ensure compatibility between different data resources and software applications and increase the efficiency and accuracy of data queries by standardizing the wide variations in terminology that exists. Successful homology modeling will depend on data integration, and the applications of these ontologies to the specific problems inherent to the modeling process.

ACKNOWLEDGMENTS

Support from the National Institute of Health is acknowledged.

PROBLEMS

1.1. a. What are usually the strongest interactions (in amplitude) in a molecule in its native conformation?
   1. bonded interactions
   2. VdW interactions
   3. electrostatics interactions
   4, interaction with the solvent

   b. When would you use homology modeling to predict the structure of a protein?
   1. when the best homologous protein with known structure found by BLAST has a *P*-value of 8 or above
   2. when the best homologous protein with known structure found by BLAST has a *P*-value of 10^{-4} or below
   3. when BLAST cannot identify a homologous protein
   4. always, as the other techniques are too complicated

   c. To build a model for the structure of a target protein using homology modeling, you need:
   1. the structure of a homologous protein (the template), and the alignment between the sequence of the target and template proteins
   2. the sequence of the target protein only
   3. a lot of imagination
   4. the structures of proteins that interact with the target protein

   d. CASP is:
   1. a series of meetings set to assess programs that predict protein–protein interactions
   2, the best method currently available for homology modeling

3. an experiment set to assess protein structure prediction techniques
4. an experiment set to assess RNA structure prediction techniques

1.2. Using the Chou and Fasman propensity values given in Table 1.2 and the prediction rules defined in Section 1.5.3.1., predict the secondary structures of the following peptide sequences:
   a. WHGCITVYWMTV
   b. CAENKLDHVRGP

1.3. The following eukaryotic DNA sequence is given to you:

   5′-CCCTTAATGCGTATCGCTCACGAGATGTTGGGCGGCTAA-3′

   a. You are told that this sequence, or its complementary, codes for one gene. Find the longest "gene," or open reading frame (ORF), corresponding to this DNA sequence; remember that there are 6 possibilities, i.e., 3 possible reading frames for one strand and 3 possible reading frames for its complementary. Transcribe this ORF into an RNA sequence.
   b. As this is a eukaryotic sequence, it may contain an intron. For simplicity, we will assume that introns always start with GU and end with AG. Identify all possible introns, and explain why their removal would result in loss of the gene.
   c. Based on question (b) just above, we know that the RNA is not spliced. Find the sequence of the "protein" it encodes.
   d. Predict the secondary structure of this "protein" using the Chou and Fassman method, based on the propensities given in Table 1.2.
   e. Can you find a single mutation at the DNA level of that gene that will modify the corresponding "protein" such that it is predicted to be fully extended (i.e., predicted to be a strand by Chou and Fassman)? (**Hint**: there are several possible answers.)

FURTHER READING

Many excellent books are available for an introduction to protein structures. Among them, I would recommend:

1. Branden C, Tooze J. 1991. *Introduction to protein structure*. New York: Garland Publishing.
2. Creighton TE. 1993. *Proteins*. New York: W.H. Freeman & Co.

In addition, I would recommend reading the following review articles on protein structure stability and protein structure prediction:

1. Taylor WR, May ACW, Brown NP, Aszodi A. 2001. Protein structure: geometry, topology and classification. *Rep Prog Phys* **64**:517–590.
2. Marti-Renom MA, Stuart AC, Fiser A, Sanchez R, Melo F, Sali A. 2000. Comparative protein structure modeling of genes and genomes. *Annu Rev Biophys Biomol Struct* **29**:291–325.
3. Bonneau R, Baker D. 2001. Ab initio protein structure prediction: progress and prospects. *Annu Rev Biophys Biomol Struct* **30**:173–189.
4. Dill KA, Bromberg S, Yue KZ, Fiebig KM, Yee DP, Thomas PD, Chan HS. 1995. Principles of protein folding—a perspective from simple exact models. *Protein Sci* **4**:561–602.

REFERENCES

1. Monod J. 1973. *Le hasard et la necessité*. Paris: Seuil.
2. Levy Y, Wolynes PG, Onuchic JN. 2004. Protein topology determines binding mechanism. *Proc Natl Acad Sci USA* **101**:511–516.
3. Plaxco KW, Simons KT, Baker D. 1998. Contact order, transition state placement and the refolding rates of single domain proteins. *J Mol Biol* **277**:985–994.
4. Alm E, Baker D. 1999. Prediction of protein-folding mechanisms from free energy landscapes derived from native structures. *Proc Natl Acad Sci USA* **96**:11305–11310.
5. Munoz V, Eaton WA. 1999. A simple model for calculating the kinetics of protein folding from three-dimensional structures. *Proc Natl Acad Sci USA* **96**:11311–11316.
6. Alm E, Morozov AV, Kortemme T, Baker D. 2002. Simple physical models connect theory and experiments in protein folding kinetics. *J Mol Biol* **322**:463–476.
7. Koehl P, Levitt M. 2002. Protein topology and stability defines the space of allowed sequences. *Proc Natl Acad Sci USA* **99**:1280–1285.
8. Smalheiser NR. 2002. Informatics and hypothesis-driven research. *EMBO Rep* **3**:702.
9. Kell DB, Oliver SG. 2003. Here is the evidence, now what is the hypothesis? The complementary role of inductive and hypothesis driven science in the post genomic era. *Bioessays* **26**:99–105.
10. Liolios K, Mavrommatis K, Tavernarakis N, Kyrpides NC. 2007. The Genomes On Line Database (GOLD) in 2007: status of genomic and metagenomic projects and their associated metadata. *Nucl Acids Res* **36**:D475–D479.
11. Bernstein FC, Koetzle TF, William G, Meyer DJ, Brice MD, Rodgers JR. 1977. The protein databank: a computer-based archival file for macromolecular structures. *J Mol Biol* **112**:535–542.
12. Berman HM, Westbrook J, Feng Z, Gilliland G, Bhat TN, Weissig H. 2000. The Protein Data Bank. *Nucl Acids Res* **28**:235–242.
13. Schulz GE, Schirmer RH. 1979. *Principles of protein structure*. New York: Springer-Verlag.
14. Cantor CR, Schimmel PR. 1980. *Biophysical chemistry: the conformation of biological macromolecules*. New York: W.H. Freeman Company.
15. Branden C, Tooze J. 1991. *Introduction to protein structure*. New York: Garland Publishing.
16. Creighton TE. 1993. *Proteins*. New York: W.H. Freeman & Co.
17. Taylor WR, May ACW, Brown NP, Aszodi A. 2001. Protein structure: geometry, topology and classification. *Rep Prog Phys* **64**:517–590.
18. Timberlake KC. 2004. *General, organic, and biological chemistry: structures of life*. San Francisco: Benjamin Cummings.
19. Brooks C, Karplus M, Pettitt M. 1988. Proteins: a theoretical perspective of dynamics, structure and thermodynamics. *Adv Chem Phys* **71**:1–259.
20. Kendrew J, Dickerson R, Strandberg B, Hart R, Davies D, Philips D. 1960. Structure of myoglobin: a three dimensional Fourier synthesis at 2 angstrom resolution. *Nature (London)* **185**:422–427.
21. Perutz M, Rossmann M, Cullis A, Muirhead G, Will G, North A. 1960. Structure of haemoglobin: a three-dimensional Fourier synthesis at 5.5 angstrom resolution, obtained by X-ray analysis. *Nature (London)* **185**:416–422.
22. Levitt M, Chothia C. 1976. Structural patterns in globular proteins. *Nature (London)* **261**:552–558.
23. Lesk AM, Chothia C. 1980. How different amino-acid sequences determine similar protein structures: the structure and evolutionary dynamics of the globins. *J Mol Biol* **136**:225–270.
24. Chothia C, Janin J. 1981. Relative orientation of close packed beta pleated sheets in proteins. *Proc Nat Acad Sci USA* **78**:4146–4150.
25. Cohen FE, Sternberg MJE, Taylor WR. 1981. Analysis of the tertiary structure of protein beta sheet sandwiches. *J Mol Biol* **148**:253–272.
26. Chothia C, Janin J. 1982. Orthogonal packing of beta pleated sheets in proteins. *Biochemistry* **21**:3955–3965.
27. Cohen FE, Sternberg MJE, Taylor WR. 1982. Analysis and prediction of the packing of aplha helices against a beta sheet in the tertiary structure of globular proteins. *J Mol Biol* **156**:821–862.
28. Chou KC. 1995. A novel approach to predicting protein structural classes in a (20-1)-D amino acid composition space. *Proteins: Struct Funct Genet* **21**:319–344.
29. Chou KC, Zhang CT. 1995. Prediction of protein structural classes. *Crit Rev Biochem Molec Biol* **30**:275–349.

30. Bahar I, Atilgan AR, Jernigan RL, Erman B. 1997. Understanding the recognition of protein structural classes by amino acid composition. *Proteins: Struct Funct Genet* **29**:172–185.
31. Liu WM, Chou KC. 1998. Prediction of protein structural classes by modified mahalanobis discriminant algorithm. *J Prot Chem* **17**:209–217.
32. Chou KC, Liu WM, Maggiora GM, Zhang CT. 1998. Prediction and classification of domain structural classes. *Proteins: Struct Funct Genet* **31**:97–103.
33. Cai YD, Li YX, Chou KC. 2000. Using neural networks for prediction of domain structural classes. *Biochim Biophys Acta* **1476**:1–2.
34. Zhou GP, Assa-Munt N. 2001. Some insights into protein structural class prediction. *Proteins: Struct Funct Genet* **44**:57–59.
35. Luo RY, Feng ZP, Liu JK. 2002. Prediction of protein structural class by amino acid and polypeptide composition. *Eur J Biochem* **269**:4219–4225.
36. Xiao X, Lin W-Z, Chou KC. 2008. Using grey dynamic modeling and pseudo amino acid composition to predict protein structural classes. *J Comput Chem* **29**:2018–2024.
37. Hutchinson EG, Thornton JM. 1993. The Greek key motif: extraction, classification and analysis. *Protein Eng* **6**:233–245.
38. Meirovitch H. 2007. Recent developments in methodologies for calculating the entropy and free energy of biological systems by computer simulation. *Curr Opin Struct Biol* **17**:181–186.
39. Dill KA, Shortle D. 1991. Denatured states of proteins. *Annu Rev Biochem* **60**:795–825.
40. Cozetto D, Tramontano A. 2005. Relationship between multiple sequence alignments and quality of protein comparative models. *Proteins: Struct Funct Genet* **58**:151–157.
41. Chothia C, Lesk A. 1986. The relation betweeen the divergence of sequence and structure in proteins. *EMBO J* **5**:823–826.
42. Flores TP, Orengo C, Moss DS, Thornton J. 1993. Comparison of conformation characteristics in structurally similar protein pairs. *Protein Sci* **2**:1811–1826.
43. Russel RB, Saqi AS, Sayle RA, Bates PA, Sternberg MJE. 1997. Recognition of analogous and homologous protein folds: analysis of sequence and structure conservation. *J Mol Biol* **269**:423–439.
44. Sauder JM, Arthur JW, Dunbrack RL. 2000. Large-scale comparison of protein sequence alignment algorithms with structure alignments. *Proteins: Struct Funct Genet* **40**:6–22.
45. Lipman DJ, Pearson WR. 1985. Rapid and sensitive protein similarity searches. *Science* **227**:1435–1441.
46. Altschul SF, Gish W, Miller W, Myers EW, Lipman DJ. 1990. Basic local alignment search tool. *J Mol Biol* **215**:403–410.
47. Pearson WR. 1995. Comparison of methods for searching protein sequence databases. *Protein Sci* **4**:1145–1160.
48. Agarwal P, States DJ. 1998. Comparative accuracy of methods for protein sequence similarity search. *Bioinformatics* **14**:40–47.
49. Brenner SE, Chothia C, Hubbard TJ. 1998. Assessing sequence comparison methods with reliable structurally identified distant evolutionary relationships. *Proc Nat Acad Sci USA* **95**:6073–6078.
50. Rost B. 1999. Twilight zone of protein sequence alignments. *Protein Eng* **12**:85–94.
51. Park J, Karplus K, Barrett C, Hughey R, Haussler D, Hubbard T, Chothia C. 1998. Sequence comparisons using multiple sequences detect three times as many remote homologues as pairwise methods. *J Mol Biol* **284**:1201–1210.
52. Altschul SF, Madden TL, Schäffer AA, Zhang J, Zhang Z, Miller W, Lipman DJ. 1997. Gapped BLAST and PSI-BLAST: a new generation of protein database search programs. *Nucl Acids Res* **25**:3389–33402.
53. Eddy SR. 1996. Hidden Markov models. *Curr Opin Struct Biol* **6**:361–365.
54. Jones DT. 1997. Progress in protein structure prediction. *Curr Opin Struct Biol* **7**:377–387.
55. Marchler-Bauer A, Bryant SH. 1997. A measure of success in fold recognition. *Trends Biochem Sci* **22**:236–240.
56. Levitt M. 1997. Competitive assessment of protein fold recognition and alignment accuracy. *Proteins: Struct Funct Genet* **Suppl 1**:92–104.
57. Godzik A. 2003. Fold recognition methods. *Methods Biochem Anal* **44**:525–546.
58. Chothia C. 1992. One thousand fold families for the molecular biologist? *Nature (London)* **357**:543.
59. Sali A, Blundell TL. 1993. Comparative protein modelling by satisfaction of spatial restraints. *J Mol Biol* **234**:779–815.

60. Sanchez R, Sali A. 1997. Evaluation of comparative protein structure modelling by MODELLER-3. *Proteins* **Suppl 1**:50–58.
61. Lemer CMR, Rooman MJ, Wodak SJ. 1995. Protein structure prediction by threading methods: evaluation of current techniques. *Proteins: Struct Funct Genet* **23**:337–355.
62. Marti-Renom MA, Stuart AC, Fiser A, Sanchez R, Melo F, Sali A. 2000. Comparative protein structure modeling of genes and genomes. *Annu Rev Biophys Biomol Struct* **29**:291–325.
63. Go N, Scheraga HA. 1970. Ring closure and local conformational deformations of chain molecules. *Macromolecules* **3**:178–187.
64. Palmer KA, Scheraga HA. 1991. Standard-geometry chains fitted to X-ray derived structures: validation of the rigid-geometry approximation, 1: chain closure through a limited search of loop conformations. *J Comput Chem* **12**:505–526.
65. Wedemeyer WJ, Scheraga HA. 1999. Exact analytical loop closure in proteins using polynomial equations. *J Comput Chem* **20**:819–844.
66. Bruccoleri RE, Karplus M. 1985. Chain closure with bond angle variations. *Macromolecules* **18**:2767–2773.
67. Moult J, James MNG. 1986. An algorithm which predicts the conformation of short lengths of chain in proteins. *J Mol Graphics* **4**:180.
68. Deane CM, Blundell TL. 2000. A novel exhaustive search algorithm for predicting the conformation of polypeptide segments in proteins. *Proteins: Struct Funct Genet* **40**:135–144.
69. Bruccoleri RE, Karplus M. 1990. Conformational sampling using high-temperature molecular dynamics. *Biopolymers* **29**:1847–1862.
70. Carlacci L, Englander SW. 1993. The Loop problem in proteins: a Monte-Carlo simulated annealing approach. *Biopolymers* **33**:1271–1286.
71. Ring CS, Cohen FE. 1994. Conformational sampling of loop structures using genetic algorithms. *Israel J Chem* **34**:245–252.
72. Zheng Q, Rosenfeld R, Vajda S, Delisi C. 1993. Loop closure via bond scaling and relaxation. *J Comput Chem* **14**:556–565.
73. Zheng Q, Rosenfeld R, Delisi C, Kyle JD. 1994. Multiple copy sampling in protein loop modeling: computational efficiency and sensitivity to dihedral angle perturbations. *Protein Sci* **3**:493–506.
74. Lavalle SM, Finn PW, Kavraki LE, Latombe JC. 2000. A ramdomized kinematics-based approach to pharmacophore-constrained conformational search and database screening. *J Comput Chem* **21**:731–747.
75. Fine RM, Wang H, Shenkin PS, Yarmush DL, Levinthal C. 1996. Predicting antibody hyper-variable loop conformations, II: minimization and molecular dynamics studies of mcp603 from many randomly generated loop conformations. *Proteins: Struct Funct Genet* **1**:342–362.
76. Canutescu AA, Dunbrack RL. 2003. Cyclic coordinate descent: a robotics algorithm for protein loop closure. *Protein Sci* **12**:963–972.
77. Jones TA, Thirup S. 1986. Using known substructures in protein model building and crystallography. *EMBO J* **5**:819–822.
78. Fidelis K, Stern PS, Bacon D, Moult J. 1994. Comparison of systematic search and database methods for constructing segments of protein-structure. *Protein Eng* **7**:953–960.
79. Kolodny R, Guibas L, Levitt M, Koehl P. 2005. Inverse kinematics in biology: the protein loop closure problem. *Int J Rob Res* **24**:151–163.
80. Ponder JW, Richards FM. 1987. Tertiary templates for proteins: use of packing criteria in the enumeration of allowed sequences for different structural classes. *J Mol Biol* **193**:775–791.
81. Dunbrack RL, Karplus M. 1994. Conformational-analysis of the backbone-dependent rotamer preferences of protein side-chains. *Nat Struct Biol* **1**:334–340.
82. Pierce NA, Winfree E. 2002. Protein design is NP-hard. *Protein Eng* **15**:779–782.
83. Chazelle B, Kingsfort C, Singh MA. 2004. A semi-definite programming approach to side-chain positioning with new rounding strategies. *INFORMS J Comput* **16**:380–392.
84. Desmet J, Maeyer MD, Hazes B, Lasters I. 1992. The dead end elimination theorem and its use in protein side-chain positioning. *Nature (London)* **356**:539–542.
85. Lasters I, Maeyer MD, Desmet J. 1995. Enhanced dead-end elimination in the search for the global minimum conformation of a collection of protein side chains. *Protein Eng* **8**:815–822.
86. Goldstein RF. 1994. Efficient rotamer elimination applied to protein side-chains and related spin glasses. *Biophys J* **66**:1335–1340.

87. Gordon DB, Mayo SL. 1998. Radical performance enhancements for combinatorial optimization algorithms based on the dead-end elimination theorem. *J Comput Chem* **19**:1505–1514.
88. Looger LL, Hellinga HW. 2001. Generalized dead-end elimination algorithms make large-scale protein side-chain structure prediction tractable: implications for protein design and structural genomics. *J Mol Biol* **307**:429–445.
89. Holm L, Sander C. 1991. Database algorithm for generating protein backbone and side-chain co-ordinates from a C-alpha trace: Application to model building and detection of co-ordinate errors. *J Mol Biol* **218**:183–194.
90. Peterson RW, Dutton PL, Wand AJ. 2004. Improved side-chain prediction accuracy using an ab initio potential energy function and a very large rotamer library. *Protein Sci* **13**:735–751.
91. Lu M, Dousis AD, Ma J. 2008. OPUS-Rota: a fast and accurate method for side-chain modeling. *Protein Sci* **17**:1576–1585.
92. Xiang Z, Honig B. 2001. Extending the accuracy limits of prediction for side-chain conformations. *J Mol Biol* **311**:421–430.
93. Samudrala R, Moult J. 1998. A graph theoretic algorithm for comparative modeling of protein structure. *J Mol Biol* **279**:298–302.
94. Canutescu AA, Shelenkov AA, Dunbrack RL. 2003. A graph theory algorithm for rapid protein side-chain prediction. *Protein Sci* **12**:2001–2014.
95. Dukka-Bahadur KC, Tomita E, Suzuki J, Akutsu T. 2005. Protein side-chain packing problem: a maximum edge-weigth clique algorithmic approach. *J Bioinfo Comput Biol* **3**:103–126.
96. Koehl P, Delarue M. 1994. Application of a self consistent mean field theory to predict protein side-chains conformation and estimate their conformational entropy. *J Mol Biol* **239**:249–275.
97. Koehl P, Delarue M. 1996. Mean-field minimization methods for biological macromolecules. *Curr Opin Struct Biol* **6**:222–226.
98. Koehl P, Delarue M. 1995. A self consistent mean field approach to simultaneous gap closure and side-chain positioning in homology modelling. *Nat Struct Biol* **2**:163–170.
99. Levitt M, Lifson S. 1969. Refinement of protein conformations using a macromolecular energy minimization procedure. *J Mol Biol* **46**:269–279.
100. Koehl P, Levitt M. 1999. A brighter future for protein structure prediction. *Nat Struct Biol* **6**:108–111.
101. Venclovas C, Zemla A, Fidelis K, Moult J. 2003. Assessment of progress over the CASP experiments. *Proteins: Struct Funct Genet* **53**:585–595.
102. Laskowski RA, Mc Arthur MW, Moss DS, Thornton J. 1993. PROCHECK: a program to check the stereochemical quality of protein structures. *J Appl Cryst* **26**:283–291.
103. Hooft RW, Vriend G, Sander C, Abola EE. 1996. Errors in protein structures. *Nature (London)* **381**:272.
104. Bowie JU, Lüthy R, Eisenberg D. 1991. A method to identify protein sequences that fold into a known three-dimensional structure. *Science* **253**:164–170.
105. Lüthy R, Bowie JU, Eisenberg D. 1992. Assessment of protein models with three-dimensional profiles. *Nature (London)* **356**:83–85.
106. Eisenberg D, Luthy R, Bowie JU. 1997. VERIFY3D, assessment of protein models with three-dimensional profiles. *Methods Enzymol* **277**:396–404.
107. Sippl MJ. 1993. Recognition of errors in three-dimensional structures of proteins. *Proteins: Struct Funct Genet* **17**:355–362.
108. Wiederstein M, Sippl MJ. 2007. ProSA-web: interactive web service for the recognition of errors in three-dimensional structures of proteins. *Nucleic Acids Res* **35**:W407–W410.
109. Pawlowski M, Gajda MJ, Matlak R, Bujnicki JM. 2008. MetaMQAP: a meta-server for the quality assessment of protein models. *BMC Bioinformatics* **9**:403.
110. Jones DT. 2001. Evaluating the potential of using fold-recognition models for molecular replacement. *Acta Cryst* **D57**:1428–1434.
111. Rossmann MG. 2001. Molecular replacement—historical background. *Acta Crystallogr D Biol Crystallogr* **57**:1360–1366.
112. Ilari A, Savino C. 2008. Protein structure determination by x-ray crystallography. *Methods Mol Biol* **452**:63–87.
113. Taylor G. 2003. The phase problem. *Acta Crystallogr D Biol Crystallogr* **59**:1881–1890.
114. Friedberg I, Jaroszewski L, Ye Y, Godzik A. 2004. The interplay of fold recognition and experimental structure determination in structural genomics. *Curr Opin Struct Biol* **14**:307–312.

115. Claude J-B, Suhre K, Notredame C, Claverie J-M, Abergel C. 2004. CaspR: a web server for automated molecular replacement using homology modeling. *Nucl Acids Res* **32**:W606–W609.
116. Giorgetti A, Raimondo D, Miele AE, Tramontano A. 2005. Evaluating the usefulness of protein structure models for molecular replacement. *Bioinformatics* **21**:72–76.
117. Qian B, Raman S, Das R, Bradley P, McCoy AJ, Read RJ, Baker D. 2007. High-resolution structure prediction and the crystallographic phase problem. *Nature (London)* **450**:259–264.
118. Topf M, Sali A. 2005. Combining electron microscopy and comparative protein structure modeling. *Curr Opin Struct Biol* **15**:578–585.
119. Zheng W, Doniach S. 2002. Protein structure prediction constrained by solution X-ray scattering data and structural homology identification. *J Mol Biol* **316**:173–187.
120. Chen SW, Pellequer JL. 2004. Identification of functionally important residues in proteins using comparative models. *Curr Med Chem* **11**:595–605.
121. Skrabanek L, Saini HK, Bader GD, Enright AJ. 2008. Computational prediction of protein–protein interactions. *Mol Biotechnol* **38**:1–17.
122. Hutchins C, Greer J. 1991. Comparative modeling of proteins in the design of novel renin inhibitors. *Crit Rev Biochem Mol Biol* **26**:77–127.
123. Hillisch A, Pineda LF, Hilgenfeld R. 2004. Utility of homology models in the drug discovery process. *Drug Discovery Today* **9**:659–669.
124. Rockey WM, Elcock AH. 2006. Structure selection for protein kinase docking and virtual screening: homology models or crystal structures? *Curr Protein Pept Sci* **7**:437–457.
125. Villoutreix BO, Renault N, Lagorce D, Sperandio O, Montes M, Miteva MA. 2007. Free resources to assist structure-based virtual ligand screening experiments. *Curr Protein Pept Sci* **8**:381–411.
126. Roessler CG, Hall BM, Anderson WJ, Ingram WM, Roberts SA, Montfort WR, Cordes MH. 2008. Transitive homology-guided structural studies lead to the discovery of Cro proteins with 40% sequence identity but different folds. *Proc Nat Acad Sci USA* **105**:2343–2348.
127. Bradley P, Misura KM, Baker D. 2005. Toward high-resolution de novo structure prediction for small proteins. *Science* **309**:1868–1871.
128. Bonneau R, Baker D. 2001. Ab initio protein structure prediction: progress and prospects. *Annu Rev Biophys Biomol Struct* **30**:173–189.
129. Hardin C, Pogorelov TV, Luthey-Schulten Z. 2002. Ab initio protein structure prediction. *Curr Opin Struct Biol* **12**:176–181.
130. Chivian D, Robertson T, Bonneau R, Baker D. 2003. Ab initio methods. *Methods Biochem Anal* **44**:547–557.
131. Jauch R, Yeo HC, Kolatkar PR, Clarke ND. 2007. Assessment of CASP7 structure predictions for template free targets. *Proteins: Struct Funct Genet* **69**(Suppl 8):57–67.
132. Dill KA, Ozkan SB, Welkl TR, Chodera JD, Voetz VA. 2007. The protein folding problem: when will it be solved? *Curr Opin Struct Biol* **17**:342–346.
133. Zhang Y. 2008. Progress and challenges in protein structure prediction. *Curr Opin Struct Biol* **18**:342–348.
134. Dill KA, Bromberg S, Yue KZ, Fiebig KM, Yee DP, Thomas PD, Chan HS. 1995. Principles of protein folding—a perspective from simple exact models. *Protein Sci* **4**:561–602.
135. Covell DG, Jernigan RL. 1990. Conformations of folded proteins in restricted space. *Biochemistry* **29**:3287–3294.
136. Park BH, Levitt M. 1995. The complexity and accuracy of discrete state models of protein structure. *J Mol Biol* **249**:493–507.
137. Lau KF, Dill K. 1989. A lattice statistical mechanics model of the conformational and sequence spaces of proteins. *Macromolecules* **22**:3986–3997.
138. Shakhnovich EI, Gutin AM. 1993. Engineering of stable and fast-folding sequences of model proteins. *Proc Natl Acad Sci USA* **90**:7195–7199.
139. Go N, Takemoti H. 1978. Resepctive roles of short- and long-range interactions in protein folding. *Proc Nat Acad Sci USA* **75**:559–563.
140. Miyazawa S, Jernigan RL. 1985. Estimation of effective interresidue contact energies from protein crystal structures: quasi-chemical approximation. *Macromolecules* **18**:534–552.
141. Chan HS, Dill K. 1989. Compact polymers. *Macromolecules* **22**:4559–4573.
142. Chan HS, Dill K. 1990. Origins of structure in globular proteins. *Proc Nat Acad Sci USA* **87**:6388–6392.

143. Karplus M, McCammon JA. 2002. Molecular dynamics simulations of biomolecules. *Nat Struct Biol* **9**:646–652.
144. Duan Y, Kollman PA. 1998. Pathways to a protein folding intermediate observed in a 1-microsecond simulation in aqueous solution. *Science* **282**:740–744.
145. Pitera JW, Swope W. 2003. Understanding folding and design: replica-exchange simulations of "Trp-cage" miniproteins. *Proc Nat Acad Sci USA* **100**:7587–7592.
146. Lei H, Wu C, Liu H, Duan Y. 2007. Folding free energy landscape of vllin headpiece subdomain from molecular dynamic simulations. *Proc Nat Acad Sci USA* **104**:4925–4930.
147. Zagrovic B, Snow CD, Shirts MR, Pande VS. 2002. Simulation of folding of a small alpha-helical protein in atomistic detail using worldwide-distributed computing. *J Mol Biol* **323**:927–937.
148. Chou PY, Fasman GD. 1974. Conformational parameters for amino-acids in helical, beta-sheet, and random coil regions calculated from proteins. *Biochemistry* **13**:211–222.
149. Garnier J, Osguthorpe D, Robson B. 1978. Analysis of the accuracy and implications of simple methods for predicting the secondary structure of globular proteins. *J Mol Biol* **120**:97–120.
150. Heringa J. 2000. Computational methods for protein secondary structure prediction using multiple sequence alignments. *Curr Protein Pept Sci* **1**:273–301.
151. Rost B. 2001. Review: protein secondary structure prediction continues to rise. *J Struct Biol* **134**:204–218.
152. Rost B, Eyrich VA. 2001. EVA: large-scale analysis of secondary structure prediction. *Proteins: Struct Funct Genet* **Suppl 5**:192–199.
153. Rost B, Sander C. 1993. Prediction of protein secondary structure at better than 70% accuracy. *J Mol Biol* **232**:584–599.
154. Montgomerie S, Sundararaj S, Gallin WJ, Wishart DS. 2006. Improving the accuracy of protein secondary structure prediction using structural alignment. *BMC Bioinformatics* **7**:301.
155. Fain B, Levitt M. 2001. A novel method for sampling alpha-helical protein backbones. *J Mol Biol* **305**:191–201.
156. Bradley P, Baker D. 2006. Improved beta-protein structure prediction by multilevel optimization of nonlocal strand pairings and local backbone conformation. *Proteins: Struct Funct Genet* **65**:922–929.
157. Wu GA, Coutsias EA, Dill KA. 2008. Iterative assembly of helical proteins by optimal hydrophobic packing. *Structure* **16**:1257–1266.
158. Orengo C, Bray J, Hubbard T, Lo Conte L, Sillitoe I. 1999. Analysis and assessment of ab initio three-dimensional prediction, secondary structure, and contacts prediction. *Proteins: Struct Funct Genet* **37**:149–170.
159. Ortiz AR, Kolinski A, Skolnick J. 1998. Native-like topology assembly of small proteins using predicted restraints in Monte Carlo folding simulations. *Proc Nat Acad Sci USA* **95**:1020–1025.
160. Rohl CA, Strauss CE, Misura KM, Baker D. 2004. Protein structure prediction using Rosetta. *Methods Enzymol* **383**:66–93.
161. Das R, Baker D. 2008. Macromolecular modeling with Rosetta. *Annu Rev Biochem* **77**:363–382.
162. Lazaridis T, Karplus M. 2000. Effective energy functions for protein structure prediction. *Currr Opin Struct Biol* **10**:139–145.
163. Huang ES, Samudrala R, Park BH. 2000. Scoring functions for ab initio protein structure prediction. *Methods Mol Biol* **143**:223–245.
164. Ngan S-C, Hung LH, Liu T, Samudrala R. 2008. Scoring functions for de novo protein structure prediction revisited. *Methods Mol Biol* **413**:243–281.
165. Roux B, Simonson T. 1999. Implicit solvent models. *Biophys Chem* **78**:1–20.
166. Koehl P. 2006. Electrostatics calculations: latest methodological advances. *Curr Opin Struct Biol* **16**:142–51.
167. Sippl M. 1990. Calculation of conformational ensembles from potentials of mean force: an approach to the knowledge-based prediction of local structures in globular proteins. *J Mol Biol* **1990**:859–883.
168. Sippl M. 1993. Boltzmann's principle, knowledge-based mean fields and protein folding: an approach to the computational determination of protein structures. *J Comput Aided Mol Des* **7**:473–501.
169. Samudrala R, Moult J. 1998. An all-atom distance dependent conditional probability discriminatory function for protein structure prediction. *J Mol Biol* **275**:895–916.
170. Moult J, Pedersen JT, Judson RS, Fidelis K. 1995. A large scale experiment to assess protein structure prediction methods. *Proteins: Struct Funct Genet* **23**:R2–R4.
171. Subbiah S, Laurents DV, Levitt M. 1993. Structural similarity of DNA-binding domains of bacteriophage repressors and the globin core. *Curr Biol* **3**:141–148.

2

MOLECULAR MODELING OF BIOMEMBRANES: A HOW-TO APPROACH

Allison N. Dickey* and Roland Faller
Department of Chemical Engineering and Materials Science, University of California Davis

2.1. INTRODUCTION TO MOLECULAR DYNAMICS

Computer Simulations have become an important complementary technique to experiment and analytical theory for scientific discoveries. Molecular Dynamics (MD) is one of the most abundant techniques of computer modeling, and is frequently used simulation methods in biomolecular applications. Its popularity may stem from its simplicity and versatile applicability. The fundamental underlying assumption of MD is that the system consists of particles that interact via the classical equations of motion, i.e., both quantum mechanical and relativistic effects are neglected. The exclusion of these effects, however, does not generally have a significant impact on the biomolecular questions being studied.

The simplest equation of motion is Newton's equation, which states that the force acting on a particle is the product of its mass and its acceleration:

$$\vec{F} = m\vec{a}\,. \tag{2.1}$$

Assuming furthermore that we have only conservative forces in our system, i.e., we neglect friction and any velocity dependent forces, we can write the force as the negative gradient of a potential function that now depends only on the particle positions

$$\vec{F}_i = -\nabla_i V(\{\vec{r}_j\})\,. \tag{2.2}$$

Address correspondence to Roland Faller, Department of Chemical Engineering and Materials Science, University of California Davis, Bainer Hall, One Shields Avenue, Davis, CA 95616, USA, 530 752-5839, 530 752-1031 (fax), <rfaller@ucdavis.edu>. *Current address: Chemical and Biological Engineering Department, Northwestern University, Evanston, IL 60208, USA.

T. Jue (ed.), *Biomedical Applications of Biophysics*,
Handbook of Modern Biophysics 3, DOI 10.1007/978-1-60327-233-9_2,
© Springer Science+Business Media, LLC 2010

The gradient here is taken with respect to the position of particle *i*. Since the acceleration is the second time derivative of particle position, a combination of Eqs. (2.1) and (2.2) leads to a second-order partial differential equation:

$$m\partial_t^2 \vec{r}_i = -\partial_{r_i} V(\{\vec{r}_j\}). \tag{2.3}$$

Even though Newton's equations are not linked to statistical or thermodynamical properties and cannot be extended to include quantum mechanical variables like Hamilton's equations, for the systems that we will discuss here Hamilton's equations reduce to the same Eq. (2.3), and so we will not go through the more complex derivation. Through the use of an integrator, MD solves the second-order PDE in Eq. (2.3) iteratively, where the positions and velocities (or momenta) of all the particles at one time point are used as the initial conditions.

An integrator is an algorithm that solves the PDE by iterative integration, and there are several varieties commonly used in MD. Here we will discuss the Verlet integrator [1] as a prototypical example. The Verlet algorithm has several advantages: it is time inversion symmetric as well as symplectic. These characteristics are required for the correct statistical mechanical behavior of the ensuing ensemble, and the interested reader is referred to several excellent books for a more detailed discussion of MD [2–4].

Most integrators are based on a Taylor expansion of the positions in time:

$$\vec{r}(t) = \sum_{i=0}^{\infty} \partial_t^i \vec{r}_{t=t_0} \frac{(t-t_0)^i}{i!}. \tag{2.4}$$

For the Verlet integrator, this expansion is performed to the third order, and the first derivative of position is velocity and the second derivative of position is acceleration. For symmetry reasons, the expansion is performed in positive and negative time. If $\Delta t = t - t_0$, the forward and backward expansions are

$$\vec{r}(t = t_0 + \Delta t) = \vec{r}(t_0) + \vec{v}(t_0)\Delta t + \frac{1}{2}\vec{a}(t_0)\Delta t^2 + \frac{1}{6}\partial_t^3 \vec{r}(t_0)\Delta t^3 + O(\Delta t^4), \tag{2.5a}$$

$$\vec{r}(t = t_0 - \Delta t) = \vec{r}(t_0) - \vec{v}(t_0)\Delta t + \frac{1}{2}\vec{a}(t_0)\Delta t^2 - \frac{1}{6}\partial_t^3 \vec{r}(t_0)\Delta t^3 + O(\Delta t^4). \tag{2.5b}$$

If the forward and the backward expansions are combined, the position Verlet algorithm (Eq. (2.6)) is obtained, and it is exact to the 4th order in the timestep Δt. The position Verlet algorithm predicts the atomic positions at the next timestep given that the atomic positions at the current and previous timesteps are known:

$$\vec{r}(t_0 + \Delta t) = 2\vec{r}(t_0) - \vec{r}(t_0 - \Delta t) + \vec{a}(t_0)\Delta t^2 + O(\Delta t^4). \tag{2.6}$$

Here the velocity is not explicitly represented but rather computed as a difference between positions.

2.2. SPECIFICS OF THE MOLECULAR MODELING OF BIOMEMBRANES

We now focus on biomembrane simulation specifics, where biological systems that span a variety of length scales have been commonly studied using MD over the last two decades [5–13]. Even so, simulation details must still be carefully selected, as it has been found that unfortunate choices in conditions or parameters can lead to unsuccessful simulations [14–16].

In an all-atom simulation, every atom is represented in the model by one interaction center. All atom simulations are very abundant and give the highest possible degree of detail (except for quantum chemical calculations). However, one downfall to a simulation having this degree of accuracy is that they are limited in both the length of simulation time as well as the system size. In biomembrane simulations, this is especially apparent for fully hydrated phospholipid bilayers. Even though the system of interest is the lipid bilayer, a considerable amount of computational effort is spent on simulating the water molecules, which are needed to accurately represent a cellular bilayer. In order to minimize the computational resources that are spent on water molecules, several approaches, such as solvent free bilayer models that use special interactions to mimic water [11,17] and coarse-grained models [18–20], have been devised that allow for the use of larger timescales. Most biomembrane simulations have been performed on free-standing bilayers in water, although a few freestanding monolayers [21–23] and supported bilayers [24,25] have been examined as well.

Biomembranes are inherently anisotropic, and this property needs to be reproduced in the model. Most simulations are performed with periodic boundary conditions, i.e., the simulation box is surrounded by replicas of itself in all directions, which leads to an infinitely extended periodic system. In order to maintain a constant pressure, the simulation volume size needs to fluctuate since pressure and volume are conjugate thermodynamic variables. However, because Newton's equations of motion conserve volume, the pressure has to be controlled via a barostat. Similar in spirit to barostats are thermostats, which modify the system temperature, and further discussion of temperature-controlling techniques can be found elsewhere. A commonly used barostat is the Berendsen, or weak-coupling, barostat [26]. This technique compares an instantaneous pressure with a predefined target pressure, and if the values differ (and they generally do), the box volume and all particle positions are rescaled according to

$$\frac{V_{\mathrm{new}}}{V_{\mathrm{old}}} = \frac{\Delta t}{\tau}\left(1 + \frac{p}{p_{\mathrm{target}}}\right). \tag{2.7}$$

The parameter τ represents the correlation time and should be chosen judiciously. A large correlation time will lead to weaker coupling than a short correlation time. Typical values are on the order of a few ps, with 1 fs being a common choice of timestep for atomistic simulations. The timestep has to be an order of magnitude shorter than the smallest characteristic time found in the system, which typically corresponds to bond or angle vibrations (on the order of 10 fs). If both a thermostat and a barostat are used, their respective correlation times should differ by an order of magnitude as well, with the pressure correlation time having a larger value. If the barostat results in box volume fluctuations, the thickness and area of a membrane become coupled. To avoid this, we independently couple the three axes to the external pressure to reproduce a tension-free bilayer. Of course, the axes (normal and lateral to the membrane) can be coupled separately to different values, which would produce an overall surface tension. For this situation we have to measure the instantaneous pressure tensor rather than the isotropic pressure.

The overall charge of a simulated system has to be zero. However, biomembrane simulations usually contain charged atoms, and electrostatic interactions differ from most other modeled interactions in that they are long ranged. A long-ranged interaction means that the integral of the potential (which we for simplicity assume to be spherically symmetric) over all space diverges:

$$\int_0^\infty V(r) r^2 dr = \infty . \tag{2.8}$$

In contrast to this, the integral for the Lennard-Jones and other short-range interactions converges. A problem with long-ranged interactions stems now from periodic boundary conditions. We cannot neglect charges outside the box. A number of ways to address this have been developed. One of the most successful is the Particle Mesh Ewald [27] technique. The main idea behind this method is that every charged interaction below a certain cutoff is calculated directly. For interactions that do not fall within this cutoff, the interactions are calculated in Fourier space. For the exact implementation we again refer to specialized literature. A second method is the reaction field technique. In this case, a charge interacts with all neighboring charges that fall within a cutoff radius at full strength. However, the charge will only feel an effective dielectric medium for electrostatic interactions whose distance separations exceed the cutoff radius. This may not be the most accurate method, however, since the variety of atom types present in biomembrane simulations leads to a wide range of dielectric values. For example, water has a dielectric constant of ~80 and the inner part of a membrane around 2-4. Hence, it is generally advisable for atomistic simulations to use PME for electrostatic calculations.

2.3. FORCE FIELDS: SIMULATION MODELS

Here we discuss some frequently used lipid bilayer models. Many simulations that retain atomistic level detail consist of 128 fully hydrated lipids, with 64 lipids per leaflet. A few studies also examine lipid bilayer structural changes that accompany a reduced level of hydration [28,29]. DPPC (dipalmitoylphosphatidylcholine) is the most abundantly studied lipid [30], and lipids that differ in the saturation or number of carbons in the acyl chains have also been well studied [30–32]. One of the most commonly used phosphatidylcholine (PC) force fields was developed by Berger et al. [33]. Force fields for non-PC lipids, such as phosphatidylserine [34], sphingomyelin [35,36], phosphatidylglycerol [31,32,37], phosphatidylethanolamine [37,38], dimyristoyltrimethylammonium propane [39], and phosphatidic acid [32], have also been published. Sterols, mainly cholesterol [36,40–42] and ergosterol [42,43] but also lanosterol [42], have been modeled. Several websites offer downloadable lipid configurations and topologies to users, and these models contain intra- and intermolecular interactions. The intramolecular bond and angle terms are typically modeled via a harmonic potential. Bonds may also be constrained using algorithms such as LINCS [44] and SHAKE [45]. The torsional degrees of freedom are most often represented by a Fourier series in order to satisfy the required 360° symmetry. Torsions are the most important part of the intramolecular potential, and they are often based on quantum chemical calculations. In some cases, such as for double bonds, special potentials, such as harmonic dihedrals, are used to help avoid unphysical local conformers. The intramolecular interactions serve a twofold purpose, where they define both the molecule geometry as well as energy differences between different local conformations.

The non-bonded Lennard-Jones interactions are modeled for atom pairs that belong to different molecules as well as for atom pairs that belong to the same molecule but are located a few bonds apart (e.g., at least 3). The Lennard-Jones potential contains two parts. A long-range attractive r^{-6} term comes from the fluctuating dipole London interactions and a short-range repulsive r^{-12} term models the Pauli repulsion. As there is no analytical form for the repulsive segment, the potential is chosen for computational convenience as r^{-12} is just the square of r^{-6} and computing the square is computationally cheap.

We would like to include here a brief discussion on the use of water models. Water is not only important in biomembrane simulations because it is an important constituent of cellular environments, but it is also a unique and interesting compound that has chemical and physical characteristics that lead to complex phase behavior. Most models are adjusted to represent liquid water at ambient conditions and do not necessarily reproduce accurate descriptions of the solid phases. This can be problematic in studies involving low temperatures where freezing becomes important. There is therefore no single "best" water model. The most commonly used water model that is employed in conjunction with standard lipid force fields is the SPC model [46,47]. In this model only the oxygen has a Lennard-Jones interaction site and the two hydrogen atoms only serve as charge sites. In general, one has to be careful to choose a water model that "matches" the chosen lipid model.

2.4. DEGREE OF DETAIL: ATOMISTIC VERSUS COARSE-GRAINED

A number of computational techniques and models have been developed to study a variety of systems of interest. Atomistic models accurately describe not only the molecular structure but also the chemical bonding, electrostatic, and van der Waals interactions. Because of the interaction scale, atomistic models employ a time step that can be as short as a tenth of the period of the fastest mode in the system [48], which often corresponds to a covalent bond or angle stretch. Atomistic models include at least every non-hydrogen atom into the system, and with the short time step atomistic simulations can model membranes that are a few tens of nanometers in size. These simulations often consist of tens of nanoseconds of data and the simulation jobs can be submitted to large-scale computing facilities. Atomistic simulations are widely used in the study of the local structure and dynamics of membranes [49–51]. They can also be used to determine how a particular component, such as sterols, affects membrane structure [52,53]. In most atomistic simulations, nonpolar hydrogens are neglected, i.e., they are subsumed into neighboring heavy atoms, which leads to a united atom description.

Since events such as self-assembly, phase transitions and phase separations occur on length and timescales beyond atomistic capabilities, coarse-grained (CG) or mesoscale models have to be applied to reach the relevant size and time periods. Reducing the degrees of freedom by combining several atoms into one effective particle and eliminating short-range dynamics are two techniques included in CG models that speed up simulations and allow access to collective phenomena [54–60]. One widely used coarse-grained model for lipids is the MARTINI model [18,61], and we refer the reader to the original literature for the exact interaction parameters between the lipids and the water molecules. We mention here only that 4–6 heavy atoms are represented by one interaction sites and that the coarse-grained parameters are chosen to reproduce important properties of lipid membranes. This model has been effectively used in the study of phase behavior [8,62] and supported membranes [25]. Vesicle fusion and formation [63], as well as hexagonal phase formation [59], have been captured at the molecular level and the phase behavior of several lipid mixtures has been semiquantitatively reproduced [58,64,65]. It has been noted that the increase in dynamics in the CG models is due to the CG molecules being "smoother" than the atomistic molecules, i.e., they exhibit less friction. In the Martini model, the dynamics are a factor of 4 faster than those of atomistic simulations and experiments, where the speed-up was determined through diffusion coefficient comparisons [18]. The phase behavior and pressure–area isotherms are reproduced semiquantitatively, i.e., within 20–30 K [66], and the model uses a reaction field for electrostatics with a dielectric constant of 20. The CG model allows the use of a time step of 40 fs and one CG-water represents 4 real waters.

Less detailed models than the Martini model exist, where water is not explicitly taken into account [11,17,67,68]. The motivation for using such models is to avoid the use of significant amounts of computational time on simulating water molecules. When examining phenomena that occur on large time and length scales, the behavior of the water molecules is usually not a primary interest of the study. To obtain the self-assembly of lipids into a fluid bilayer, the normal Lennard-Jones interaction model needs to be modified, and as an example we briefly discuss a model proposed by Cooke et. al. [11,69,70], where interactions between tail beads are represented as

$$V_{\mathrm{att}}(r) = -\varepsilon, \quad r < r_c, \tag{2.9a}$$

$$V_{\mathrm{att}}(r) = -\varepsilon \cos^2 \frac{\pi(r - r_c)}{2w_c}, \quad r_c < r < r_c + w_c, \tag{2.9b}$$

$$V_{\mathrm{att}}(r) = 0, \quad r > r_c + w_c. \tag{2.9c}$$

The interactions between the head beads as well as between the head and tail beads follows a purely repulsive version of Lennard-Jones that we obtain by cutting off the potential in its minimum and shifting it to zero at that point (WCA potential) [71].

The model uses a tunable long-range weakly varying attractive potential that reproduces a fluid bilayer with properties that are commensurate with experimentally measured values [11].

Further coarse-graining leads to the realm of two-dimensional models, where, for example, one leaflet of a membrane is modeled using hard disks with the sole parameter being the excluded volume (or, more precisely, area). This type of model has been used to study a dipalmitoyl phosphatidylethanolamine (DPPE) and ganglioside GM_1 mixture [72]. A circular excluded area of 45 Å^2 was used for DPPE and an area of 65 Å^2 was used for GM_1 and these values were based on experimental pressure–area isotherms for each lipid [73]. GM_1 however, in low to intermediate density mixtures with DPPE does not strongly change the overall area per molecule. Hence, a minimum packing area of 40 Å^2 per molecule was used for GM_1 in conjunction with DPPE molecules, which leads to the peculiar situation of a binary hard-disk fluid having a cross-interaction radius that is not the average of the self-interaction radii.

2.5. VISUALIZATIONS

It is always a good idea to visualize a simulation system, from either output that is generated during a run or once the simulation has completed. Most simulation packages can write atom coordinates into a protein databank file that can be easily visualized with a number of software packages [74,75]. Visualizing a simulation system, especially in the initial equilibration phase, can serve as a check that the membrane configuration remains intact. The images also provide a convenient method for closely examining the interactions between molecules in a particular region of the system. Figure 2.1 shows images from different regions of a POPA (palmitoyl oleyl phosphatidic acid) lipid bilayer [32].

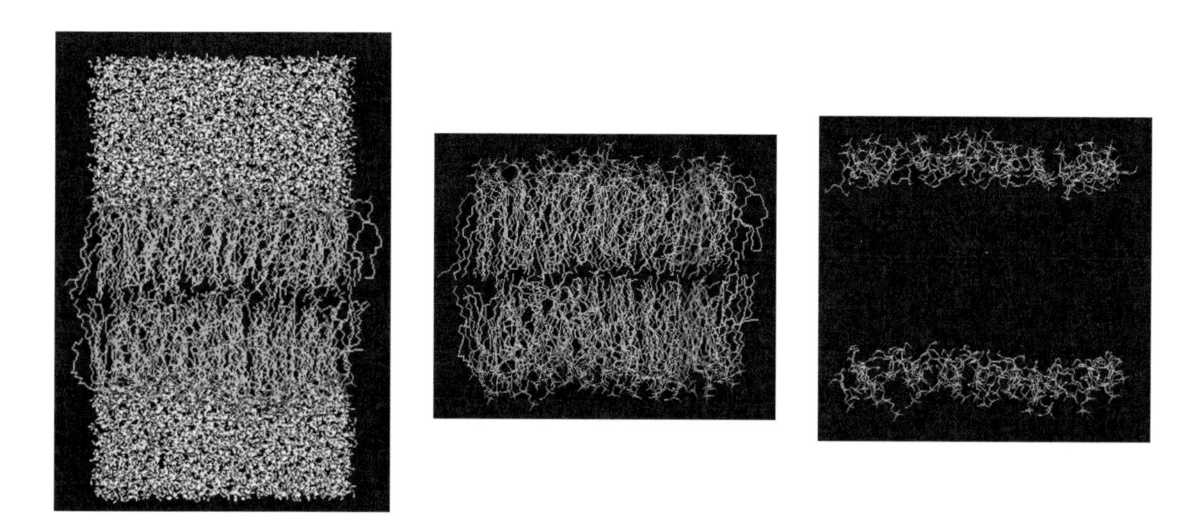

Figure 2.1. Visualization of a POPA lipid bilayer [32]. Left: The system contains 128 lipids with 5443 water molecules. Middle: The POPA lipids without the water molecules. Right: Only the POPA headgroups are shown. Please visit http://extras.springer.com/ to view a high-resolution full-color version of this illustration.

2.6. AREA PER MOLECULE AND THICKNESS, COMPARISON TO X-RAY DATA

In homogeneous simulations, the simplest thermodynamic property to calculate is the overall density since we know both the mass and volume of the simulation box for all time steps. The area per molecule and bilayer thickness are two often-calculated membrane quantities, and their values can be extracted from the simulation box volume. In our simulations the area per molecule is determined by dividing the product of the x and y dimensions of a simulation cell by the number of lipids per leaflet. One can also determine the area per molecule by measuring the membrane thickness under the assumption of constant volume per lipid [76].

To determine the membrane thickness, the density profile along the bilayer normal (usually the z-axis), must first be calculated. The density profile can be tabulated by dividing the bilayer into equidistant slabs along the bilayer normal. Each atom is assigned to a slab, and this ensures that we know again both the mass and the volume of each slab. Density profiles are useful in determining how the locations of different groups (water molecules, lipid headgroups, etc.) vary along the bilayer normal. Because of the short simulation timescales, lipid flip-flop is a rarely witnessed event [77] and hence the density profiles for the top and bottom leaflets of a bilayer can be calculated separately. By looking at the individual leaflet density profiles, one can examine interdigitation, which occurs when lipids from separate leaflets intertwine. This phenomenon has been studied extensively in the context of membrane–alcohol interactions [43,78,79].

The electron density profile can also be determined, and it is a useful simulation measurement because it can be related to X-ray scattering or X-ray reflectivity data since the X-ray profile is essentially a one-dimensional Fourier transform of the electron density [80]. An example of an electron density profile for a POPA lipid bilayer is shown in Figure 2.2. To compare simulation data with that of neutron scattering, an atomic scattering length for all atoms needs to be known so that a scattering length density profile can be assembled.

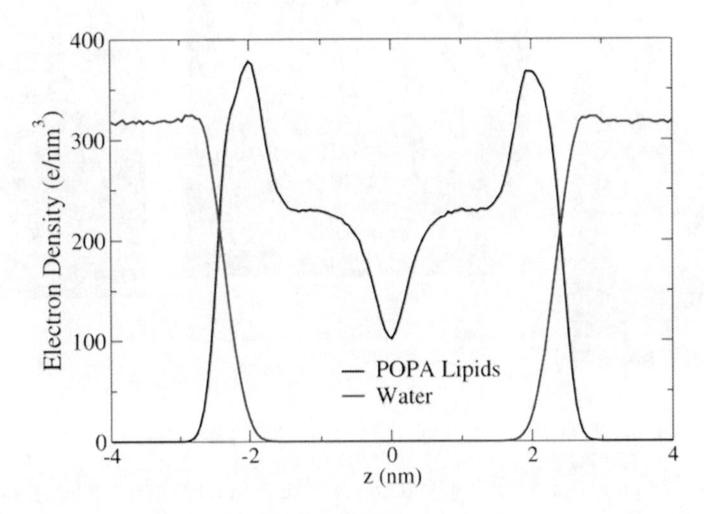

Figure 2.2. Electron density profiles of the POPA lipid bilayer from Figure 2.1. The bilayer center corresponds to a value of z = 0 nm. Please visit http://extras.springer.com/ to view a high-resolution full-color version of this illustration.

2.7. ORDER PARAMETERS AND OTHER SINGLE-LIPID PROPERTIES

Another experimentally relevant parameter is the chain order parameter. This parameter connects the degree of chain ordering to the bilayer normal and, in contrast to the properties previously discussed, the order parameters can be defined for a single lipid. The order parameter is a useful measurement because it can be compared with the experimental deuterium order parameter, which can be determined using nuclear magnetic resonance spectroscopy. Since most atomistic simulations use a united atom representation to model the hydrocarbon chains (even less description is included in coarse-grained systems), hydrogen atoms are not explicitly represented and the C–H bonds have to be reconstructed assuming a tetrahedral geometry of the CH_2 groups. The order parameter is defined as

$$S_{CD} = 0.5\langle 3\cos^2\Theta_{CD} - 1\rangle, \tag{2.10}$$

where Θ_{CD} is the angle between the CD-bond and the bilayer normal in experiments, and in simulations the CD-bond is replaced by the CH-bond. This quantity cannot be measured directly in experiments; however, there is a recurrent formulation that allows the calculation of S_{CC} from S_{CD} order parameters. For the C_n groups, the deuterium order parameter for the nth carbon in computer simulations can be calculated by

$$-S^n_{CD} = \frac{2}{3}S^n_{xx} + \frac{1}{3}S^n_{yy}.$$

Here $S_{JJ} = \langle \cos\theta_j \cos\theta_j - \delta_{jj}\rangle$ and ($j = x, y, z$) with $\cos\theta_j = \hat{u}_j \hat{u}_z$, $\hat{u}_j$ is the unit vector for the jth molecular axis in the bilayer, and $\hat{u}_z$ is the unit vector in the z direction (average bilayer normal). The order parameters are normally defined for all saturated carbons that have two neighboring carbon atoms. For DPPC the order parameters can therefore be calculated for atoms C_2 through C_{15} (see Fig. 2.3 for numbering). The order parameters for the two hydrocarbon chains are normally analyzed separately, even for DPPC, whose tails each have 16 carbon atoms, because the distance between the two hydrocarbon chains and the water/bilayer interface are not equivalent, For saturated bonds the order parameter is a measure of the spatial restriction of the motion of the C–H vector [81] and is proportional to the deuterium quadrupolar splittings [82] in NMR measurements. For unsaturated bonds (i.e., a double bond) the chain kinks and, although the order parameter can in principle be calculated (nothing in the formula is undefined), it is not usually performed because the local geometry of the atom is not tetrahedral anymore and the values are typically significantly lower than for saturated carbons.

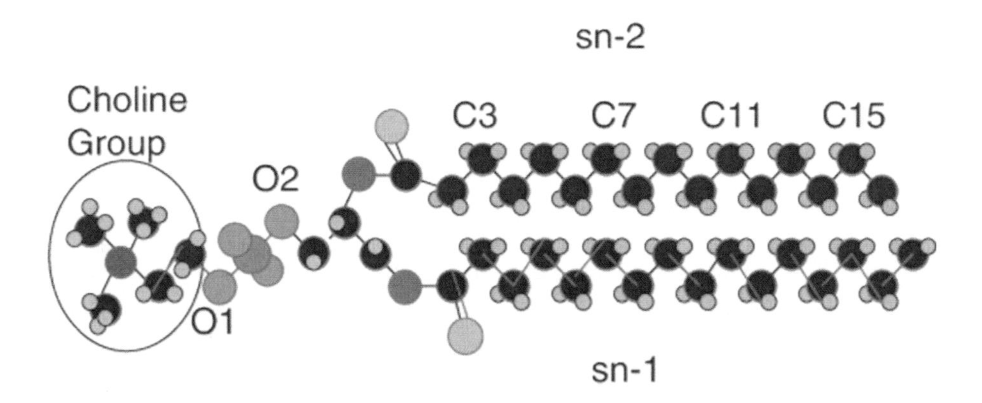

Figure 2.3. An example of how the lipid acyl chains are numbered in calculating the order parameter. The POPA lipid shown here differs from the more commonly studied DPPC lipid in that the DPPC choline group is replaced with a hydrogen atom and the sn-2 tail contains a double bond and 18 atoms. Please visit http://extras.springer.com/ to view a high-resolution full-color version of this illustration.

There are a number of other properties that describe the intra- and intermolecular conformations of individual lipids. One property is the tilt of the molecule or of the headgroup. To calculate the tilt of the overall molecule, one can define a unit vector that spans from the glycerol group to a specified atom in one of the hydrocarbon chains. To calculate the tilt of the headgroup one can define a unit vector that spans from the choline group to the phosphate group. We can calculate the tilt of the headgroup by measuring the angle of this vector with that of the bilayer normal. The scalar product of the unit vectors (the bilayer normal and the vector of interest) is the cosine of the tilt angle [83]. Another simple way to characterize the structure of a lipid is to calculate its end-to-end distance, i.e., the distance from the headgroup to the end of the tails. This gives a crude estimation of the overall order and phase. For example, the gel phase lipids are normally more ordered than liquid phase lipids and hence have a longer length.

2.8. RADIAL DISTRIBUTION FUNCTIONS

One parameter that can be used to characterize the structure of a molecular modeling system in detail is a radial distribution function (RDF). An RDF provides additional information about the membrane morphology and the structure that is complementary to the density profiles. The RDF ($g_{AB}(r)$) between particles of type A and B is defined as

$$g_{AB} = \frac{\langle \rho_B(r) \rangle}{\langle \rho_B \rangle_{\text{local}}} = \frac{1}{\langle \rho_B \rangle_{\text{local}}} = \frac{1}{N_A \sum_{i,j} N_A N_B} \frac{\delta(r_{i,j} - r)}{4\pi r^2}, \tag{2.11}$$

with $\langle \rho_B(r) \rangle$ being the particle density of type B at a distance r around particle A, and $\langle \rho_B \rangle_{\text{local}}$ the particle density of type B averaged over all spheres around particles A with radius r. Radial distribution functions have two appealing properties. First, they quantitatively describe how many neighbors of a certain type are found around a given atom, and thus they characterize the local neighborhood. Second, the RDFs are a three-dimensional Fourier transform of the static structure factor, which can be measured using either X-rays or neutrons.

For membranes we often only calculate a two-dimensional RDF in order to characterize the in-plane lipid neighborhood. An illustration of an RDF for the POPA example system is shown in Figure 2.4.

2.9. HYDROGEN BONDING AND ADVANCED STATIC ANALYSIS

Since we know the positions and the momenta of all particles at all times in an MD simulation, we have access to all system information. This becomes useful when we want to calculate properties that cannot (or only very indirectly) be determined using experiments. One such example is a study of the hydrogen bonds that exist within a system. A caveat here is that the analysis and the properties are only as good as the model. Hence, for hydrogen bonding studies in atomistic simulations, this means that the hydrogen bond definitions are based on geometry rather than explicit bonding between hydrogen atoms and neighboring polar atoms since that would require the use of quantum degrees of freedom. A typical criterion that we use for hydrogen

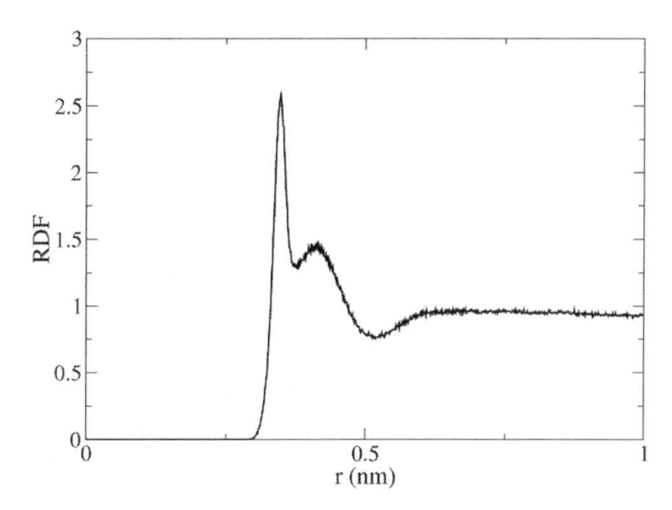

Figure 2.4. The RDF between the water oxygen atoms and the phosphate atom in a POPA lipid bilayer. The figure shows that there are two particularly favorable positions for the water molecules about the phosphate atom. The first location is represented by the sharp peak, which occurs when the water oxygen atom and the phosphate atom are located 0.35 nm apart. The second favorable location occurs at a distance of 0.42 nm and this peak has a smaller g_{AB} value and is much broader than the first peak, indicating that the interaction between the water and the phosphate atom at $r = 0.42$ nm is not as strong as at $r = 0.35$ nm.

bond existence is that the distance between the hydrogen atom and the hydrogen bond acceptor be less than 3.5 Å and the angle between the hydrogen atom, hydrogen bond donor, and hydrogen bond acceptor be less than 30° [84]. In order to determine the hydrogen bond donor and acceptor pairs, one can assume that an OH or NH group is a good donor and a bare oxygen or nitrogen atom is a good acceptor [77]. Thus, in phospholipid membranes common examples of hydrogen bond acceptors are the oxygen atoms in the phosphate group or the ester groups. To determine the lifetime of a hydrogen bond, one can define a function A that is equal to 1 if a hydrogen bond (in one defined pair) exists and 0 if it does not exist. Based on these values, the hydrogen bond lifetime can be calculated using the function

$$C(t) = \frac{\langle A(t)A(0)\rangle}{\langle A(t)A(t)\rangle}, \tag{2.12}$$

as defined by Luzar and Chandler [85]. Even if a bond does not exist continuously between time 0 and time t, the bond will still be included in the correlation function for the time periods where the bond does exist. In order to calculate the correlation time, we integrate this correlation function:

$$\tau = \int_0^\infty C(t)\,dt\,. \tag{2.13}$$

2.10. PRESSURE AND PRESSURE PROFILES

An important thermodynamic property is pressure, which, as discussed earlier, can be kept constant in the three Cartesian directions through the use of a barostat. In an anisotropic medium pressure is not a scalar but a second-rank tensor that is defined in terms of forces, velocities, and positions. When only pair forces exist, the overall pressure is defined as

$$\underline{\underline{P}} = \sum_{\text{pairs}} \vec{F} \otimes \vec{r} + \frac{1}{2} m\vec{v} \otimes \vec{v}. \quad (2.14)$$

If we cannot define all forces based on pair interactions, then we have to be careful about how the origin of the coordinate system is defined when using periodic boundary conditions as the pressure can depend explicitly on this choice. For many applications, the overall pressure calculation, even as a tensor, is not accurate enough, and hence a more localized pressure calculation via formulations such as the Irving-Kirkwood equation are necessary [86]. We do not discuss these methods here but note that one can obtain pressure as a function of position or more often as a function of the position along only the bilayer normal. This is mainly for statistical reasons since pressure can have significantly fluctuating values and good statistics are needed to obtain reliable data. The surface tension can be incorporated into the coupling scheme [77,87] as

$$\int_{Z_1}^{Z_2} \frac{P_T(Z)dZ}{Z_2 - Z_1} = p_{\text{ref}} - \frac{\gamma}{Z_2 - Z_1}, \quad (2.15)$$

with p_{ref} being the chosen reference pressure. An example of a lateral pressure profile is shown in Figure 2.8 for a phosphatidylglycerol system [32].

2.11. TWO-DIMENSIONAL DIFFUSION

Previously, we discussed static and thermodynamic properties of membrane lipids. Since MD gives the user access to particle momenta, we can monitor dynamical properties as well. The lipid mobility is generally examined through the lateral diffusion coefficient, which can be calculated from the slope of the mean-square displacement (MSD) via the Einstein equation:

$$D = \lim_{t\to\infty} \frac{\langle (r(t) - r(0))^2 \rangle}{2dt}, \quad (2.16)$$

where d is the dimensionality of the system (i.e., $d = 2$ for lateral diffusion in the membrane plane) and $r(t)$ and $r(0)$ are the coordinates of the lipid molecules at times t and 0. An average is calculated for all particles of interest, and over time as a "running-time average" such that an interval of length Δt can be realized from $t = 0$ to $t = \Delta t$, or from $t = t_0$ to $t = t_0 + \Delta t$. Hence, every interval that corresponds to a particular Δt is included in the average. Therefore, after equilibration, all time points are equivalent as reference points. If we have 1000 time steps, this leads to 999 datapoints per particle for a $\Delta t = 1$ versus the existence of only 1 datapoint having a Δt of 999. Thus, the accuracy of the MSD decreases with increasing Δt. As the diffusion coefficient should be calculated from the long time limit of the MSD slope, the accuracy of datapoints from regions corresponding to long Δt values may be questionable. For very short Δt values, the MSD will increase with the square of time according to the equations of motion $\Delta r \propto \Delta t^2$. Sometimes at intermediate Δt values the dynamical regimes may be subdiffusive, resulting in only long Δt values for calculating diffusive behavior. However, as mentioned above, at long Δt values the MSD may deviate from linearity due to the deteriorated statistical quality. MSDs are regularly plotted double logarithmically in order to easily distinguish the different dynamical regimes; any dynamic regime with algebraic time dependence $\Delta r \propto \Delta t^{\alpha}$ will then appear as a line with the exponent α equal to the slope. In the case of lipid flip-flop, undulations, or protrusions, one-dimensional MSDs along the bilayer normal are mainly of interest, whereas for water the 3D MSD is usually generated. Figure 2.5 shows how a POPA two-dimensional diffusional coefficient can be derived from the MSD.

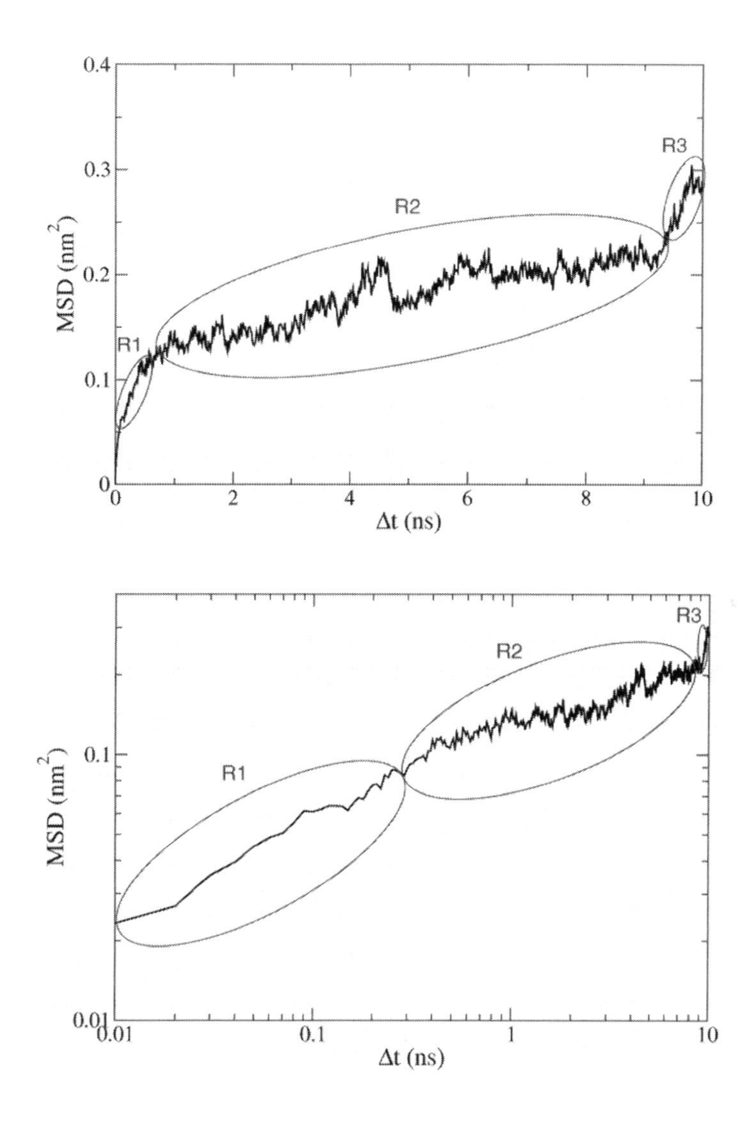

Figure 2.5. A) The lateral mean-squared displacement MSD (in x and y) for POPA. The curve is quadratic for region 1 (R1) since there are no collisions between molecules for small Δt values. The diffusion coefficient can be calculated from the slope of the curve in region 2 (R2) using equation (2.16). The data from region 3 (R3) is not used in calculating the diffusion coefficient because of poor statistics. B) The same data in a log-log plot of the MSD.

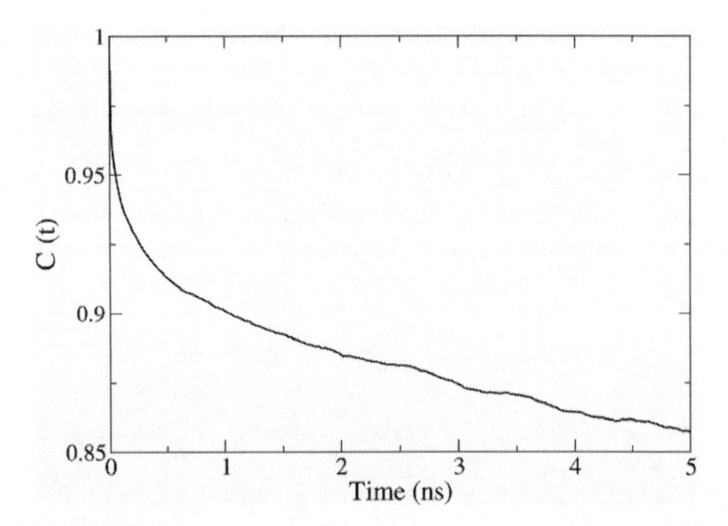

Figure 2.6. An example of a rotational correlation function for the POPA headgroup. Because POPA has such a small headgroup, the vector for this figure spanned only across the phosphate group, from oxygen atom O1 to atom O2 (see Fig. 2.3). For statistical reasons, the correlation function is only calculated for the first 5 ns of a 10 ns trajectory and we can see from the figure that a significantly longer simulation time is required for a complete POPA headgroup relaxation.

2.12. REORIENTATIONS AND NMR

The reorientation of a molecule (e.g., water) and the lipid rotational correlation are also dynamical properties that can be obtained from MD simulations. These parameters require that we either define a plane or a unit vector with which to calculate the reorientation. Since a plane is represented by its normal vector, here we discuss only vector reorientation. The rotational correlation function is calculated using the autocorrelation function for the unit vector **V**. Figure 2.6 shows an example of a function, the reorientation of the vector

$$C(t) = \langle V(t)V(0) \rangle . \tag{2.17}$$

Other definitions based on higher polynomials can be used as well. The rotational relaxation time τ can again be calculated from the integral of the autocorrelation function (see Eq. (2.13)). This correlation time is related to the T_1 time in NMR experiments [88] as

$$\frac{1}{T_1} = \frac{\hbar^2 \gamma_C^2 \gamma_H^2}{10 r_{\mathrm{CH}}^2} [J(\omega_H - \omega_C) + 3J(\omega_C) + 6J(\omega_H + \omega_C)] . \tag{2.18}$$

We restrict ourselves here to CH vectors, but others are equivalent; γ are gyromagnetic ratios of the respective nuclei and ω are the Larmor frequencies. while r is the distance between the nuclei and the J functions are the Fourier transforms of the correlation functions:

$$J(\omega) = \int_{-\infty}^{\infty} C(t) e^{i\omega t} dt . \tag{2.19}$$

We can e.g., use a vector that connects the two tails that measures the in-plane rotation of the lipids since this vector is roughly perpendicular to the long axis of the lipid. Since this vector is nearly perpendicular to the bilayer normal, it decays at long times to a value close to zero. It cannot de-correlate completely since we do not find transbilayer flip-flop on timescales reachable in atomistic simulations. Thus, the correlation function gets stuck at a small nonzero value. The reorientation of the lipid headgroups can be determined by defining a vector that for example spans from the phosphate atom to an atom in the amine or choline groups. This rotation is typically much faster [8] than the rotation of the lipid tails. The rotational correlation vectors of the POPA lipid headgroups are shown in Figure 2.6.

2.13. DYNAMICS OF INDIVIDUAL MOLECULES: CORRELATIONS OF DISTRIBUTION FUNCTIONS

One additional benefit of computer simulations is that the static and dynamical properties of individual molecules can be captured. This selectivity cannot be matched using experiments where typically only average distributions are determined. In studying individual dynamical properties, we essentially do exactly the same calculations that we would perform for an average property, but we restrict ourselves to a small subset. The extreme case of this is a subset with only one member. Another possibility is that we compare spatially or temporally differentiated regions of the bilayer. Far from a phase transition, one typically finds that the average molecule behavior is a reasonable approximation for the individual molecule behavior, i.e., the molecules behave in an essentially identical manner. However, heterogeneities in the system that stem from domain or density differences may be misleading when interpreting individual molecule dynamics, especially when the distribution of lipid behaviors is non-Gaussian and there are qualitative differences between classes. An example is shown in Figure 2.7, where we calculate the rotational correlation function for a simulation of POPA lipids around a protein [89,90]. Each curve displays a set of lipids that is located a different distance from the protein. This is an example of how different observables can be correlated, where the POPA lipid headgroup rotational correlation function value is dependent upon a two-dimensional RDF. Another example could involve classifying lipids into groups of more and less highly ordered lipids based on the individual order parameter values. This lipid separation would then allow additional dynamical properties to be analyzed for each class separately [91]. One word of caution here is that if we investigate the correlation function or the mean-squared displacements of individual lipids, the statistics are very weak and noisy. Thus, an integration of an orientation correlation function to determine a reorientation time or the differentiation of a mean-squared displacement to determine a diffusion coefficient is unreliable.

2.14. INTERACTIONS WITH SMALL MOLECULES

Here, as an example application, we discuss how small molecules modify the behavior of a phospholipid bilayer. Understanding how membranes interact with small molecules is of tremendous biological importance as the cell membrane serves as a barrier between the extracellular environment and the intracellular contents. Therefore, a number of molecular simulations

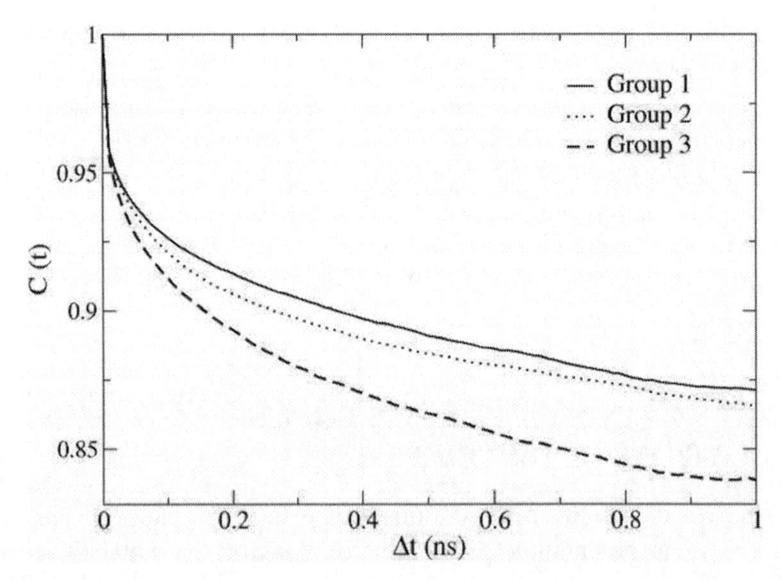

Figure 2.7. This figure shows the rotational correlation function of the POPA lipid head group (vector O7 – O10) for three groups of lipids, where the lipids are sorted into groups based on their lateral root mean squared distance (RMSD) from a transmembrane protein. Group RMSD specifications: Group 1 (<1.0 nm), Group 2 (1.0 nm ≤ 2.0 nm) , and Group 3 (>2.0 nm) [89].

have been performed on systems that contain lipid bilayers and small molecules [7,9,43,77,78, 92,93]. Among the small molecules that have been studied are sugars and alcohols. Sugar molecules are nutrients for living organisms; under the proper conditions they can serve as cryoprotectants [94–96]. In particular, trehalose, which is a disaccharide of glucose, has been found to be very effective in this respect. Recently, it has been shown that the molecular mechanism underlying this cryoprotective effect is a result of hydrogen bonding between the trehalose molecules and the bilayer headgroups [7]. The sugar can replace some of the hydrogen bonds that normally form between the water molecules and the bilayer headgroups and thereby stabilize the fragile bilayer structure. Stabilization here means that the bilayer is able to withstand harsher environmental conditions in the presence of small molecules when compared with a single-lipid bilayer in water. Experimentally, it has been shown that trehalose prevents lipids from undergoing a phase transition under cooling, i.e., it shifts the main-phase transition temperature significantly [97]. Simulations have in general been able to corroborate these effects [92,98].

It has been found experimentally that alcohol molecules have the opposite effect of trehalose and destabilize model membranes [99,100]. It has been observed that upon the addition of alcohol molecules the lipid bilayer becomes thinner and the area per molecule increases. One application for alcohol/membrane experiments is that of stuck fermentations in the wine industry [101,102]. In a stuck fermentation, the yeast cells do not convert all available sugar molecules into alcohol but stop at an incomplete stage. It has been proposed that the underlying mechanism of stuck fermentations is an alcohol-triggered structural transition in the membrane that results in conformational changes to transmembrane proteins that render them dysfunctional [101,103]. Aside from wine production, increased sugar conversion would also be beneficial in the production of ethanol as a component in biofuels. Furthermore, alcohols have been used as model anesthetic molecules where it is has been proposed that anesthetic molecules may alter the lateral pressure profile of lipids, again resulting in conformational changes to trans-membrane proteins [104,105]. An example of a lateral pressure profile is shown in Figure 2.8 for a phosphatidylglycerol system.

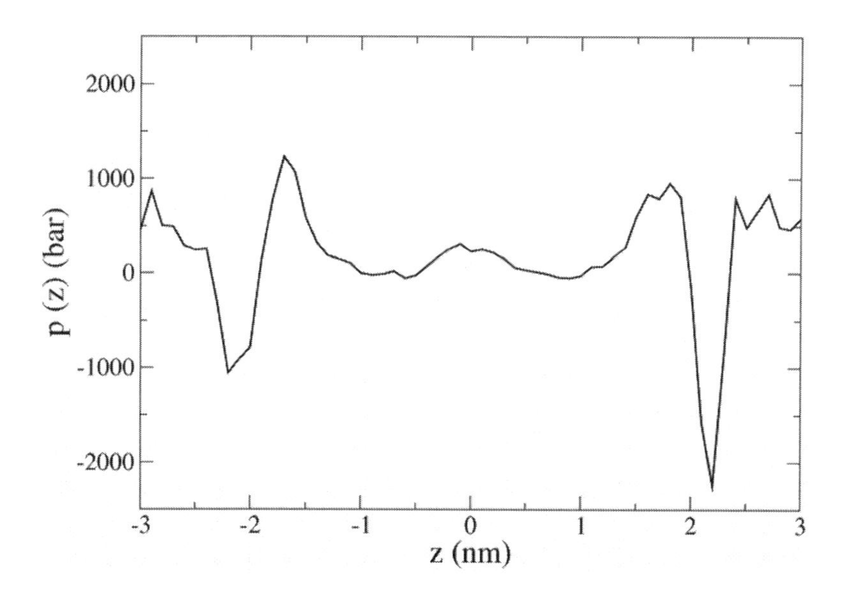

Figure 2.8. Lateral Pressure Profile for a 128 lipid POPG bilayer[32]. The asymmetry of the figure is due to statistical sampling deficiencies. We can clearly identify the headgroup water interfaces by the large peaks.

2.15. SUMMARY

It is clear that biomolecular modeling in general and molecular dynamics of biomembranes in particular is an efficient and useful tool to understand lipid bilayer systems. No other technique allows us to access all atom positions directly and offers in that way an unprecedented and unrivalled degree of detail. The question of modeling biomolecules is therefore no longer a "Why" or "If" but a "How." In this chapter we could only give a brief overview of the fundamentals of the technique. We could not discuss the statistical mechanical fundamentals of molecular dynamics. There are, however, many excellent books on that. We rather wanted to deliver a brief introduction that can be rapidly implemented by the non-expert user as molecular modeling starts to become a tool that is not only used by experts who devote their full time and often whole career to it. But due to the abundance and reliability of modern software packages, it can often be used as a tool for a short side project without going into all the depths.

Nonetheless, as with any other technique there are several traps that need to be avoided. It is a danger of modern software packages that they provide default values for almost all switches and numbers to be chosen in a simulation such that the user is tempted not to think about settings at all. We wanted therefore to explain the meaning of the most important of these often elusive settings that can fail simulations. The failure of a simulation is often not very obvious; if it crashes and the software aborts we know we did something wrong. But often simulations do not crash but are completely meaningless because the software is used in an inadequate manner; and for computers we must keep one thing in mind: they do *exactly* what they are programmed to do, not what we think we programmed them to do. So the two main ideas we wanted to convey here are that nobody should be dissuaded from using molecular modeling because of a fear of its complexity and that there are a few points about which one really has to be careful.

In general, simulations in many cases offer a very good visual understanding of a system where other techniques may be more precise. The strength of a simulation is most often in its interplay with other, mainly experimental, techniques. Simulations complement experiments by providing access to the mechanisms on a molecular level that give rise to experimental observables. The numbers themselves may often be only qualitative, but a mechanistic understanding is an invaluable asset.

ACKNOWLEDGMENTS

We would like to thank our former and current collaborators for many stimulating discussions. Funding was partially provided by NIH (NIH-NIGMS Grant Number T32-GM08799) and NSF (Grant Number CBET 0506602).

PROBLEMS

Simulations are a useful method for studying how molecule location and orientation are influenced by the positions of neighboring molecules and the local environment. If you happen to have access to a Unix/Linux computer and can download the MD simulation program Gromacs (http://www.gromacs.org/), complete Exercise 2.1 to simulate a solvated lipid bilayer and analyze the resulting water molecule orientations. If you do not have such access, use the resulting simulation data from Exercise 2.1 (available for download at http://www.chms.ucdavis.edu/research/web/faller/downloads.html) and proceed with the data analysis portion of the problem.

Preparing the Simulation Program

Download Gromacs. In this exercise, the syntax for the input will correspond to Gromacs version 3.3.1.

We are interested in comparing the molecular ordering between water molecules that are located close to a lipid bilayer surface and bulk-like water molecules.

Follow the steps below to generate a hydrated lipid bilayer

a. Make a new directory and name it lipid
b. Go to http://moose.bio.ucalgary.ca/index.php?page=Structures_and_Topologies and download the *dppc128.pdb file*. Save this file in your lipid directory
c. Download the *dppc.itp* and *lipid.itp* files into your directory
d. Download the *example2.top* file and rename it topol.top
e. Copy the *grompp.mdp* file from /share/tutor/water into your directory

f. Use the command *editconf –f dppc128.pdb –o conf.gro* to convert your configuration file into a .gro file format
g. Change the number of time steps in the *grompp.mdp* file from 10,000 to 250,000
h. Run the *grompp* and *mdrun* commands. This will generate a trajectory that is 0.5 ns in length since the time step in the *grompp.mdp* file is 2 fs.
i. When the simulation is complete you will see a confout.gro file,
j. An index file needs to be created to specify which atom in the water molecule is of interest. We will choose the oxygen atom, which is labeled OW.
k. Type the command *make_ndx –f conf.gro –o index.ndx*. A prompt will open and ask you to specify the atom of interest. Type *2 & a OW* to specify the oxygen atom in water. Then enter *q* to exit.
l. Calculate the radial distribution function with *g_rdf –n index.ndx*. You will be asked to specify two groups. Since we are interested in the water oxygen atoms, choose "*3*" twice. The results are in rdf.xvg.

Water Box

a. Make a new folder and name it waterbox
b. Copy the *topol.top* and *grompp.mdp* files from the lipid directory into the waterbox directory.
c. We need to make a new configuration file that contains only water molecules. Type *genbox -cs -box 6.41840 6.44350 2.67 –o conf.gro*. This creates a box of 3652 water molecules, where the x and the y box lengths are the same as those in the lipid simulation *conf.gro* file.
d. In the *topol.top* file, delete the line that says "DPPC 128" and change the number of water molecules to 3652.
e. Run the *grompp* and *mdrun* commands.
f. Make a new index file with *make_ndx –f conf.gro –o index.ndx*. A prompt will open and ask you to specify the atom of interest. Type *1 & a OW* to specify the oxygen atom in water. Then enter *q* to exit.
g. Calculate the radial distribution function with *g_rdf –n index.ndx*. You will be asked to specify two groups. Since we are interested in the water oxygen atoms, choose "2" twice.

Exercise 2.1

Plot the resulting radial distribution functions (RDFs) on the same graph. Discuss the differences between the resulting RDFs from the two systems. Are the water molecules more or less ordered near the surface? How would the RDF differ if we simulated water molecules near a smooth surface rather than a bilayer? Based on the definition of an RDF, why do the RDF values approach 1 with an increase in molecular distance?

You can now repeat the steps for the POPA layer discussed in this chapter in detail. The necessary configuration/topology files can be found at:

<http://www.chms.ucdavis.edu/research/web/faller/downloads.html>.

FURTHER STUDY

Molecular Dynamics and Other Modeling Techniques in General

Allen MP, Tildesley DJ. 1987. *Computer simulation of liquids*. Oxford: Clarendon Press.

Frenkel D, Smit B. 1996. *Understanding molecular simulation: from basic algorithms to applications*. San Diego: Academic Press.

Leach A. 1997. *Molecular modelling: principles and applications*. Essex: Addison Wesley.

Specific Software Packages and Force Fields

Gromos/Gromacs

van Gunsteren WF, Billeter SR, Eising AA, Hünenberger P, Krüger AE, Mark WRP, Scott AE, Tironi IG. 1996. *Biomolecular simulation: the GROMOS manual and user guide*. Zürich: Vdf.

Lindahl E, Hess B, van der Spoel D. 2001. GROMACS 3.0: a package for molecular simulation and trajectory analysis. *J Mol Model* **7**(8):306–317.

Hess B, Kutzner C, van der Spoel D, Lindahl E. 2008. GROMACS 4: algorithms for highly efficient, load-balanced, and scalable molecular simulation. *J Chem Theor Comput* **4**(3):435–447.

http://www.gromacs.org

Charmm

Brooks BR, Bruccoleri RE, Olafson DJ, States DJ, Swaminathan S, Karplus M. 1983. CHARMM: a program for macromolecular energy, minimization, and dynamics calculations. J Comput Chem **4**:187–217.

MacKerel Jr AD, Brooks III CL, Nilsson L, Roux B, Won Y, Karplus M. 1998. CHARMM: the energy function and its parameterization with an overview of the program. In *The Encyclopedia of Computational Chemistry*, Vol. 1, pp. 271–277. Ed. PvR Schleyer. Chichester: John Wiley & Sons.

http://www.charmm.org

Amber

Cornell WD, Cieplak P, Bayly CI, Gould IR, Merz KM, Ferguson DM, Spellmeyer DC, Fox T, Caldwell JW, Kollman PA. 1995. A second generation force field for the simulation of proteins, nucleic acids, and organic molecules. *J Am Chem Soc* **117**(19):5179–5197.

http://amber.scripps.edu/

REFERENCES

1. Verlet L. 1967. Computer 'experiments' on classical fluids, I: thermodynamical properties of Lennard-Jones molecules. *Phys Rev* **159**:98–103.
2. Allen MP, Tildesley DJ. 1987. *Computer simulation of liquids*. Oxford: Clarendon Press.
3. Frenkel D, Smit B. 1996. Understanding molecular simulation: from basic algorithms to applications. San Diego: Academic Press.
4. Leach A. 1997. Molecular modelling: principles and applications. Essex: Addison Wesley.
5. Egberts E, Marrink S-J, Berendsen HJC. 1994. Molecular dynamics simulation of a phospholipid membrane. *Eur Biophys J* **22**:423–436.
6. Tieleman DP, Marrink SJ, Berendsen HJC. 1997. A computer perspective of membranes: molecular dynamics studies of lipid bilayer systems. *Biochim Biophys Acta* **1331**:235–270.
7. Sum AK, Faller R, de Pablo JJ. 2003. Molecular simulation study of phospholipid bilayers and insights of the interactions with disaccharides. *Biophys J* **85**:2830–2844.

8. Faller R, Marrink S-J. 2004. Simulation of domain formation in mixed DLPC–DSPC lipid bilayers. *Langmuir* **20**:7686–7693.
9. Lee BW, Faller R, Sum AK, Vattulainen I, Patra M, Karttunen M. 2004. Structural effects of small molecules on phospholipid bilayers investigated by molecular simulations. *Fluid Phase Equilib* **225**:63–68.
10. Switzer J, Bennun S, Longo ML, Palazoglu A, Faller R. 2006. Karhunen-Loeve analysis for pattern description in phase separated lipid bilayer systems. *J Chem Phys* **124**:234906.
11. Cooke IR, Kremer K, Deserno M. 2005. Tunable generic model for fluid bilayer membranes. *Phys Rev E* **72**:011506.
12. Moore PB, Lopez CF, Klein ML. 2001. Dynamical properties of a hydrated lipid bilayer from a multinanosecond molecular dynamics simulation. *Biophys J* **81**:2484–2494.
13. Lopez CF, Nielsen SO, Moore PB, Shelley JC, Klein ML. 2002. Self-assembly of a phospholipid Langmuir monolayer using coarse-grained molecular dynamics simulations. *J Phys Condens Matter* **14**:9431–9444.
14. Patra M, Karttunen M, Hyvönen MT, Falck E, Vattulainen I. 2004. Lipid bilayers driven to a wrong plane in molecular dynamics simulations by truncation of long-range electrostatic interactions. *J Phys Chem B* **108**:4485–4494.
15. Feller SE, Pastor RW, Rojnuckarin A, Bogusz S, Brooks BR. 1996. Effect of electrostatic force truncation on interfacial and transport properties of water. *J Phys Chem* **100**:17011–17020.
16. Anezo C, de Vries AH, Holtje H, Tieleman DP, Marrink SJ. 2003. Methodological issues in lipid bilayer simulations. *J Phys Chem B* **107**:9424–9433.
17. Farago O. 2003. "Water-free" computer model for fluid bilayer membranes. *J Chem Phys* **119**:596–605.
18. Marrink SJ, de Vries AH, Mark AE. 2004. Coarse grained model for semi-quantitative lipid simulation. *J Phys Chem B* **108**:750–760.
19. Stevens MJ. 2004. Coarse-grained simulations of lipid bilayers. *J Chem Phys* **121**:11942–1198.
20. Shelley JC, Shelley MY, Reeder RC, Bandyopadhyay S, Klein ML. 2001. A coarse grain model for phospholipid simulations. *J Phys Chem B* **105**:4464–4470.
21. Kaznessis YN, Kim ST, Larson RG. 2002. Simulations of zwitterionic and anionic phospholipid monolayers. *Biophys J* **82**:1731–1742.
22. Alper HE, Bassolino D, Stouch TR. 1993. Computer simulation of a phospholipid monolayer-water system: the influence of long-range forces on water structure and dynamics. *J Chem Phys* **98**:9798–9807.
23. Feller SE, Zhang YH, Pastor RW. 1995. Computer-simulation of liquid/liquid interfaces, II: surface-tension area dependence of a bilayer and a monolayer. *J Chem Phys* **103**:10267–10276.
24. Heine DR, Rammohan AR, Balakrishnan J. 2007. Atomistic simulations of the interaction between lipid bilayers and substrates. *Mol Simul* **33**:391–397.
25. Xing C, Faller R. 2008. Interactions of lipid bilayers with supports: a coarse-grained molecular simulation study. *J Phys Chem B* **112**:7086–7094.
26. Berendsen HJC, Postma JPM, van Gunsteren WF, Dinola A, Haak JR. 1984. Molecular dynamics with coupling to an external bath. *J Chem Phys* **81**:3684–3690.
27. Essman U, Perela L, Berkowitz ML, Darden HLT, Pedersen LG. 1995. A smooth particle mesh Ewald method. *J Chem Phys* **103**:8577–8592.
28. Mashl RJ, Scott HL, Subramaniam S, Jakobsson E. 2001. Molecular simulation of dioleoylphosphatidylcholine lipid bilayers at differing levels of hydration. *Biophys J* **81**:3005–3015.
29. Feller SE, Yin D, Pastor RW, MacKerel Jr. AD. 1997. Molecular dynamics simulation of unsaturated lipid bilayers at low hydration: parameterization and comparison with diffraction studies. *Biophys J* **73**:2269–2279.
30. Tieleman DP, Berendsen HJC. 1996. Molecular dynamics simulations of fully hydrated dipalmitoylphosphatidylcholine bilayer with different macroscopic boundary conditions and parameters. *J Chem Phys* **105**:4871–4880.
31. Balali-Mood K, Harroun TA, Bradshaw JP. 2003. Molecular dynamics simulations of a mixed DOPC/DOPG bilayer. *Eur Phys J E* **12**:S135–S140.
32. Dickey A, Faller R. 2008. Examining the contributions of lipid shape and headgroup charge on bilayer behavior. *Biophys J* **95**:2636–2646.
33. Berger O, Edholm O, Jahnig F. 1997. Molecular dynamics simulations of a fluid bilayer of dipalmitoylphosphatidylcholine at full hydration, constant pressure, and constant temperature. *Biophys J* **72**:2002–2013.
34. Pandit SA, Bostick D, Berkowitz ML. 2003. Mixed bilayer containing dipalmitoylphosphatidylcholine and dipalmitoylphosphatidylserine: lipid complexation, ion binding, and electrostatics. *Biophys J* **85**:3120–3131.

35. Niemela P, Hyvonen MT, Vattulainen I. 2004. Structure and dynamics of sphingomyelin bilayer: insight gained through systematic comparison to phosphatidylcholine. *Biophys J* **87**:2976–2989.
36. Pandit SA, Jakobsson E, Scott HL. 2004. Simulation of the early stages of nano-domain formation in mixed bilayers of sphingomyelin, cholesterol, and dioleylphosphatidylcholine. *Biophys J* **87**:3312–3322.
37. Murzyn K, Rog T, Pasienkiewicz-Gierula M. 2005. Phosphatidylethanolamine–phosphatidylglycerol bilayer as a model of the inner bacterial membrane. *Biophys J* **88**:1091–1103.
38. Leekumjorn S, Sum AK. 2006. Molecular simulation study of structural and dynamic properties of mixed DPPC/DPPE Bilayers. *Biophys J* **90**:3951–3965.
39. Gurtovenko A, Patra M, Karttunen M, Vattulainen I. 2004. Cationic DMPC/DMTAP lipid bilayers: molecular dynamics study. *Biophys J* **86**:3461–3472.
40. Falck E, Patra M, Karttunen M, Hyvonen MT, Vattulainen I. 2004. Lessons of slicing membranes: interplay of packing, free area and lateral diffusion in phospholipid/cholesterol bilayers. *Biophys J* **87**:1076–1091.
41. Pandit SA, Bostick D, Berkowitz ML. 2004. Complexation of phosphatidylcholine lipids with cholesterol. *Biophys J* **86**:1345–1356.
42. Smondryev AM, Berkowitz ML. 2001. Molecular dynamics simulation of the structure of dimyristolphosphatidylcholine bilayers with cholesterol, ergosterol and lanosterol. *Biophys J* **80**:1649–1658.
43. Dickey AN, Yim W-S, Faller R. 2009. Using ergosterol to mitigate the deleterious effects of ethanol on bilayer structure. *J Phys Chem B.* **113**:2388–2397.
44. Hess B, Bekker H, Berendsen HJC, Fraaije JGEM. 1997. LINCS: a linear constraint solver for molecular simulations. *J Comput Chem* **18**:1463–1472.
45. Müller-Plathe F, Brown D. 1991. Multi-colour algorithms in molecular simulation: vectorisation and parallelisation of internal forces and constraints. *Comput Phys Commun* **64**:7–14.
46. Berendsen HJC, Postma JPM, van Gunsteren WF, Hermans J. 1981. Interaction models for water in relation to protein hydration. In *Intermolecular forces*, pp. 331–342. Ed. B Pullman. Dordrecht: Reidel.
47. Berendsen HJC, Grigera JR, Straatsma TP. 1987. The missing term in effective pair potentials. *J Phys Chem* **91**:6269–6271.
48. Mackerel AD. 2004. Empirical force fields for biological macromolecules: overview and issues. *J Comput Chem* **25**:1584–1604.
49. Aman K, Lindahl E, Edholm O, Hakansson P, Westlund PO. 2003. Structure and dynamics of interfacial water in an L-alpha phase lipid bilayer from molecular dynamics simulations. *Biophys J* **84**:102–115.
50. Pasenkiewicz-Gierula M, Takaoka Y, Miyagawa H, Kitamura K, Kusumi A. 1999. Charge pairing of headgroups in phosphatidylcholine membranes: a molecular dynamics simulation study. *Biophys J* **76**:1228–1240.
51. Tieleman DP, Berendsen HJC. 1996. Molecular dynamics simulations of a fully hydrated dipalmitoyl phosphatidylcholine bilayer with different macroscopic boundary conditions and parameters. *J Chem Phys* **105**:4871–4880.
52. Hofsass C, Lindahl E, Edholm O. 2003. Molecular dynamics simulations of phospholipid bilayers with cholesterol. *Biophys J* **84**:2192–2206.
53. Tu KC, Klein ML, Tobias DJ. 1998. Constant-pressure molecular dynamics investigation of cholesterol effects in a dipalmitoylphosphatidylcholine bilayer. *Biophys J* **75**:2147–2156.
54. Muller M, Katsov K, Schick M. 2006. Biological and synthetic membranes: what can be learned from a coarse-grained description? *Phys Rep* **434**:113–176.
55. Shelley JC, Shelley MY, Reeder RC, Bandyopadhyay S, Moore PB, Klein ML. 2001. Simulations of phospholipids using a coarse grain model. *J Phys Chem B* **105**:9785–9792.
56. Shelley JC, Shelley MY, Reeder RC, Bandyopadhyay S, Klein ML. 2001. A coarse grain model for phospholipid simulations. *J Phys Chem B* **105**:4464–4470.
57. Nielsen SO, Lopez CF, Srinivas G, Klein ML. 2004. Coarse grain models and the computer simulation of soft materials. *J Phys Condens Matter* **16**:R481–R512.
58. Marrink SJ, Risselada J, Mark AE. 2005. Simulation of gel phase formation and melting in lipid bilayers using a coarse grained model. *Chem Phys Lipids* **135**:223–244.
59. Marrink SJ, Mark AE. 2004. Molecular view of hexagonal phase formation in phospholipid membranes. *Biophys J* **87**:3894–3900.
60. Lopez CF, Moore PB, Shelley JC, Shelley MY, Klein ML. 2002. Computer simulation studies of biomembranes using a coarse grain model. *Comput Phys Commun* **147**:1–6.

61. Marrink SJ, Risselada HJ, Yefimov S, Tieleman DP, de Vries AH. 2007. The MARTINI force field: coarse grained model for biomolecular simulations. *J Phys Chem B* **111**:7812–7824.
62. Bennun SV, Longo ML, Faller R. 2007. Phase and mixing behavior in two-component lipid bilayers: a molecular dynamics study in DLPC/DSPC mixtures. *J Phys Chem B* **111**:9504–9512.
63. Marrink SJ, Mark AE. 2003. Molecular dynamics simulation of the formation, structure, and dynamics of small phospholipid vesicles. *J Am Chem Soc* **125**:15233–15242.
64. Faller R, Marrink SJ. 2004. Simulation of domain formation in DLPC–DSPC mixed bilayers. *Langmuir* **20**:7686–7693.
65. Bennun SV, Longo, M., Faller R. Molecular scale structure in fluid-gel patterned bilayers: stability of interfaces and transmembrane distribution. To be submitted.
66. Bennun S, Longo ML, Faller R. 2007. The molecular scale structure in fluid-gel patterned bilayers: stability of interfaces and transmembrane distribution. *Langmuir* **23**:12465–12468.
67. Brannigan G, Brown FLH. 2004. Solvent-free simulations of fluid membrane bilayers. *J Chem Phys* **120**:1059–1071.
68. Noguchi H, Takasu M. 2001. Self-assembly of amphiphiles into vesicles: a Brownian dynamics simulation. *Phys Rev E* **64**:041913.
69. Reynwar BJ, Illya G, Harmandaris VA, Müller MM, Kremer K, Deserno M. 2007. Aggregation and vesiculation of membrane proteins by curvature-mediated interactions. *Nature* **447**:461–467.
70. Harmandaris VA, Deserno M. 2006. A novel method for measuring the bending rigidity of model lipid membranes by simulating tethers. *J Chem Phys* **125**:204905.
71. Weeks JD, Chandler D, Andersen HC. 1971. Role of repulsive forces in determining the equilibrium structure of simple liquids. *J Chem Phys* **54**:5237–5247.
72. Faller R, Kuhl TL. 2003. Modeling the binding of cholera-toxin to a lipid membrane by a non-additive two-dimensional hard disk model. *Soft Mater* **1**:343–352.
73. Majewski J, Kuhl TL, Kjaer K, Smith GS. 2001. Packing of ganglioside-phospholipid monolayers: an x-ray diffraction and reflectivity study. *Biophys J* **81**:2707–2715.
74. Humphrey W, Dalke A, Schulten K. 1996. VMD—visual molecular dynamics. *J Mol Graphics* **14**:33–38.
75. Sayle RA, Milner-White EJ. 1995. RASMOL: biomolecular graphics for all. *Trends Biochem Sci* **20**:374–376.
76. Pandit SA, Chiu S-W, Jakobsson E, Grama A, Scott HL. 2007. Cholesterol surrogates: a comparison of cholesterol and 16:0 ceramide in POPC bilayers. *Biophys J* **92**:920–927.
77. Dickey AN, Faller R. 2007. How alcohol chain-length and concentration modulate hydrogen bond formation in a lipid bilayer. *Biophys J* **92**:2366–2376.
78. Dickey AN, Faller R. 2005. Investigating interactions of biomembranes and alcohols: a multiscale approach. *J Polym Sci B* **43**:1025–1032.
79. Kranenburg M, Venturoli M, Smit B. 2003. Phase behavior and induced interdigitation in bilayers studied with dissipative particle dynamics. *J Phys Chem B* **107**:11491–11501.
80. Miller CE, Majewski J, Gog T, Kuhl TL. 2005. Characterization of biological thin films at the solid–liquid interface by x-ray reflectivity. *Phys Rev Lett* **94**:238104.
81. Tieleman DP, Marrink SJ, Berendsen HJC. 1997. A computer perspective of membranes: molecular dynamics studies of lipid bilayer systems. *Biochim Biophys Acta* **1331**:235–270.
82. Davis JH. 1983. The description of membrane lipid conformation, order and dynamics by 2H-NMR. *Biochim Biophys Acta* **737**:117–171.
83. Aittoniemi J, Róg T, Niemela P, Pasenkiewicz-Gierula M, Vattulainen I, Karttunen M. 2006. Sterol tilt: major determinant of sterol ordering capability in lipid membranes. *J Phys Chem B Lett* **110**:25562–25564.
84. Lindahl E, Hess B, van der Spoel D. 2001. GROMACS 3.0: a package for molecular simulation and trajectory analysis. *J Mol Model* **7**:306–317.
85. Luzar A, Chandler D. 1996. Hydrogen-bond kinetics in liquid water. *Nature* **379**:55–57.
86. Ollila S, Hyvonen MT, Vattulainen I. 2007. Polyunsaturation in lipid membranes: dynamic properties and lateral pressure profiles. *J Phys Chem B* **111**:3139–3150.
87. Chiu SW, Clark M, Balaji V, Subramaniam S, Scott HL, Jakobsson E. 1995. Incorporation of surface tension into molecular dynamics simulation of an interface: a fluid phase lipid bilayer membrane. *Biophys J* **69**:1230–1245.
88. Schmidt-Rohr K, Spiess HW. 1994. *Multidimensional solid state NMR and polymers.* New York: Academic Press.

89. Dickey AN, Faller R. 2008. Behavioral differences between phosphatidic acid and phosphatidylcholine in the presence of the nicotinic acetylcholine receptor. *Biophys J* **95**:2636–2646.
90. Dickey AN. 2008. How bilayer composition affects the stability of a model yeast membrane and the behavior of lipids surrounding the nicotinic acetylcholine receptor. PhD dissertation. University of California Davis.
91. Wong BY, Faller R. 2007. Phase behavior and dynamic heterogeneities in lipids: a coarse-grained simulation study of DPPC–DPPE mixtures. *Biochim Biophys Acta* **1768**:620–627.
92. Sum AK, Faller R, de Pablo JJ. 2003. Molecular simulation study of phospholipid bilayers and insights of the interactions with disaccharides. *Biophys J* **85**:2830–2844.
93. Patra M, Salonen E, Terama E, Vattulainen I, Faller R, Lee BW, Holopainen J, Karttunen M. 2006. Under the influence of alcohol: the effect of ethanol and methanol on lipid bilayers. *Biophys J* **90**:1121–1135.
94. Crowe JH, Crowe LM, Carpenter JF, Rudolph AS, Wistrom CA, Spargo BJ, Anchordoguy TJ. 1988. Interactions of sugars with membranes. *Biochim Biophys Acta* **947**:367–384.
95. Crowe JH, Crowe LM, Carpenter JF, Wistrom CA. 1987. Stabilization of dry phospholipid bilayers and proteins by sugar. *Biochem J* **242**:1–10.
96. Crowe JH, Crowe LM, Oliver AE, Tsvetkova N, Wolkers W, Tablin F. 2001. The trehalose myth revisited: introduction to a symposium on stabilization of cells in the dry state. *Cryobiology* **43**:89–105.
97. Crowe JH, Crowe LM, Chapman D. 1984. Preservation of membranes in anhydrobiotic organisms: the role of trehalose. *Science* **223**:701–703.
98. Pereira CS, Lins RD, Chandrasekhar I, Carlos L, Freitas G, Hünenberger PH. 2004. Interaction of the disaccharide trehalose with a phospholipid bilayer: a molecular dynamics study. *Biophys J* **86**:2273–2285.
99. Ly HV, Block DE, Longo ML. 2002. Interfacial tension effect of ethanol on lipid bilayer rigidity, stability, and area/molecule: a micropipet aspiration approach. *Langmuir* **18**:8988–8995.
100. Ly HV, Longo ML. 2004. The influence of short-chain alcohols on interfacial tension, mechanical properties, area/molecule, and permeability of fluid lipid bilayers. *Biophys J* **87**:1013–1033.
101. Bisson LF, Block DE. 2002. Ethanol tolerance in Saccharomyces. In *Biodiversity and biotechnology of wine yeasts*, pp. 85–98. Ed. M Ciani. Trivandrum, India: Research Signpost.
102. Cramer AC, Vlassides S, Block DE. 2002. Kinetic model for nitrogen-limited wine fermentations. *Biotechnol Bioeng* **77**:49–60.
103. Bisson LF. 1999. Stuck and sluggish fermentations. *Am J Enol Vitic* **50**:107–119.
104. Cantor RS. 1997. The lateral pressure profile in membranes: a physical mechanism of general anesthesia. *Biochemistry* **36**:2339–2344.
105. Terama E, Ollila OH, Salonen E, Rowat AC, Trandum C, Westh P, Patra M, Karttunen M, Vattulainen I. 2008. Influence of ethanol on lipid membranes: from lateral pressure profiles to dynamics and partitioning. *J Phys Chem B* **112**:4131–4139.

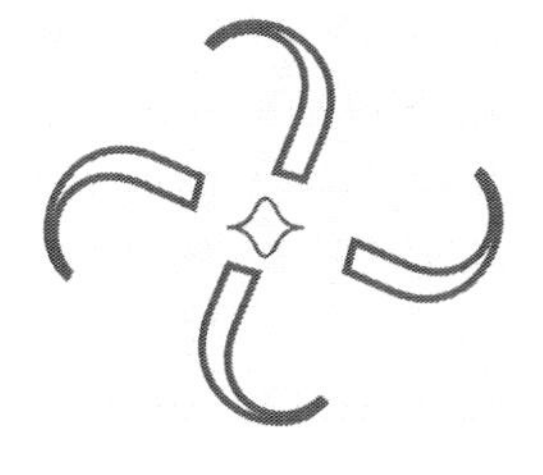

3

Introduction to Electron Paramagnetic Resonance Spectroscopy

Marcin Brynda
Department of Chemistry, University of California Davis

3.1. INTRODUCTION

The acronym EPR describes a spectroscopic technique known as electron paramagnetic resonance. In several, especially older, textbooks and research papers the alternative names ESR (electron spin resonance) or EMR (electron magnetic resonance) can be sometimes encountered. The development of EPR spectroscopy was slowed during the few decades following its invention due to the unavailability of microwave components for higher frequencies. Today, however, EPR technologies are enjoying very rapid development in all areas of application. Such regained interest is related to the fact that EPR not only proved to be an extremely powerful spectroscopic technique but to an increasing amount of related techniques that were born around EPR methodology. Recent developments in the area of instrumentation followed by the commercialization of high-end EPR spectrometers have attracted many new adepts who benefit from exploring this technique in their specific fields of research.

EPR is in many ways very similar to nuclear magnetic resonance (NMR); however, the energy required for flipping the spins provided in the form of electromagnetic radiation (microwaves) in this case is absorbed not by nuclei as in NMR, but by molecules, ions, or atoms possessing unpaired electron spins. One of the most important aspects of EPR is its broad applicability, once again similar to NMR. The method can be used to study everything from nanoparticles and small molecules to complicated biological systems, and even animal and human tissues. A large number of materials possess unpaired electrons, including semiconductors,

Address correspondence to Marcin Brynda, PhD, Department of Chemistry, University of California Davis, One Shields Avenue, Davis, CA 95616-8635, USA, 530 754-4141, 530 752-8998 (fax), <mabrynda@ucdavis.edu>.

T. Jue (ed.), *Biomedical Applications of Biophysics*,
Handbook of Modern Biophysics 3, DOI 10.1007/978-1-60327-233-9_3,
© Springer Science+Business Media, LLC 2010

polymers, and free radicals. Some solids exhibit "electron holes" that can also be observed by EPR, while others host paramagnetic impurities such as open shell metals or stable radicals. Inclusion of impurities into silicon, titanium oxide, or cadmium/selenium conglomerates result in "paramagnetically" doped materials that exhibit a broad range of interesting electronic and chromophoric properties. On the other hand, EPR can be used to probe metal centers in biologically relevant enzymes or free radicals that are formed by radiation. Many of these paramagnetic species play central roles in biological reactions and are important elements of living organisms. The unpaired electrons are also present in key intermediates of many processes such as photosynthesis, oxidation, catalysis, and polymerization reactions.

The EPR technique is truly interdisciplinary, lying at the interface of chemistry, physics, and biology, but EPR techniques are not limited to these scientific disciplines. They are also useful in other popular fields such as geochemistry, food industry, material/polymer science, medicine, pharmacology, and many others. The power of EPR lies in the fact that in a way similar to NMR it can yield direct structural information at microscopic and macroscopic scales. From details about the specific orientation of the particular amino acid residues in a enzymatic molecule to the mapping of the oxidative stress in cancerous tissues at a scale of a living organism, EPR yields a tremendous amount of information. Often, information unavailable from other spectroscopies, such as X-ray or molecular spectroscopy, can be obtained directly from EPR. In addition, the interaction of the unpaired electron with nearby magnetic nuclei yields details of the atomic scale environment of the electron.

3.2. HISTORICAL BACKGROUND

With the intense scientific research during the Second World War, a number of new experimental techniques emerged. Several months after Purcell, Torrey and Pound observed the first NMR transition in paraffin wax at Stanford University in 1944, Russian physicist Yevgeny K. Zavoisky started to explore a new technique on his laboratory-built EPR spectrometer at the Kazan Institute of Physics. His first experiment was performed at a relatively low field (50 G) and low frequency (130 MHz) and resulted in an EPR spectrum of a hydrated copper chloride complex shown in Figure 3.1 (depicted in its original state). But already a year later, the first experiments were conducted at much higher fields close to 4000 G. At this magnetic field, the resonant frequency of the unpaired electron is close to 9 GHz and lies within the so-called EPR X-band. Since microwave technology at these frequencies was developed for use in radar equipment during WW II, for many years X-band became the most common frequency used by EPR spectroscopists around the world. Early on it became evident that in a way similar to NMR (where the chemical shifts of nuclei can be better resolved with increasing magnetic field strength) the improved resolution at higher fields might render EPR much more attractive. The EPR analog of a chemical shift, σ, encountered in NMR is the so-called g factor. The free electron value of g is 2.0023, and most of the organic free radicals exhibit g values close to the free electron value. However, g is very sensitive to the electronic environment of the radical, and its deviation from the free electron value can yield detailed information on the immediate chemical surrounding of the unpaired electron. Accessing higher EPR frequencies would allow increased resolution of the g tensor. After early attempts with higher frequencies (25 and 37.5 GHz), the

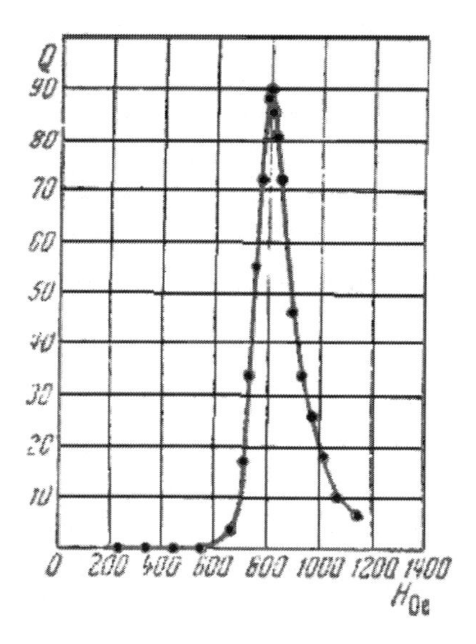

Figure 3.1. First EPR spectrum of $CuCl_2{\cdot}2H_2O$ observed by Y. Zavoisky at a frequency of 133 MHz. Note that the strength of the magnetic field is reported in Oersted (Oe).

lack of adequate electronics precluded development of the technique for almost three decades. It was only in the 1970s that the field of high-frequency EPR regained interest, and very rapid advances were made. However, in the meantime many new techniques built around EPR emerged. Some of them will be briefly discussed in the following paragraphs.

- Stern and Gerlach (1921–1922) — quantization of electron magnetic momentum
- Pauli (1924) — nuclear moments connected with hyperfine structure in atomic spectra
- Uhlenbeck and Goudsmidt (1925) — electron magnetic moment linked with the spin of the electron
- Breit and Rabi (1931) — energy levels for the H atom; link between nuclear and electron spin angular momentum
- Rabi, Zacharias, Millman, and Kusch (1938) — first experimental observation of magnetic resonance
- Zavoisky (1945) — first EPR spectrum of $CuCl_2{\cdot}2H_2O$ at 4.76 mT and a frequency of 133 MHz

3.3. BASIC EQUATIONS

As in the case of NMR, the EPR technique relies on the fact that electron spins in the external magnetic filed act as small magnets and can be manipulated by the electromagnetic fields of the incident microwaves. In quantum mechanics, the electron possesses an internal degree of freedom called "spin" with a quantum number S. Its operator is denoted $\hat{S}$, and it is a vector $[S_x,S_y,S_z]$ corresponding to an angular momentum operator. The magnitude of the angular momentum is given by $(S(S + 1))^{1/2}$, where $(S(S + 1))$ are the eigenvalues of S. The eigenvalues for the projection of the angular momentum on the specific axis (we will use Z as principal axis) are M_S. In a static magnetic field B_0 (along the Z axis), an electron will have a state of lowest energy when its magnetic moment μ_e, is aligned with the magnetic field and a state of highest energy when μ_e is aligned against the magnetic field. The two states are labeled by the projection of the electron spin, M_S, on the direction of the magnetic field. Since electrons are fermions (spin ½ particles), the parallel state is designated as $M_S = -½$ and the antiparallel state is $M_S = +½$.

The magnetic moment of an electron is directly proportional to the angular momentum μ_e:

$$\mu_e = -g_e \cdot \mu_B \cdot S, \tag{3.1}$$

where the Bohr magnethon is expressed as $\mu_B = e\hbar / 2m_e$ and g_e is a correction factor equal to 2.0023 and is intrinsic to a free e^-.

Some nuclei also possess spin values greater than zero (nuclear spin). The angular momentum operator for a nucleus is I, and its magnitude is $I(I+1))^{½}$, where $(I(I+1))$ are the eigenvalues of I. The eigenvalues for their projection are M_I. The magnetic moment of nucleus is expressed as

$$\mu_N = g_N \cdot \mu_N \cdot I, \tag{3.2}$$

where

$$\mu_N = \frac{e_p \hbar}{2m_p}. \tag{3.3}$$

The proportionality constant γ is called the gyromagnetic ratio and can be written as

$$\gamma = g\frac{q}{2m} = g\frac{\beta}{\hbar}, \tag{3.4}$$

with the β factor is equal to

$$\beta = \frac{q\hbar}{2m}. \tag{3.5}$$

The component μ_z of the spin magnetic moment of an electron along the direction of the magnetic field B_0 applied along z is

$$\mu_{Z_e} = \gamma_e \hbar M_S = -g_e \beta_e M_S. \tag{3.6}$$

Similarly, for the magnetic moment of nucleus

$$\mu_{Z_N} = \gamma_N \hbar M_I = -g_N \beta_N M_I. \tag{3.7}$$

In a static magnetic field, the classical energy of a magnetic moment is

$$E = -\mu \cdot B_0. \tag{3.8}$$

In a quantum mechanical system we replace μ by an appropriate operator, and we obtain the Hamiltonian H for the free electron:

$$H = -g_e \mu_B S \cdot B . \tag{3.9}$$

With the field aligned with the Z direction of our spin, we can reduce the scalar product $S \cdot B$ to

$$H = -g_e \mu_B S_Z B_0 . \tag{3.10}$$

In a static magnetic field B_0, electron spin is aligned with Z, and the quantum mechanics stipulates that only two values are permitted for an electron ($S_Z = \pm\hbar/2$), which means that the electron magnetic moment can only assume two projections, parallel or antiparallel, with the corresponding energies

$$E_+ = -\frac{1}{2} g_e \mu_B B_0 \text{ and } E_- = +\frac{1}{2} g_e \mu_B B_0 . \tag{3.11}$$

The energy difference between these two states equals $\Delta E = E_+ - E_-$. If the Bohr frequency condition for a radiation of quanta of energy is satisfied,

$$\Delta E = h\nu , \tag{3.12}$$

the corresponding transition will occur:

$$\Delta E = h\nu = g_e \mu_B B_0 . \tag{3.13}$$

Equation (3.13) is the basic expression for the resonance condition in EPR. The physical meaning of this equation is as follows. Since the g value is an intrinsic property of each paramagnetic system, at a given (and constant) frequency of the spectrometer ν, the EPR transition will occur at the magnetic field corresponding to B_0, and will be a characteristic spectral signature of this system.

3.4. SPIN HAMILTONIAN

The concept of a spin Hamiltonian comes from a quantum mechanical description of the molecular system. However, for a given molecule (or ensemble of molecules), a complete Hamiltonian can be quite complex. This is because it has to include the space and spin coordinates of all electrons and nuclei of the system. To overcome this complexity, the magnetic resonance phenomena can be described by its simplified form, the so-called spin Hamiltonian. The main simplification rests on taking the average value for molecular and electron coordinates and express them for a given spin state of the system.

Let us consider a simple paramagnet with an electronic spin S and a nuclear spin I placed in a homogenous magnetic field B_0. The complete EPR spin Hamiltonian of such a system can be written as

$$H = \beta_e \cdot S \cdot \overline{g} \cdot B_0 - g_N \cdot \beta_N \cdot B_0 \cdot I + S \cdot T \cdot I + I \cdot P \cdot I + S \cdot D \cdot S . \tag{3.14}$$

Different terms of this Hamiltonian have the following meaning:

$\beta_e \cdot S \cdot g \cdot B_0$ — represents the electronic Zeeman interaction (interaction between the magnetic momentum of the electronic spin and the magnetic field).

$\overline{g}$ — differs from the free electron g value ($g_e = 2.0023$) and is a consequence of the contribution of the orbital magnetic momentum; g is defined as

$$g = 1 + \frac{S(S+1) - L(L+1) + J(J+1)}{2J(J+1)}, \tag{3.15}$$

where L is the orbital angular momentum and J is the electron angular momentum quantum number. For most organic radicals $L \sim 0$ and J represents the spin quantum number S. In this case, g is close to the free electron value, g_e. However, for paramagnetic transition metals, S and L can be quite large, making the corresponding g values very different from g_e. It has to be stressed out that g is in fact a two dimensional matrix. Its expression depends on the choice of the axis system of (3.16). In the random axis system g' is represented by a 3 × 3 matrix with nine g_{ij} values, but it is always possible to find an axis system that will diagonalize the g tensor. In this case, the g matrix is simplified and has only diagonal elements:

$$g' = \begin{bmatrix} g'_{xx} & g'_{xy} & g'_{xz} \\ g'_{yx} & g'_{yy} & g''_{yz} \\ g'_{zx} & g'_{zy} & g'_{zz} \end{bmatrix}, \quad g = \begin{bmatrix} g_{xx} & 0 & 0 \\ 0 & g_{yy} & 0 \\ 0 & 0 & g_{zz} \end{bmatrix}. \tag{3.16}$$

Strongly anisotropic g tensors influence the shape of the EPR spectrum; some examples are discussed in more details in the next paragraph.

$g_N \cdot \beta_N \cdot B_0 \cdot I$ — nuclear Zeeman interaction, which represents the interaction between the magnetic momentum of the nuclear spin and the static magnetic field. The nuclear Zeeman interaction is usually very small compared to the electron Zeeman interaction owing to the large difference in particles masses.

$S \cdot T \cdot I$ — represents the hyperfine interaction (the interaction between the magnetic momentum of the nuclear spin and the magnetic momentum of the electronic spin). T (in the same way as $\overline{g}$) is a tensor, which it is possible to decompose into an isotropic part and an anisotropic part:

$$S \cdot T \cdot I = A_{\text{iso}} \cdot I \cdot S + I \cdot \tau_{\text{aniso}} \cdot S \quad \text{with} \quad \text{Tr}(\tau_{\text{aniso}}) = 0. \tag{3.17}$$

The isotropic part results from the Fermi contact interaction, and is given by

$$A_{\text{iso}} = \frac{8\pi}{3} g_e \beta_e g_N \beta_N (\Psi_{(0)})^2 \cdot I \cdot S \,. \tag{3.18}$$

This interaction is thus directly a function of the density of probability to find an unpaired electron at the nucleus. The anisotropic contribution is of purely dipolar character and is described by the tensor

$$\tau_{\text{aniso}} = -g_e \beta_e g_N \beta_N \begin{bmatrix} \left\langle \frac{r^2 - 3x^2}{r^5} \right\rangle & \left\langle \frac{-3xy}{r^5} \right\rangle & \left\langle \frac{-3xz}{r^5} \right\rangle \\ \left\langle \frac{-3xy}{r^5} \right\rangle & \left\langle \frac{r^2 - 3y^2}{r^5} \right\rangle & \left\langle \frac{-3yz}{r^5} \right\rangle \\ \left\langle \frac{-3xz}{r^5} \right\rangle & \left\langle \frac{-3yz}{r^5} \right\rangle & \left\langle \frac{-3z^2}{r^5} \right\rangle \end{bmatrix}, \tag{3.19}$$

where r represents the nucleus–electron distance and x, y, and z the coordinates of the electron. The trace of $\tau = 0$ and, consequently, when T is diagonalized, $\tau_i = T_i - A_{\text{iso}}$. The hyperfine interaction is also a direct measure of the shape of the molecular orbital in which resides the unpaired electron. For the s-type orbitals the hyperfine term is perfectly spherical, for p- or d-orbitals τ is rhombic or axial.

$I \cdot P \cdot I$ — quadrupolar interaction results from the interaction between the quadrupolar electric moment of the nucleus with the electric field gradient created by the surrounding electrons and occurs for nuclei with nuclear spin >1/2.

$S \cdot D \cdot S$ — Zero-Field Splitting (ZFS) results from quadrupolar electronic interactions, in a way similar to that observed with nuclear quadrupolar interaction. ZFS removes the spin degeneracy for systems with $S > ½$, even in the absence of an applied magnetic field. In the same manner as g, the D tensor comports isotropic and anisotropic parts:

$$H_{\text{ZFS}} = D\left[S_z^2 - \frac{S(S+1)}{3}\right] + E\left(S_x^2 - S_y^2\right). \tag{3.20}$$

For odd-electron systems, axial ZFS (D parameter) results in so-called Kramer's doublets (two energy levels), which are split by the rhombic ZFS (E parameter).

The energy levels of a given system described by such a spin Hamiltonian can be obtained by diagonalization of the matrix, whose elements are calculated using basis functions represented by the wavefunctions of electronic and nuclear spins. To illustrate the notion of the energy levels derived from the spin Hamiltonian in the case of a simple organic radical, energy levels obtained for a system with one electronic spin S = 1/2 and one nuclear spin I = 1/2 are schematically represented in Figure 3.2. Due to the magnetic field, the degenerate electronic levels are split into $M_S = +½$ and $M_S = –½$ levels. Additional splitting arises from the hyperfine interaction of the magnetic momentum of the electron and the nuclear spin ($M_I = +½$ or $M_I = –½$). The two allowed EPR transitions ($\Delta M_S = \pm 1$, $\Delta M_I = 0$) can occur between $M_S = –½$, $M_I = –½$ and $M_S = +½$, $M_I = –½$ and between $M_S = –½$, $M_I = +½$ and $M_S = +½$, $M_I = +½$ (thin arrows). Note that transitions between nuclear levels of the same electron spin manifold (e.g., $M_S = –½$) can also occur (bold gray arrows). Such an experiment is known as ENDOR (Electron Nuclear Double Resonance) and will be discussed in more detail below in Section 3.5.2.

For a more practical aspect when analyzing the EPR spectra, it is convenient to use "stick diagrams" that account for hyperfine splittings arising from different nuclei (note that for the moment we can disregard the anisotropy of the g and A tensors; such a situation corresponds to an EPR spectrum taken in solution). In the stick diagram (in a way analogous to Pascal's triangle) the vertical line is a single transition between two energy levels, the horizontal spacing between two lines represents the hyperfine coupling, and the number arising from the summation

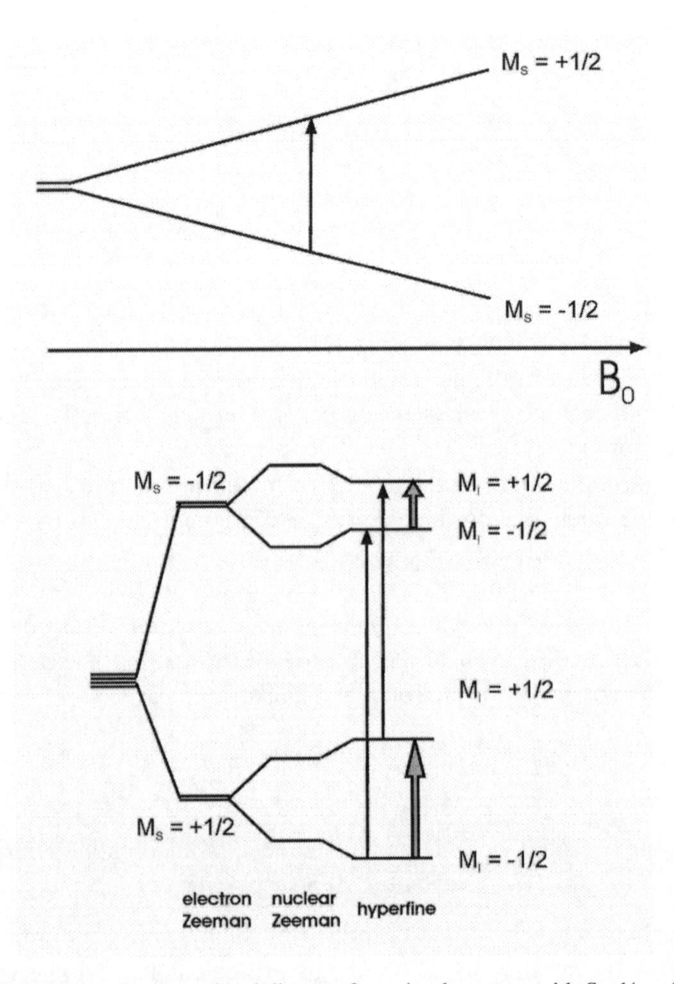

Figure 3.2. Schematic energy level diagram for a simple system with $S = ½$ and $I = ½$.

of all the single lines in a base row gives the EPR line intensity. The corresponding stick diagram for the energy diagram in Figure 3.2 is shown in Figure 3.3. The examples in Figure 3.4 represent stick diagrams with the corresponding EPR spectra for similar systems with one single unpaired electron (S = 1/2) and (A) three equivalent nuclei of spin I = 1/2 or (B) seven equivalent nuclei of spin I = 1/2. The respective numbers in the stick diagrams stand for the relative intensity of the resulting lines.

In a typical case, different nuclei will exhibit different hyperfine splitting. Let us illustrate this with the following example. The hypothetical phenol-type radical $M^{\bullet}$ (Fig. 3.5; note that "•" symbol is used to specify that M bears an unpaired electron), possesses 4 nuclei with nuclear spin I = 1/2 that can interact with the unpaired electron. Fluorine nucleus F_1 has the largest hyperfine coupling, followed by two equivalent protons, H_2 and H_3, and finally by the proton H_3, which has the smallest hyperfine coupling. To simplify, we assume that R does not possess any magnetic nuclei. The stick diagram and the resulting spectrum consist of 12 lines with an alternating 1:1:2:2:1:1 intensity pattern.

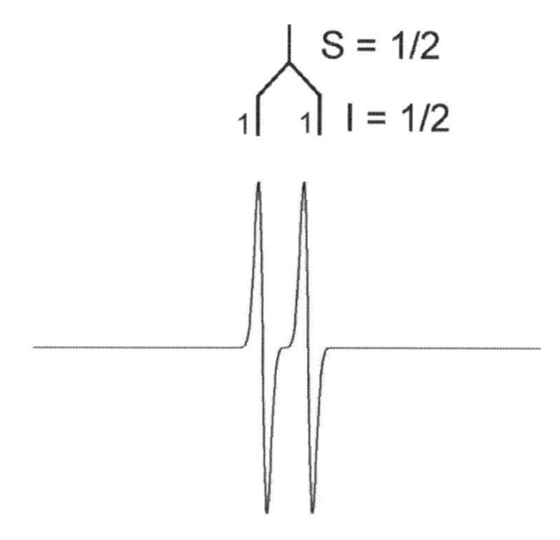

Figure 3.3. Stick diagram and the corresponding spectrum for a simple system with $S = ½$ and $I = ½$.

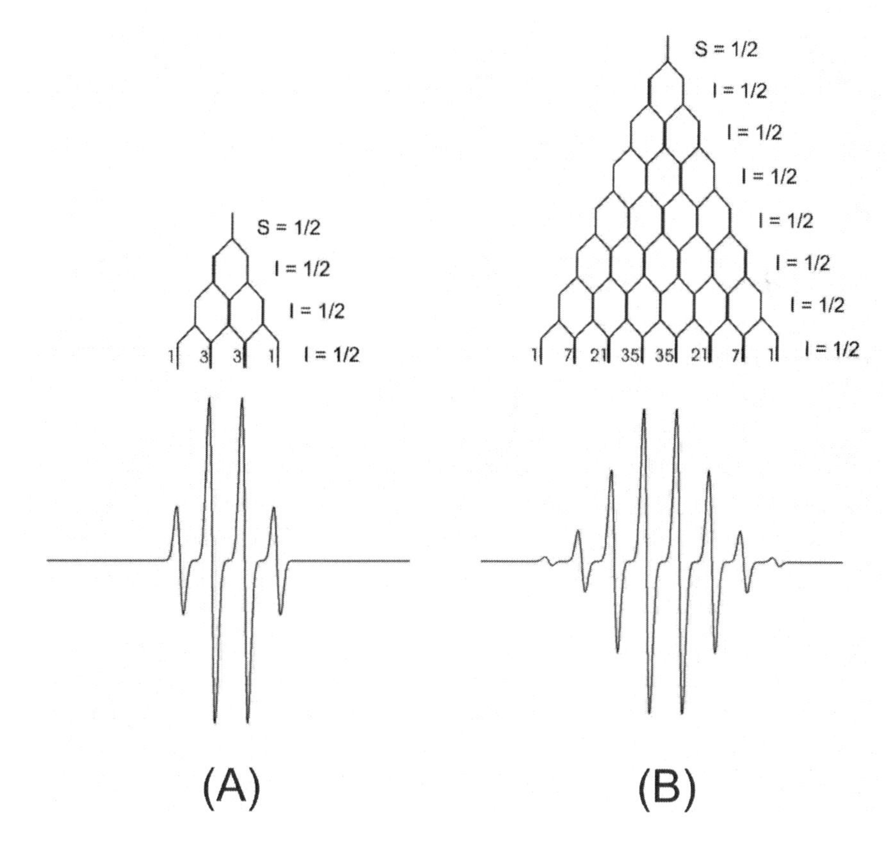

Figure 3.4. Stick diagrams and the corresponding spectra for $S = 1/2$ and three (A) or seven (B) equivalent nuclei of spin $I = 1/2$.

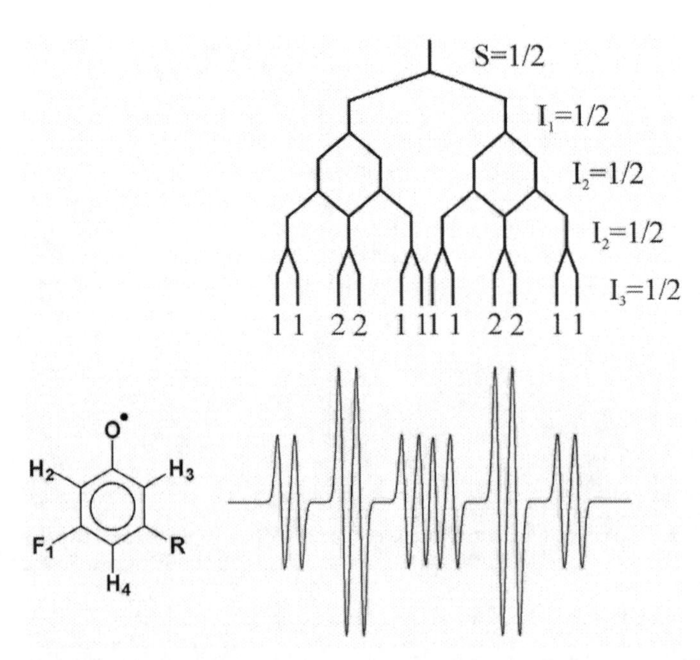

Figure 3.5. Stick diagram and the corresponding spectrum for $S = 1/2$, and four inequivalent nuclei of spin $I = 1/2$.

It is useful to remember that in many cases the isotopic substitution can greatly help in the interpretation of an EPR spectrum. One of the simplest and most common practices is the substitution of the protons (H) with deuterons (D). The isotropic hyperfine coupling constant for D is approximately 6 times smaller than for the corresponding H. The exact H/D A_{iso} ratio can be obtained from the ratio of the magnetic moments of the proton and the deuteron.

The situation is slightly more complicated for the paramagnetic species comporting metals. In contrast to the organic radicals, in transition-metal ions the unpaired electron resides in d-orbitals. Such a situation results in an important spin–orbit coupling or, in other words, in a large orbit angular momentum that couples to the magnetic moment of the electron. The very direct consequence of the strong spin–orbit coupling is a highly marked anisotropy of the g tensor. The principal elements of the g tensor for transition-metal-containing molecules undergo important deviation from the electron free value (2.0023). In addition, a secondary consequence of a large spin–orbit coupling is the presence of zero-field splitting (ZFS).

The presence of ligands around the metal center causes (to the first approximation) an electric field, which will influence the way that the valence electrons will be distributed among the d-orbitals. This is at the base of the so-called ligand-field theory (a simplified version of this theory is known as the crystal field theory). Depending on the "strength" of the ligand (the ability of donating or accepting the π electrons to or from the metal ion) the metal electrons can be either in a low- or high-spin configuration. The high-spin configuration is found in cases where the ligand field strength is weak and thus the spin pairing energy exceeds that of placing an electron in a high-energy level. This corresponds to a situation where as many unpaired electrons as possible reside in the empty d-orbitals, and therefore all the valence d-electrons remain

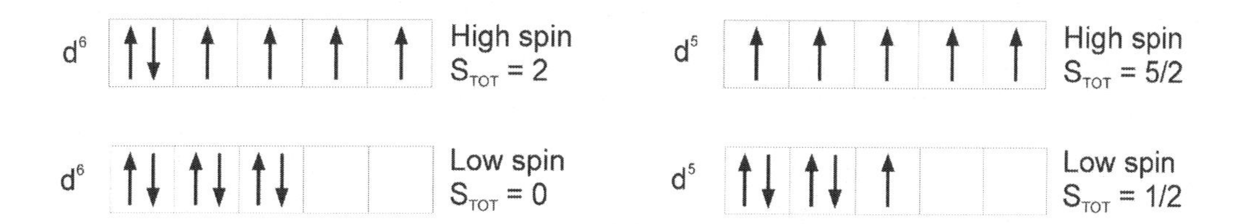

Figure 3.6. Schematic representation of the 3d shell with valence electrons in case of low- and high-spin d^6 and d^5 species.

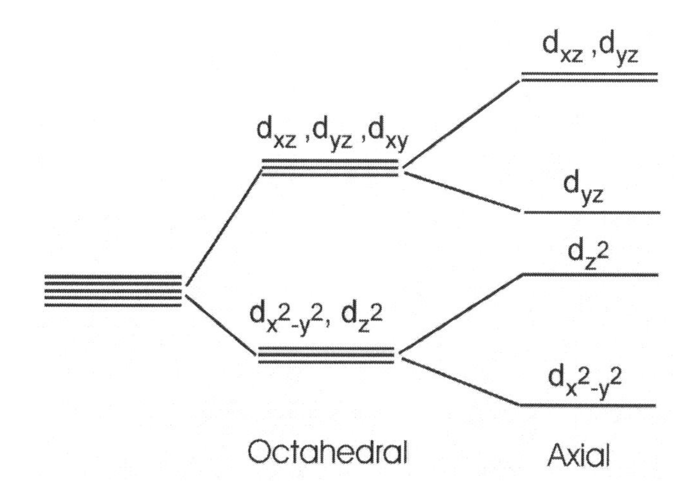

Figure 3.7. Schematic representation of the splitting of d-orbitals in the octahedral and axial ligand fields.

unpaired. In the low-spin configuration, the electrons are paired (one spin-up and one spin-down electron) in each of the d-orbitals, and this results in a configuration where all the spins are being paired (total S, $S_T = 0$) or in a configuration where only one electron remains unpaired (Fig. 3.6).

In case of the free transition metal ions, the presence of ligands is also responsible for lifting the degeneracy of the electronic levels corresponding to the d-orbitals. Figure 3.7 depicts a schematic diagram of the energy levels for the transition metal ions in octahedral and axial ligand fields. Lowering of the symmetry results in additional splitting of the levels and a more complicated EPR spectrum.

To illustrate this concept with a simple example, let us consider two systems: one with two unpaired electrons (total electron spin $S_T = 1$) and another with four unpaired electrons (total electron spin $S_T = 2$). For $S_T = 1$, three possible values of M_S are implicated: $M_S = -1$, 0 and +1. For the sake of simplicity we will consider only two cases: (A) complete absence of ZFS and consequently $D = 0$, and (B) large ZFS with $D \gg 0$. Figure 3.8 illustrates simplified energy levels for both situations. In the absence of ZFS, the three levels remain degenerate in the absence of the magnetic field. At fields where the energy difference corresponds to the quanta of the energy carried by the microwave, the transitions occur (the length of the small

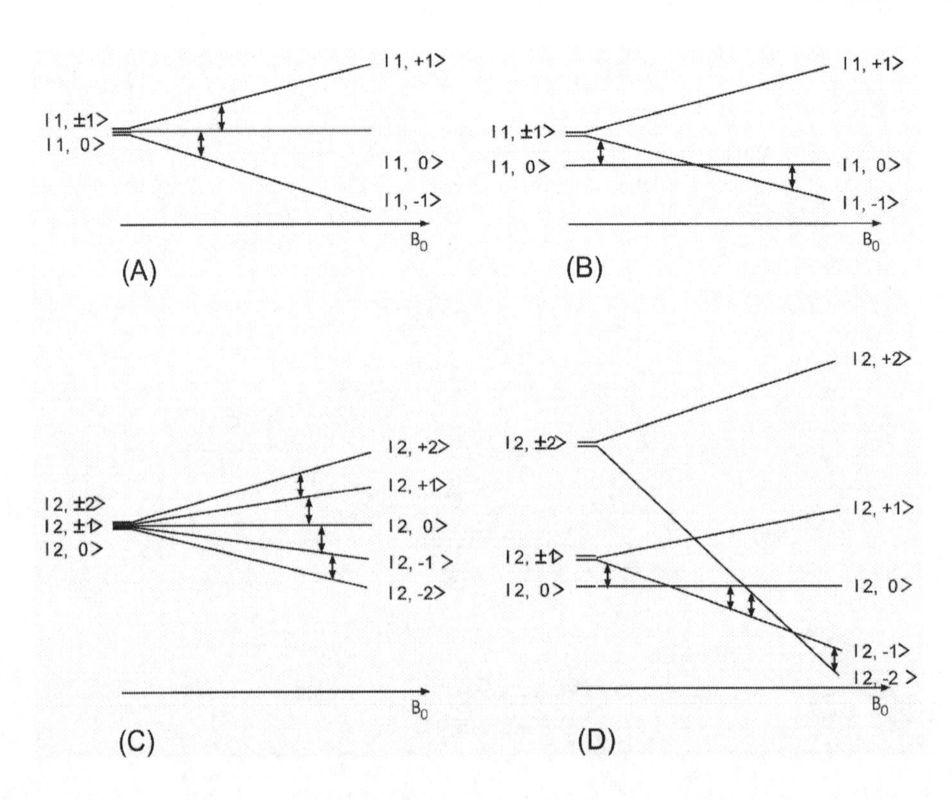

Figure 3.8. Energy levels for $S = 1$, with (A) $D = 0$ and (B) $D >> 0$ and $S = 2$, with (C) $D = 0$ and (D) $D >> 0$.

arrows represents the energy of the microwave). In this case, two closely positioned transitions result. In case of a large ZFS, however, at zero field the energy level corresponding to $M_S = 0$ lies below the energy levels of the $M_S = +1$ and $M_S = -1$ levels. But since the levels are crossed, the transitions are now shifted. The first transition appears at low magnetic field, while the second transition will occur at a much higher magnetic field. Analogous treatment applies for the $S_T = 2$ case. In the absence of ZFS, four transitions close to each other will appear in a small magnetic field window. For large ZFS, the same four transitions are spread over a large interval of the magnetic field. One consequence of such "spread" of the possible EPR transitions is the fact that they can appear at magnetic fields far outside the range operated by the spectrometer magnet. In such case some paramagnetic species can appear as "EPR silent" at the given frequency of the spectrometer. In order to perform EPR experiments on such species, high-field and high-frequency EPR, where fields as high as 30 T can be achieved, have to be used. High-field EPR is particularly useful in studying the transition metal complexes with large ZFS parameters.

3.5. CONTINUOUS WAVE (CW) VERSUS PULSED EPR TECHNIQUES

Most commercial EPR instruments were, until the last decade, operating in so-called Continuous Wave (CW) mode, where the microwave field B_1 is continuously sent to the cavity hosting the sample with the constant magnitude of oscillations. In CW mode, the static magnetic field B_0 is slowly scanned, while the B_1 of a given (and precisely locked) frequency is delivered to the sample. The signal received back from the sample by the detector results from absorption of the microwave radiation by the sample at the magnetic field where the transition occurs. This absorption signal is recorded and plotted to give the actual EPR spectrum. However, other techniques where the excitation amplitude of B_1 is time dependent are also possible. These are commonly grouped under a broad designation of pulsed EPR. We will not present here any pulsed EPR theory but instead will give the important concepts that should allow a phenomenological understanding of the pulsed EPR experiments.

The first pulsed EPR experiments were due to the pioneering work by W.B. Mims from Bell Laboratories, who used short pulses of microwave (MW) radiation to manipulate the spins. In a way analogous to the pulsed NMR experiments, the application of a suitable sequence of MW pulses can yield information that is usually "hidden" or unavailable from simple CW-EPR spectroscopy.

In the classical picture, the spin packet in the static magnetic field B_0 will precess around the quantization axis (Z // B_0) with a constant frequency, called the Larmor frequency, ν_L. The resulting total magnetization vector, M_Z, points along the z axis (Fig. 3.9). If a microwave pulse is applied along the Y axis, the spin package will start to rotate in the XZ plane. By selecting a microwave pulse of appropriate length, it is possible to rotate the spin package by 90 or 180° degrees. In our case, after such a pulse, magnetization vector M_Z will be along X (90° pulse) or –Z (180° pulse).

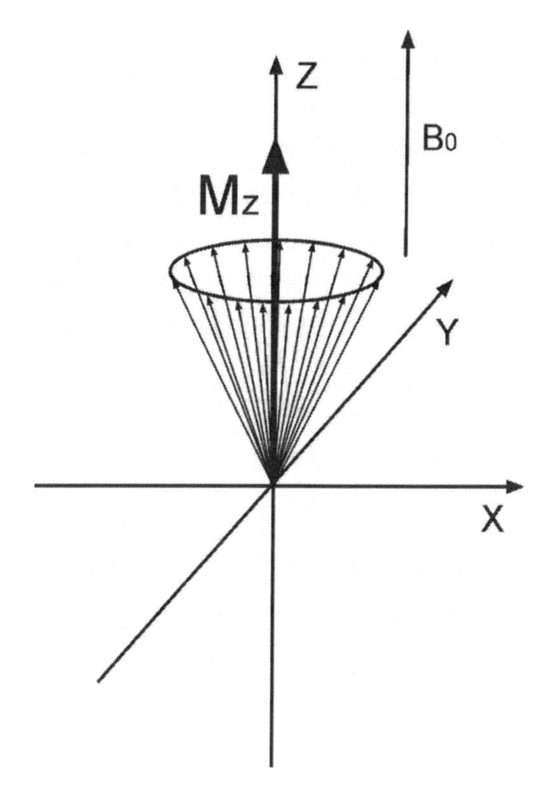

Figure 3.9. The total magnetization vector M_Z and the spin packet processing around B_0 with the Larmor frequency ν_L.

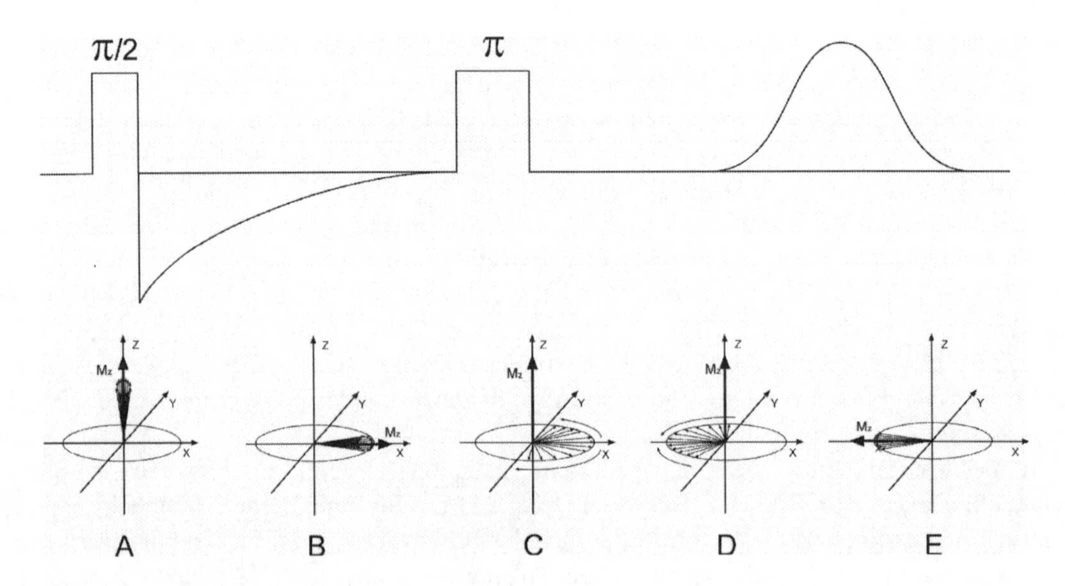

Figure 3.10. Pictorial representation of the total magnetization vector M_z during a Hahn spin-echo sequence.

If an appropriate detector is positioned along the X axis,[1] one can monitor the intensity of M_Z. The simplest pulsed EPR experiment would consist of a 90° pulse, followed by observation of the decay of the magnetization vector. This would result in a Free Induction Decay (FID) type pattern, in a way analogous to NMR that can be Fourier transformed to yield the actual EPR spectrum. However, it is not practical for EPR, especially at higher magnetic fields, because in order to excite a spin packet over a large frequency band very short and powerful pulses have to be employed.

By contrast, a simple pulse sequence that combines two pulses (the so-called Hahn sequence) allows one to easily perform an EPR experiment. The spin packet behavior accompanying the two pulses is pictorially represented in Figure 3.10. Before any pulse is applied, the spins are processing with Larmor frequency around Z (A). After the first $\pi/2$ pulse, the spin packet comes along X (B). After the pulse is finished, the spins continue to rotate around Z with different angular speed. After a certain time τ, the spins are spread in the XY plane (C). Now the second pulse (π) is applied and all the spins again rotated by 180° (D). The spins continue to rotate now with the same velocity around Z, but this time in the opposite direction (D). After the same time τ, the spins recombine (the magnitude of M_Z increases) and the so called spin-echo is observed along $-X$ (E).

The pulsed EPR spectrum can be obtained in a way similar to the CW technique. The static magnetic field is scanned, and at each step of the magnetic field scan the same pulse sequence is repeated, while the magnitude of the resulting spin-echo is recorded. This type of experiment is called Electron Spin Echo Detected EPR (ESE-EPR).

As mentioned in earlier, the pulsed techniques have almost an unlimited potential and can be used for many exciting experiments. The appearance of various pulse sequences has resulted in an armory of EPR-related techniques that can be applied for study of specific interactions. Some of the most common techniques are briefly mentioned here.

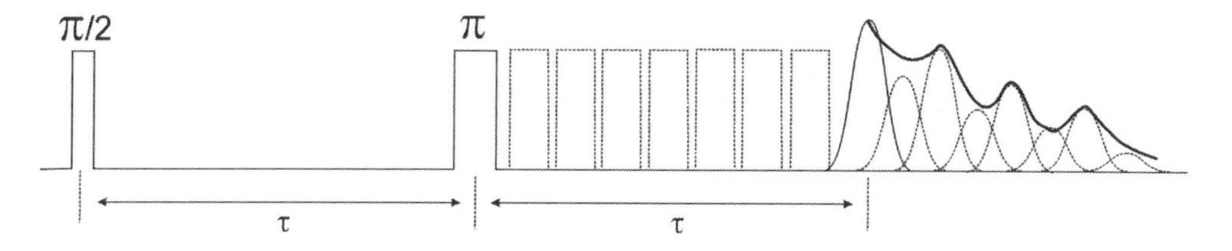

Figure 3.11. Schematic representation of the pulse sequence for the ESEEM experiment.

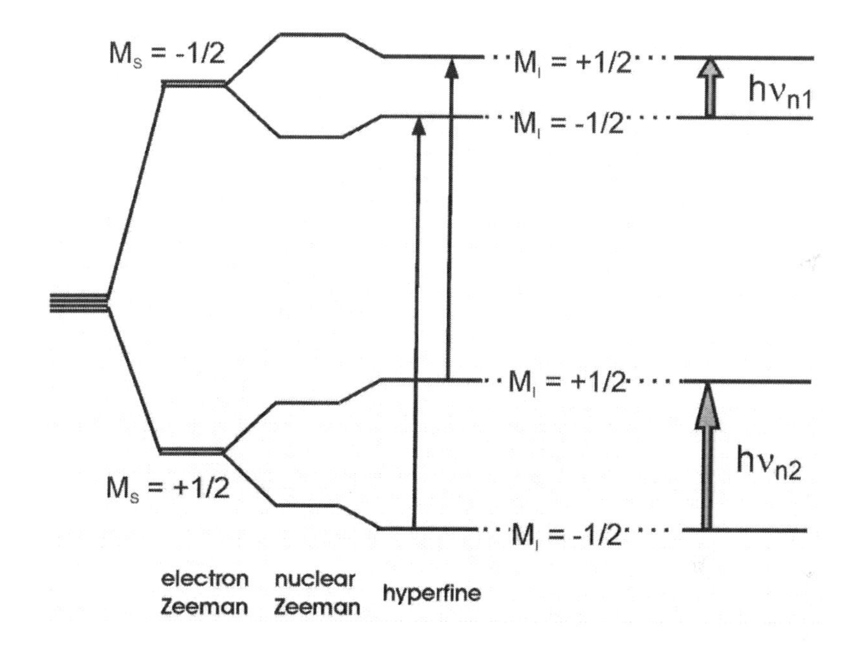

Figure 3.12. Schematic representation of the energy levels for a simple system with $S = ½$ and $I = ½$.

3.5.1. Electron Spin Echo Envelope Modulation (ESEEM)

In the ESEEM technique, the amplitude of the spin-echo signal is recorded as a function of the time interval τ between two (or three) pulses (Fig. 3.11). Variation of the amplitude of the echo at different τ results in an echo envelope that exhibits modulation. The frequency and amplitude of these modulations are related to the hyperfine or/and quadrupolar coupling of nearby nuclei. ESEEM is an extremely useful technique in probing the radial distribution of magnetic nuclei around the paramagnetic center.

3.5.2. Electron Nuclear Double Resonance (ENDOR)

As mentioned in Section 3.4, the transition between nuclear levels can also be exploited in an EPR experiment. If we look again at the schematic energy levels for an $S = 1/2$ and $I = 1/2$ system depicted in Figure 3.12, it is apparent that the energy of the NMR transitions is much smaller as compared to the analogous EPR transitions occurring at the same static magnetic field. In fact, the ratio of the frequencies, which will drive the nuclear versus electronic transitions, is mainly dictated by the ratio of the Bohr electron magneton versus nuclear magneton, and it is more than 1:1800. For a proton, as an example, at the operating frequency of the EPR spectrometer close to 10 GHz the Larmor frequency, ν_L^H for the nuclear transition will be ~15 MHz. Such an experiment (which is also possible in CW mode) is called Electron Nuclear Double Resonance (ENDOR), and it is in fact an NMR experiment, detected by EPR.

In a typical ENDOR setup, a microwave with fixed frequency is used to partially saturate electronic Zeeman transitions of the paramagnet and to monitor the intensity of the resulting EPR signal, while a strong radiofrequency is used to drive the NMR transitions. In this way, a particular nuclear transition rate is enhanced and the information about specific nuclei is transferred to the electrons via the electron–nuclear hyperfine interaction (interaction between the unpaired electron and the surrounding magnetic nuclei). The nuclear transition increases the induced relaxation rate of the electronic transition, which in turn changes the EPR signal intensity of the latter. In a simplified picture, ENDOR translates the NMR of nuclei surrounding the paramagnetic center into an EPR-detectable signal. In many cases, the hyperfine couplings that are overlapping, making their assignment difficult or are broadened in the conventional EPR line, can be detected and measured by ENDOR with a precision much higher than that in a classical EPR experiment.

The ENDOR and ESEEM techniques are complementary to EPR and are of great importance in determining structural features of the paramagnetic systems. In particular, important information about the geometries (distances and orientations) of the atoms or molecules surrounding the paramagnetic centers can be obtained from ENDOR experiments performed on biological samples with proteins and enzymes containing paramagnetic metal ions. For example, the hyperfine interaction with nuclei such as protons, deuterons, nitrogens, or phosphorus atoms (which all exhibit nuclear spin $I > 0$) can be detected by ENDOR and can yield a tremendous amount of information about the ligation of the paramagnetic metal center by the surrounding protein.

Pulsed ENDOR techniques offer several advantages over the CW-ENDOR techniques. Briefly, the CW-ENDOR technique requires a very delicate control of both temperature and the delivered microwave power. This is related to the fact that the intensity of the signal in the CW-ENDOR depends on the subtle balance between electron and nuclear spin–spin and spin–lattice relaxation rates in a critical way. This tight dependence unfortunately restricts the use of this method to a very narrow temperature range. It is beyond the scope of this introductory chapter to provide a detailed explanation of the various relaxation mechanisms, but the interested reader can find a complete discussion of these mechanisms in the monographs by Wertz & Bolton and Schweiger & Jeschke (see the section on Further Study).

The two widely used pulsed ENDOR sequences are due to W.B. Mims and E.R. Davis, and they are consequently named of their inventors (Fig. 3.13). In general, Mims ENDOR is used for the more weakly coupled nuclei (smaller hyperfine coupling constant A), while Davies ENDOR is preferred for the more strongly coupled systems. For a simple $S = 1/2$, $I = 1/2$ system (as in Fig. 3.12), two cases can occur, depending on the magnitude of the hyperfine coupling versus the Larmor frequency of a given nucleus at fixed spectrometer frequency. In the

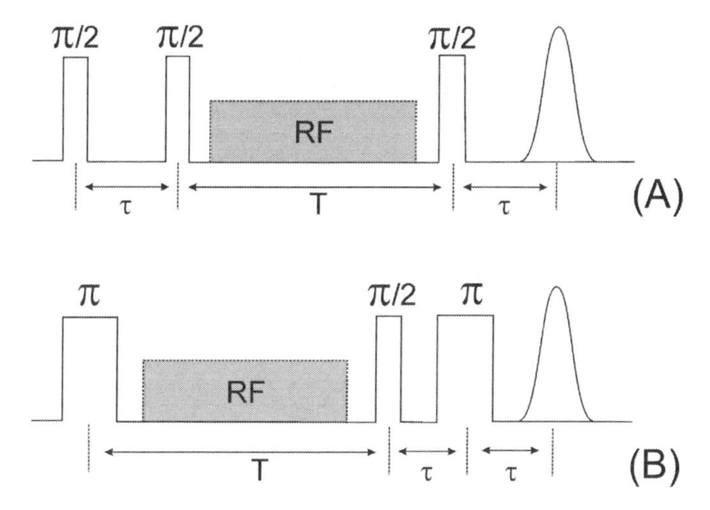

Figure 3.13. Schematic representation of the pulse sequences for the pulsed Mims ENDOR (A) and Davis ENDOR (B) experiments.

weak coupling case ($\nu_L > A/2$) the two peaks corresponding to ν_{n1} and ν_{n2} will be centered around ν_L and spaced by A. However in the strong coupling case ($\nu_L < A/2$), these peaks will be centered around $A/2$ and spaced by $2\nu_L$ (Fig. 3.14).

To illustrate the ability of the ENDOR technique to obtain information, which is not directly accessible by classical EPR, in Figure 3.15 are shown the CW-EPR spectrum and the corresponding ENDOR transitions of the radical form of canthaxanthin. This interesting biological molecule belongs to the larger class of molecules called carotenoids, which are important phytochemicals found in living organisms. Carotenoids also play a key role in photosynthesis, where they participate in energy and electron transfer in Photosystem II (PS II). As apparent from Figure 3.15A, the CW-EPR spectrum of the canthaxanthin radical cation exhibits only a broad, featureless peak. However, the corresponding ENDOR spectrum in Figure 3.15 shows three sets of peaks corresponding to the three classes of protons found in canthaxanthin. This example shows that by using the ENDOR technique it is possible to detect hyperfine couplings that are broadened in the conventional CW-EPR spectrum.

3.5.3. 2D Techniques

Many exciting 2D EPR techniques also exist. The most popular is an extension of ESEEM, called 2D-HYSCORE (Hyperfine Sublevel Correlation Spectroscopy). In this technique two nuclear Larmor frequencies associated with a particular hyperfine coupling are correlated, allowing one to distinguish between otherwise unresolved hyperfine couplings.

3.6. EPR INSTRUMENTATION

The schematic drawing of a simplest CW-EPR spectrometer is presented in Figure 3.16. The spectrometer consists of a magnet (source of a homogenous magnetic field), a microwave

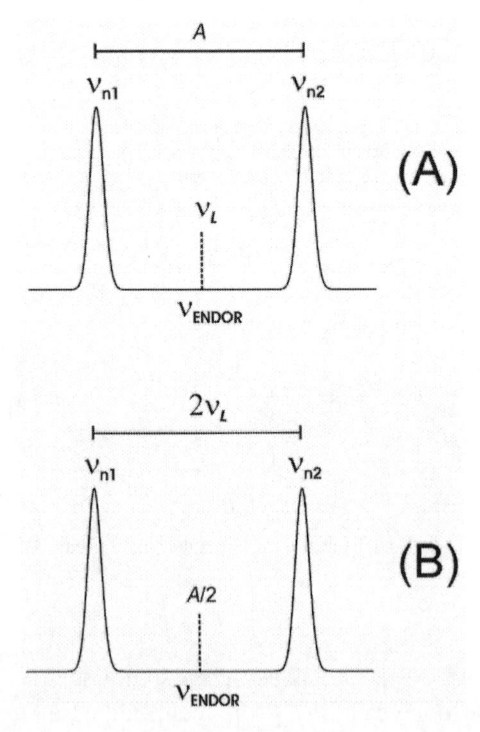

Figure 3.14. Schematic representation of the simulated ENDOR spectra for (A) the weak coupling case ($\nu_L > A/2$); ν_{n1} and ν_{n2} are centered around ν_L and spaced by A; and (B) for the strong coupling case ($\nu_L < A/2$); ν_{n1} and ν_{n2} are centered around $A/2$ and spaced by $2\nu_L$.

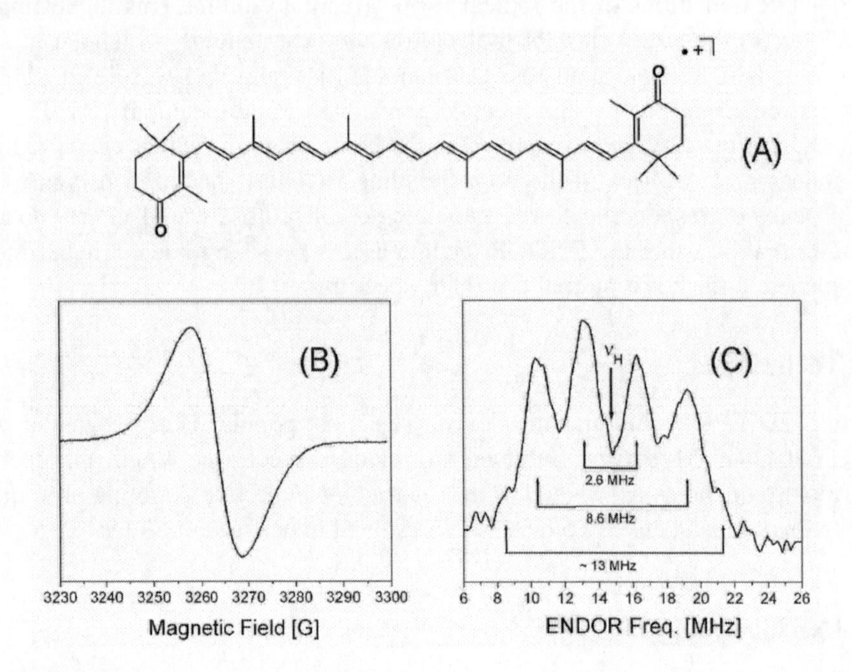

Figure 3.15. The structure of the Canthaxanthin radical cation (A) and the corresponding CW-EPR (B) and ENDOR (C) spectra.

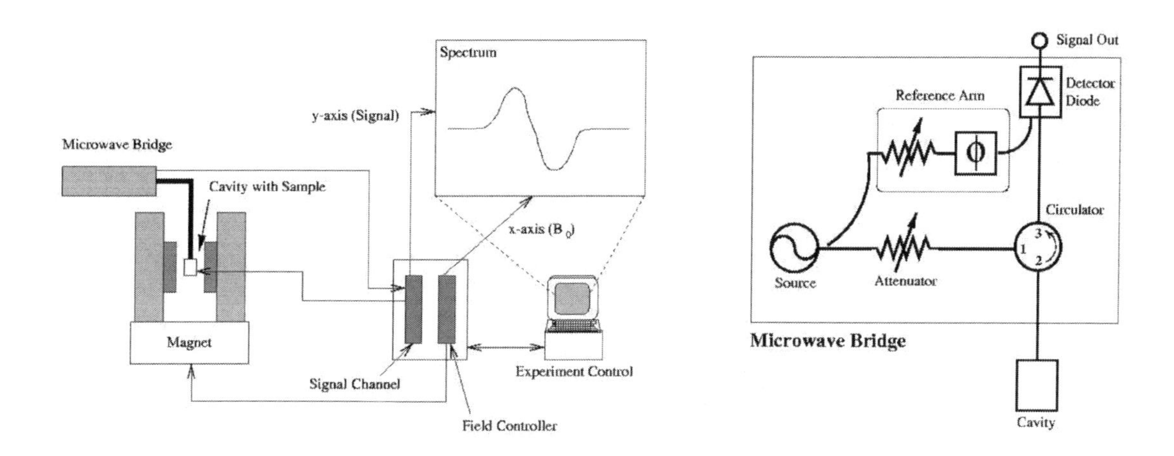

Figure 3.16. Basic scheme of a CW-EPR spectrometer. Courtesy of Dr. Josef Granwehr, ETH Zürich.

bridge (source of a microwave radiation), a cavity where the sample is placed, and the detection system. The sample is positioned in the cavity in the homogenous static magnetic field B_0. The B_0 is controlled by changing the electromagnet current via the field controller. In the bridge, the microwave source delivers a microwave radiation, which is transported via a working arm to the cavity, when at the resonant field microwave absorption by the sample occurs. A small amount of the microwave power is also sent via the reference arm. After absorption of a part of the microwave power by the sample, the remaining power is reflected and goes trough the circulator to the mixer (or detector diode). By setting the appropriate phase between the traveling microwave in both arms (working arm and reference arm) the difference signal coming from the mixer is converted into a DC voltage, which can be easily detected with a boxcar integrator or a digital oscilloscope.

For the CW experiments, the static magnetic field B_0 is also modulated with a low frequency (usually 100 kHz) to increase the signal-to-noise (S/N) ratio. Field modulation is performed during the magnetic field scan, and while going through the sample resonance the amplitude of the modulated pattern of the absorption signal is recorded, which results in a "derivative" shape of the EPR signal (Fig. 3.17). In a pulsed EPR instrument, a very similar setup is used. The main difference resides in the microwave bridge, which is the microwave power source and delivers a sequence of short (10–500 ns) low-amplitude pulses. These pulses are first amplified with a TWT (Traveling Wave Transceiver) amplifier, before being sent to the cavity. As in the CW spectrometer, the losses in the reflected microwave power resulting from absorption by the sample are then detected, giving rise to the EPR signal.

One of the most important parts of the spectrometer is the resonator, which hosts the paramagnetic sample. A broad range of different types and designs for EPR resonators exists, and we show here only an example of a commonly used rectangular cavity, depicted schematically in Figure 3.18. The microwave field B_1 (not shown) is perpendicular to the static magnetic field B_0. For CW mode an external modulation of the B_0 field is necessary, and therefore two modulation coils that deliver the modulating magnetic field parallel to B_0 are placed on both sides of the

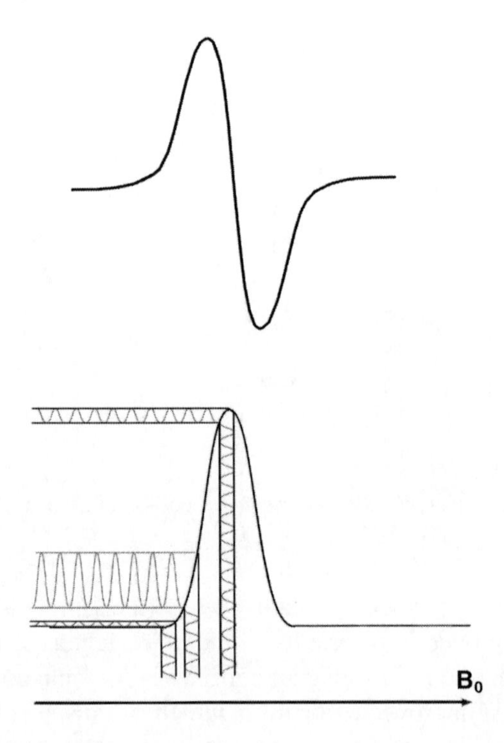

Figure 3.17. Schematic representation of the field modulation concept used in the CW-EPR spectrometer.

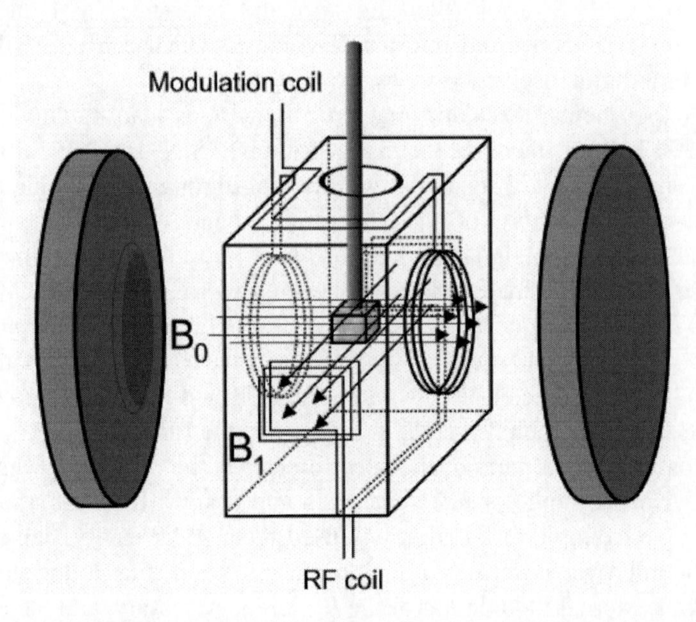

Figure 3.18. A schematic drawing of a rectangular EPR cavity with modulation coils for the modulation of the static magnetic field and with the RF coils used to generate the radiofrequency field for the ENDOR experiments.

cavity. For the double-resonance experiments such as ENDOR, an additional coil that generates the radiofrequency field B_2 (perpendicular to B_0) necessary to drive the NMR transitions is placed on the cavity walls, or directly inside the resonant part.

As highlighted in Section 3.2, contrary to NMR, the fast evolution of EPR became possible only within last two decades. This is mainly due to two factors: (i) the availability of electronic devices for frequencies much higher than the conventional X- or Q-bands, and (ii) the development of very fast electronics, such as ultrafast electronic switches or analog-to-digital converters. Such fast-performing devices are not only critical in obtaining very short and intense pulses, necessary for proper excitation of the spins, but are also required to ensure ultrafast data acquisition. Currently a number of commercial instruments operating in both pulsed and CW mode are available, ranging from very low frequencies (300 MHz, L-band) to very high frequencies of 95 GHz (W-band).

3.7. EPR MEASUREMENTS IN SOLUTION AND SOLID STATE

Paramagnetic species can be studied both in solution and in solid state. Single crystals, glasses, powders, polymers, low-temperature frozen solutions are examples of solid-state EPR samples. Some paramagnetic species occur naturally (e.g., Cu^{2+} or VO^{2+} complexes), but others have to be generated by various physicochemical methods (e.g., chemical reactions, UV, X-ray or γ irradiation, laser-flash photolysis, in-situ voltammetry). Many of the biological samples are aqueous solutions of proteins that have to be frozen at low temperature to prevent enzymatic turnover or chemical reactions. Some proteins can also be crystallized, yielding single crystals suitable for EPR studies.

The shape of the EPR spectra obtained from the same sample but acquired in different phases (e.g., solution versus powder) is usually very different and results from different arrangements of the spins at the molecular level. The differences between solution, powder, and single-crystal spectra can be rationalized by considering an ensemble of spins in a closed system that represent the sample. In Figure 3.19 such an ensemble of spins is represented: small ellip-

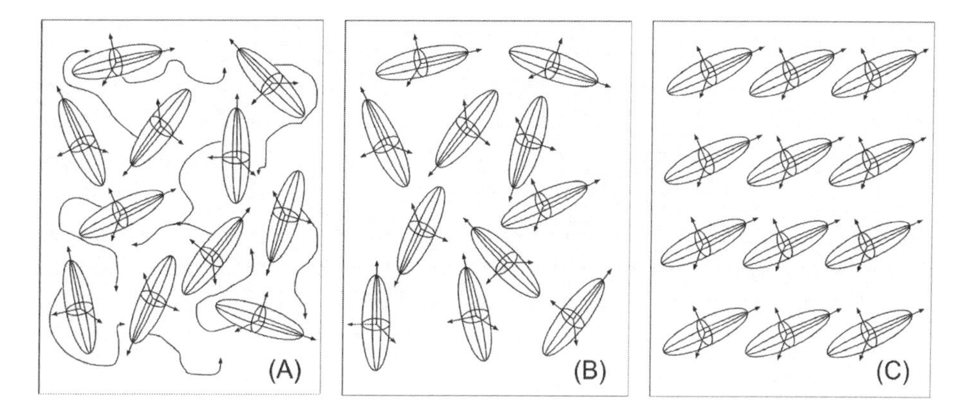

Figure 3.19. Schematic representation of the molecular anisotropy concept in different phases: (A) solution, (B) frozen solution or powder, (C) single crystal.

soids represent individual molecules (or individual spins), and the previously discussed interactions result in the anisotropy of the hyperfine and/or *g* tensors. In solution (Fig. 3.19A) the molecules undergo constant random motion on the timescale of the EPR measurement, which results in a time-averaged spectrum. The situation is similar in powder or frozen solution: the molecules are disordered; however, since no motion is present, they remain in the same position over the timescale of the experiment (Fig. 3.19B). Finally, in single crystals (or other types of oriented samples, e.g., oriented biological membranes) each molecule occupies the exact same position in the crystal cell, so that the macroscopic symmetry properties of the crystal reflect the same microscopic properties at the molecular scale.

The main difference between an EPR spectrum recorded in solution and one recorded in a solid sample is the absence of anisotropic terms for both dipolar coupling and *g* terms for the solution spectrum. The loss of anisotropy is due to the fast motion of the randomly oriented spins (Fig. 3.19A), so that the dipolar coupling and the *g* terms (which both depend on the orientation of the paramagnet in the static magnetic field) are averaged. Such a situation results in an averaged value for *g* (g_{ave}) and in a complete absence of dipolar hyperfine coupling, which vanishes ($\tau_{aniso} = 0$). Therefore, in liquid samples *g* reduces to g_{ave}, τ_{aniso} is averaged to zero, and the isotropic Fermi contact term of the hyperfine coupling, a_{iso}, is the only remaining interaction, making the spectra particularly easy to interpret. In the solid state, both a_{iso} and τ_{aniso} contribute to the spectral shape. Although the solid-state experiments result in more complicated spectra, they have a clear advantage of carrying additional (and usually interesting!) information on the system that can be extracted from analysis of the spectral patterns.

The differences between the spectral patterns obtained from different phases carrying the same paramagnetic species (solution, frozen solution, or single crystal) can be understood by analyzing the simple example illustrated in Figure 3.20. In this figure, the simulated single-crystal spectral patterns of a hypothetical paramagnetic molecule with an unpaired electron centered on the phosphorus atom (^{31}P nuclear spin is one half; $I = 1/2$) are presented. The hyperfine coupling of the unpaired electron with the phosphorus nucleus gives rise to a simple spectrum consisting of two lines (a doublet) separated by *A* (*A* is the value of the hyperfine coupling of the electron with the ^{31}P nucleus). In Figure 3.20C this spectrum is plotted for each orientation of the crystal, rotated in three perpendicular planes in steps of 10°. In solution, the same doublet with a constant separation between two lines (constant hyperfine coupling) would always be observed, independent of the orientation of the sample in the static magnetic field (Fig. 3.20A). However, if this same radical is constrained into a solid matrix, the situation is quite different. In a single crystal, each of the individual paramagnetic molecules occupies exactly the same position in the same orientation, so that the orientation of all the anisotropic ellipsoids (Fig. 3.19C) is the same for the entire crystal. If such a crystal is now placed in the cavity of the spectrometer, the resulting spectrum will depend on the orientation of the crystal in the static magnetic field B_0. By rotating the crystal in three perpendicular planes, it is possible to record the corresponding EPR transitions for each orientation of the crystal. Such collection of the field-dependent spectra results in a pattern that is commonly called “angular variation” or “field variation.” A fitting procedure can be used to reconstruct various (usually the hyperfine and *g*) tensors from such a pattern and determine the exact orientation of these tensors in the molecular frame. The situation is slightly more complicated in a frozen solution. Although all the molecules are immobilized, in contrast to the situation encountered in a single crystal, one is in the

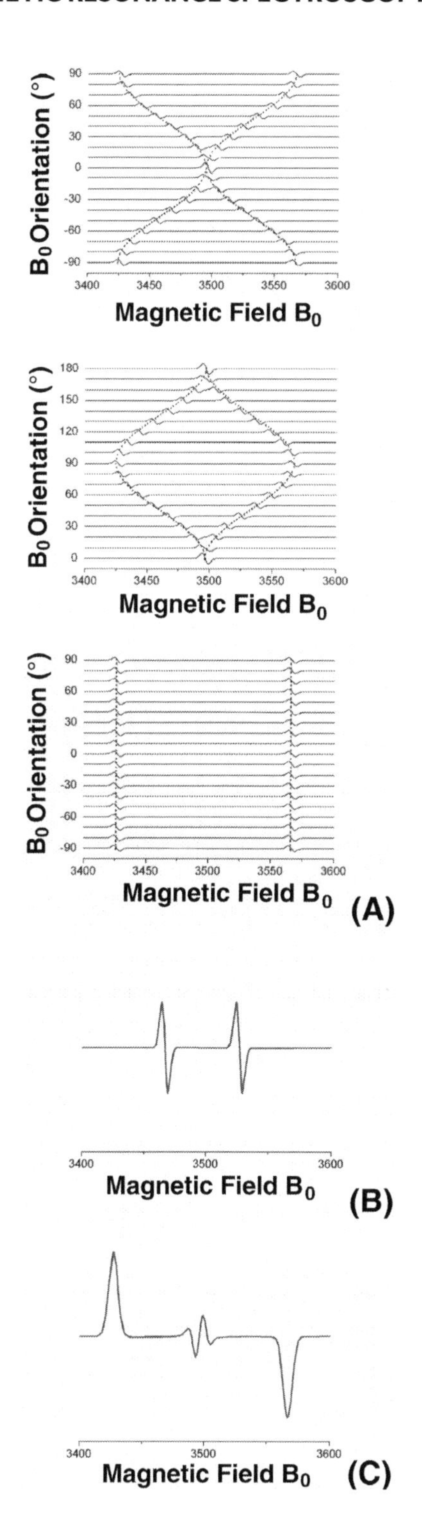

Figure 3.20. Angular variations of a single crystal of paramagnet with $S = 1/2$ and $I = 1/2$ (A). The bottom pictures show the simulated EPR spectra of the same paramagnetic species in solution (B) and in frozen solution (C).

presence of a simultaneous distribution of all possible orientations of the paramagnets. This is similar to looking simultaneously at an infinite number of "single-crystal type" spectra for each possible orientation of the crystal. The resulting spectrum of the frozen solution is in fact a sum of all such orientation-dependent spectra, and it is usually called a powder pattern (note that the same spectrum should be theoretically obtained by crushing very finely a single crystal and performing an EPR experiment on a powder sample obtained in such a way). An interesting observation can be made concerning the intensity of the peaks in the powder pattern. The outer peaks are clearly the most intense. In fact, if one analyzes the single-crystal spectra above, it is easy to notice that the spectral patterns with the largest hyperfine coupling are statistically the most abundant (all the orientations in the third rotation plane!). The contrary is true for single-crystal orientations where the hyperfine coupling is the smallest. These give rise to the weak intensity of the inner peaks in the powder pattern.

3.8. TIME-RESOLVED EPR

Only some radicals are stable on the timescale of an EPR experiment. Many paramagnetic species act as intermediates, which makes them highly reactive and usually difficult to trap. This is especially true for short-lived radicals in biological systems. Several techniques can be used to "stabilize" and investigate such short-lived species. These involve ultrafast freezing, specific mutations that can "quench" an enzymatic turnover, spin trapping, etc. However, some of the radicals have lifetimes so short that it is impossible to carry the radical for the time necessary to perform an EPR experiment. In this case, an alternative method is employed that consists in creating the paramagnetic species directly in situ, inside the EPR cavity. If paramagnetic species are generated inside the cavity of the spectrometer, it is possible to observe the kinetics of their appearance and decay by using so-called time-resolved EPR. Most popular are the experiments wherein a short laser flash is used (Fig. 3.21) to generate either an unpaired electron (this is a consequence of homolytic scission of a chemical bond, e.g., extraction of $H^\bullet$), or an excited triplet state, by activating a system with enough energy. Most of the common studies are focusing on the decay behavior of short-lived radicals, as well as on the formation of secondary paramagnetic species. Many interesting EPR experiments are also used to probe the Proton-Coupled

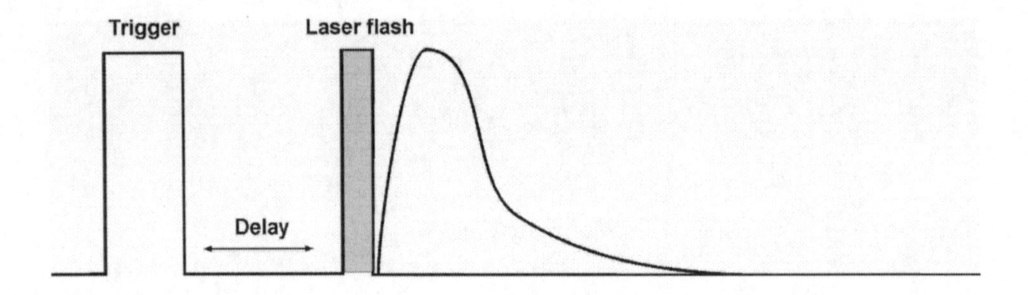

Figure 3.21. Schematic representation of the timing for the time-resolved EPR experiment. The EPR signal is observed directly after the short laser flash.

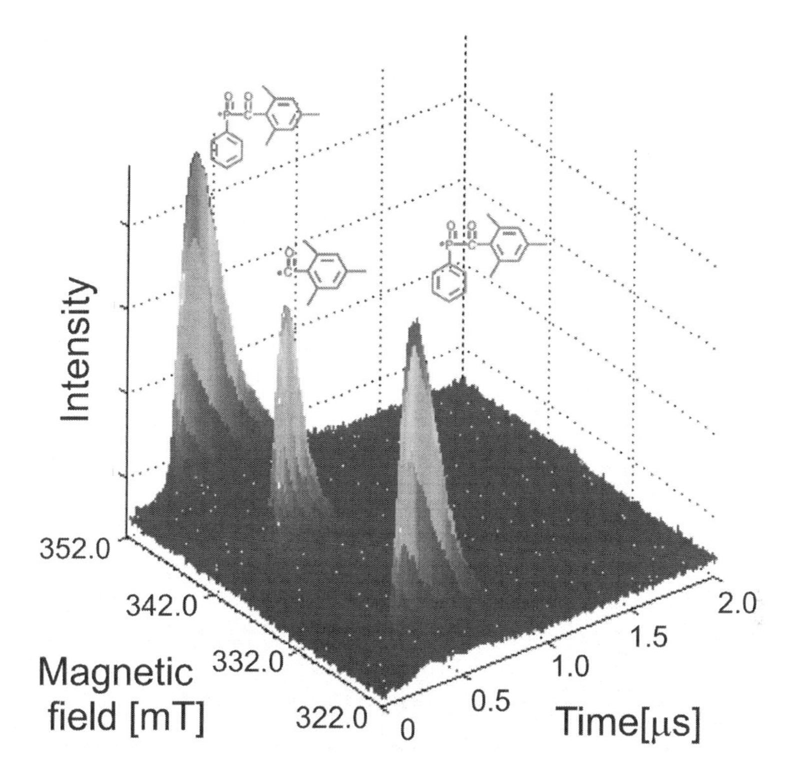

Figure 3.22. Time-resolved EPR spectrum of a $(CH_3)_2COH$ radical pair on a milliseconds timescale. Adapted from [1].

Electron Transfer (PCET) reactions that are largely present in many biological systems. The example below shows a collection of microseconds timescale time-resolved EPR spectra of phosphorus- and carbon-centered radicals produced by laser-flash photolysis of bis(2,4,6-trimethylbenzoyl) phenylphosphine oxide (Fig. 3.22).

3.9. DYNAMIC PROCESSES OBSERVED BY EPR

Many biological systems are in constant motion, at both the molecular and cellular levels. Folding and unfolding of proteins, amino acid librations, or even simple methyl group rotation are examples of such dynamic behavior. From a very general point of view, a repeated dynamic process in chemistry can be seen as a suite of discrete events or, in other words, can be associated with an exchange between two (or more) distinct configurations. The rate of such an exchange will vary with the temperature. Thus, rotation of a paramagnetic molecular fragment around a simple bond, for example, can be associated with successive exchange between a few distinct positions around the axis of rotation. In this case, one can simulate the EPR spectra resulting from such an exchange between different configurations of the paramagnetic fragments. The rate of exchange will vary with the temperature and the chemical environment. To get an understanding of what the effects of the molecular reorientation in EPR are, let us consider the simplest case, where a paramagnetic fragment can adopt two distinct configurations. In the absence of any motion (e.g., at a very low temperature), and assuming that the spectra corresponding to the different configurations are different enough to be distinct, three types of situations can be considered (for simplicity we will consider only the hyperfine and Zeeman interactions):

i. The EPR signals resulting from each configuration are distinct due to the difference in the hyperfine interaction. This effect is a consequence of two nonequivalent orientations of the orbital containing the unpaired electron. (Fig. 3.23A). This case characterizes systems with strong anisotropy of the hyperfine tensor.
ii. The EPR signals resulting from each configuration are distinct due to the difference in *g* values associated with each of the two configurations. This case characterizes systems with strong anisotropy of the electronic Zeeman effect (Fig. 3.23B).
iii. The two situations described previously appear simultaneously (Fig. 3.23C).

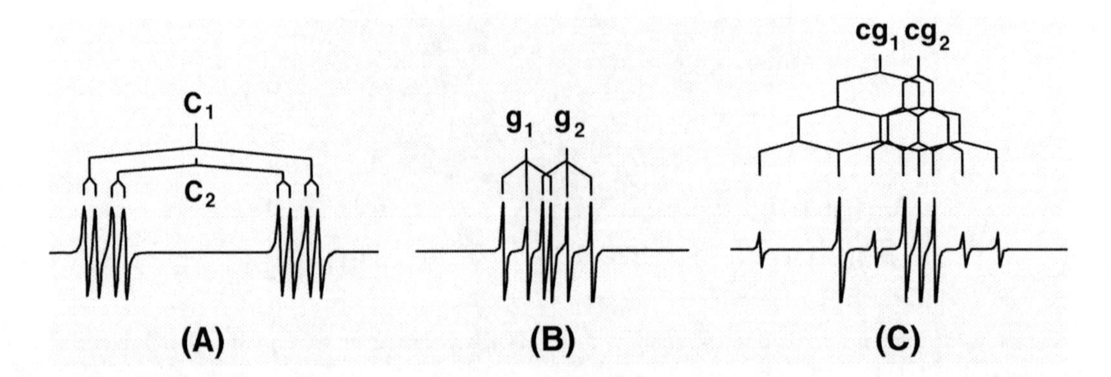

Figure 3.23. Schematic representation of the possible hyperfine and *g* tensor conformations arising from two distinct configurations of the same radical and leading to significant differences in the resulting spectral patterns.

When the temperature is increased, the exchange process will start to occur and an "average signal" between the signals associated with the two configurations will be observed. This averaging process will modify the spectrum obtained at low temperature. The more the signals of the two isolated configurations are distinct in the frozen conformations (the more the anisotropies of the *g* or hyperfine tensors are marked), the more different will be the "average" spectrum obtained at higher temperature from the "frozen conformation" spectrum. In Figure 3.24 are presented the variable-temperature experimental spectra of a phosphinyl radical (analogous to an $S = 1/2$, $I = 1/2$ system) with three distinct conformations (corresponding to three doublets with different A values) trapped at very low temperature (110 K). As the temperature is raised, the motion of the phosphinyl radical results in "jumps" between the three conformations, leading to an average spectrum with only one doublet, observed at room temperature (300 K).

A very useful technique employed in biology and biochemistry uses so-called spin labels. Spin labels are usually small paramagnetic organic molecules (e.g., nitroxides). These labels are easily functionalized to be able to bind to larger biological entities, which themselves do not exhibit any paramagnetism and, as such, cannot be directly studied by EPR. Spin labels are used

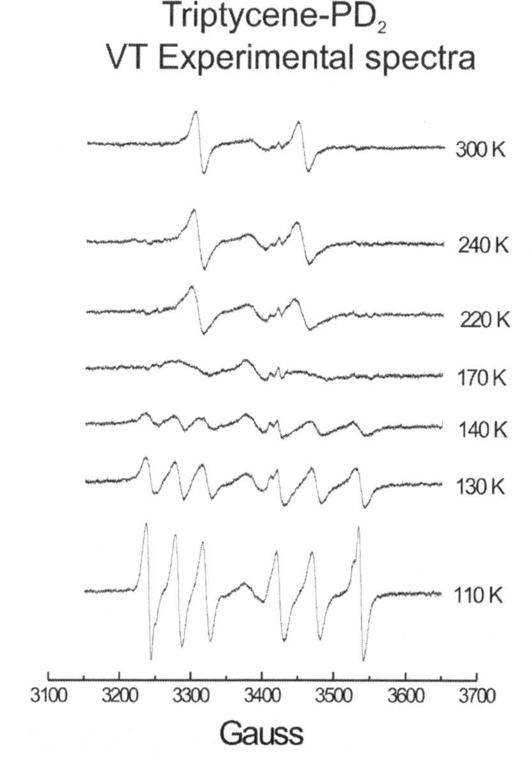

Figure 3.24. Variable-temperature EPR spectra of a phosphinyl radical with three distinct configurations trapped at low temperature (110 K).

for probing the dynamic processes that occur in large systems and are commonly involved in studying the motion, folding, and unfolding of proteins and biomolecules such as proteins or DNA. By using site-directed spin-labeling techniques it is possible to place the spin labels in the desired parts of the enzyme or membrane and to monitor a specific motion within the desired part of a large molecular system. Conformational information about DNA, for example, can be obtained by simulating the EPR spectra of the spin labels attached to the DNA helices. Complementary techniques that rely on the measurements of the dipolar coupling between two paramagnetic centers (e.g., two spin labels) are also used to precisely estimate the distances between two distant parts of the protein. Figure 3.25 shows the temperature-dependent EPR spectra recorded on the protein with specifically labeled sites to which a spin label was attached. A clear change in the line shape of the spectra, which translates into dynamical behavior, is observed at different temperatures.

3.10. MULTIFREQUENCY EXPERIMENTS

As mentioned in the introduction, most of the EPR studies were done for historical reasons in the microwave frequency region close to 9 GHz (X-band). But other frequencies (higher or

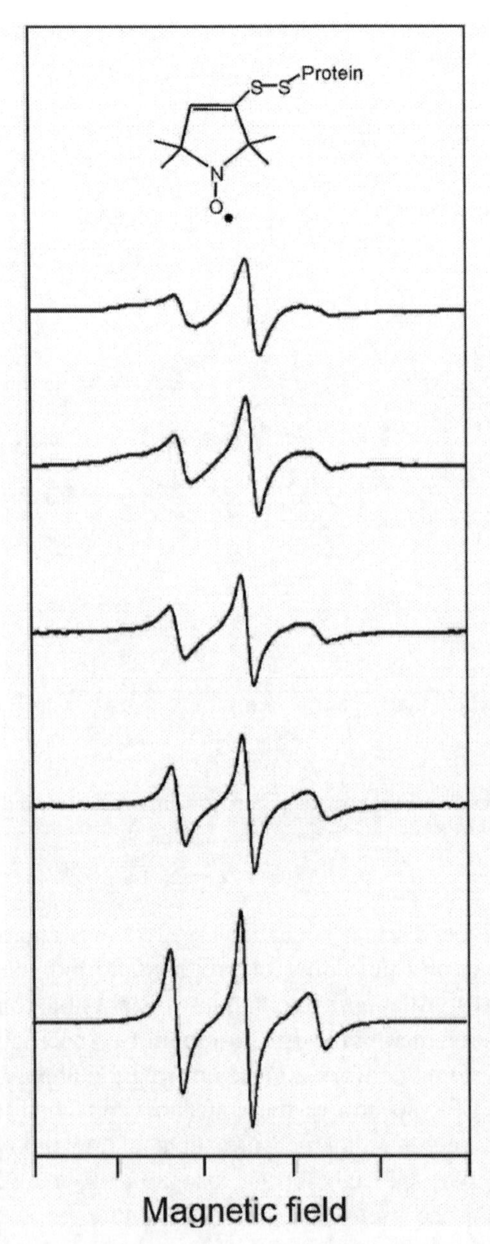

Figure 3.25. Variable temperature EPR spectrum of a full-length mouse prion protein with the attached spin STTMS spin label. Adapted from [2].

lower) are also in use, and the most popular among these are probably the Q- and W-bands. In Table 3.1 are reported some of the EPR bands used for studies of biological systems with the corresponding magnetic fields and microwave frequencies.

Table 3.1. Commonly Used EPR Frequency Bands with the Corresponding Resonant Magnetic Fields in T for $g = 2$ (1 T = 10000 G)

Band	Frequency	Resonant field (for $g = 2$)
L-band	1.5 GHz	0.054 T
S-band	3.2 GHz	0.11 T
X-band	9 GHz	0.34 T
Q-band	35 GHz	1.25 T
W-band	95 GHz	3.4 T
D-band	130 GHz	4.6 T
H-band	250 GHz	8.9 T

EPR experiments performed at several different frequencies are the most advantageous. Higher excitation frequencies require smaller samples (which is an economically important factor for many biological systems) and at the same time provide higher sensitivity. Yet the most important benefit is that acquisition of the EPR spectra of the same sample at more than one frequency can simplify interpretation of the spectral features. For example, it is common for X-band spectra to have broad and unresolved features that result from superposition of the hyperfine couplings and/or g values or overlap of spectra arising from different species present in the sample. To better understand this concept, let us examine again the EPR spin Hamiltonian in more detail:

$$H = \beta_e \cdot S \cdot \overline{g} \cdot B_0 - g_N \cdot \beta_N \cdot B_0 \cdot I + S \cdot T \cdot I + I \cdot P \cdot I + S \cdot D \cdot S\,.$$

The contributing terms can be divided into field-independent terms (they do not depend on B_0) and field-dependent terms. Field-independent terms include hyperfine and quadrupolar interactions. The other terms depend on the magnetic field, however, not in the same manner. Since this dependence of the different spin Hamiltonian terms is not uniform, by selecting an appropriate magnetic field range it is possible to enhance or reduce the contribution of one or another Hamiltonian term into the spectrum. Thus, by targeting specific Hamiltonian terms, the desired information can be selectively "extracted" from the system. The use of high magnetic fields is especially appealing because of enhanced resolution of the g tensor and quantization of the Zeeman terms, which results in simplified first-order spectra. This can be illustrated with the following example.

Let us consider again a simple $S = 1/2$ system interacting with two nuclei of $I = 1/2$ and with a slightly anisotropic g tensor. If we perform our EPR experiment at say the X-, Q-, W-. and D-bands, resonance will occur at four distinct magnetic fields, associated with four different frequencies of the microwave radiation (Fig. 3.26). Since the hyperfine interaction is field independent, only the part associated with the electron Zeeman interaction will be modified. That means that if we have a system with an anisotropic g tensor, which is unresolved at low field (the three principal values of the g tensor are too close to each other and appear at almost the same field in the EPR spectrum), it will be possible to resolve this spectral overlap (and therefore make a spectrum less complicated and extract more parameters) if we perform an experiment at higher frequency. In Figure 3.27 are shown the simulated multifrequncy EPR spectra of a hypothetical organic radical R composed of two nuclei with $I = ½$.

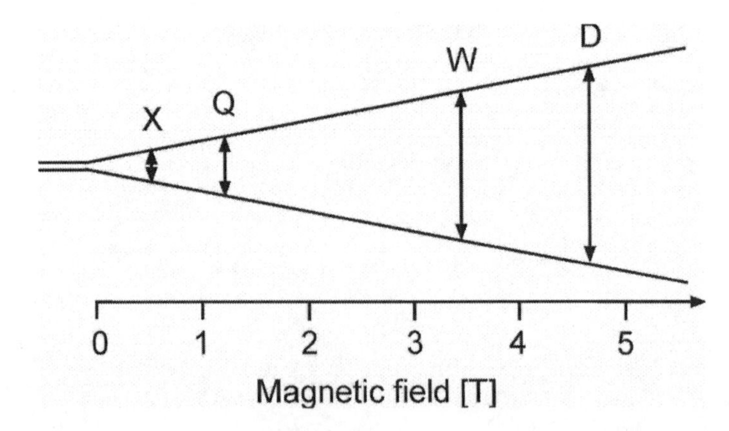

Figure 3.26. EPR transitions at different magnetic field corresponding to the common EPR frequency bands.

It is easy to notice that at the two higher frequencies not only are the three values of the *g* tensor fully resolved but also hyperfine coupling is much more clearly assigned due to less overlap of spectral lines (note, however, that this is not always the case!).

3.11. QUANTUM CHEMISTRY CALCULATIONS

Modern spectroscopy is assisted by a plethora of theoretical methods that use Quantum Mechanical (QM) calculations for description of the electronic structure of molecular systems. Some of these techniques, such as Density Functional Theory, are now available in commercial software and are easily accessible, even for people with a very rudimentary knowledge in the field. These techniques can provide useful information not only about the electronic structure of the paramagnetic species, but can be used to compute important (from a spectroscopic point of view) magnetic resonance parameters. Computed observables can then be correlated to the experimental data obtained from the EPR (or other) techniques and provide an additional tool for elucidation of the structural and electronic characteristics of the experimental system. Currently available programs can be used to calculate, *inter alia*,

a. Hyperfine tensors: Fermi term and dipolar interaction
b. Quadrupolar tensors
c. *g* tensors
d. Zero-Field Splitting (ZFS) tensors
e. Conformational changes that occur in molecular system undergoing creation of a radical
f. Rotational barriers, mechanistic aspects, energetics, etc.

The theoretical foundations of such calculations are beyond the scope of this short introductory text. Very briefly, in Density Functional Theory (DFT) one can access the total electronic density ρ of a given system. In a simplified way, ρ corresponds to an electronic cloud distributed

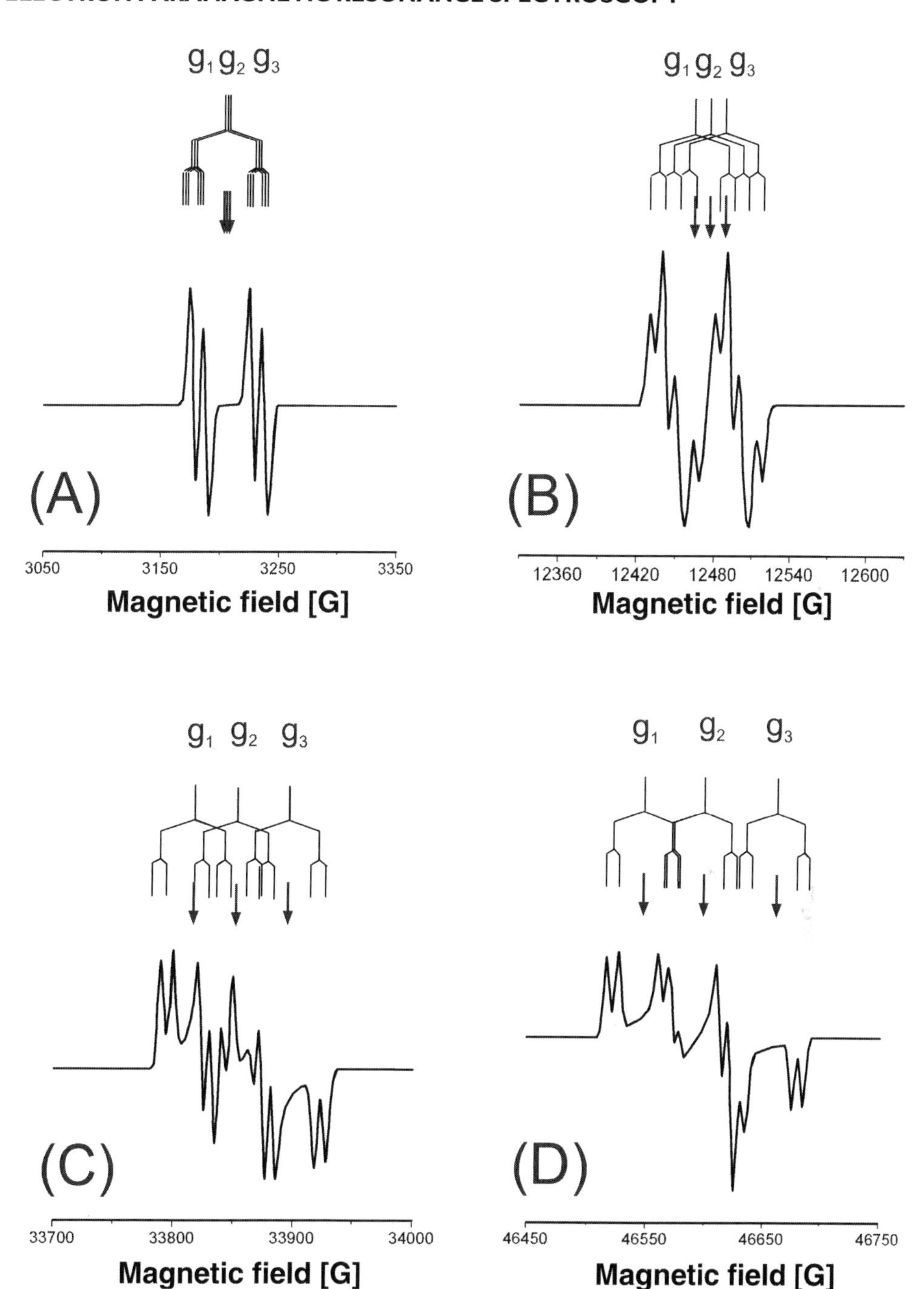

Figure 3.27. Simulated EPR spectra of radical R observed with four spectrometers operating in different EPR frequency bands: (A) X-Band, (B) Q-Band, (C) W-Band, (D) D-Band.

around the nuclei in a molecular system. The knowledge of this precise distribution allows one to derive many important observables for the given set of nuclear coordinates. For example, for hyperfine coupling, knowing ρ allows the following quantities to be directly calculated:

$$\text{Isotropic part:} \quad a_{\text{iso}} = \left(\frac{8\pi}{3}\right) g_e \beta_e g_N \beta_N \rho(N) \ , \tag{3.21}$$

$$\text{Anisotropic part:} \quad \tau_{ij} = \frac{1}{2} g_e \beta_e g_N \beta_n \left(\frac{r_{kN}^2 \rho_{ij} - 3 r_{kN,i} r_{kN,j}}{r_{kN}^5} \right) . \tag{3.22}$$

Computation of other quantities (e.g., the g tensor) requires much more sophisticated mathematical tools. However, the availability of calculations of the EPR or NMR parameters is reduced for rather small systems (~100 atoms). For large biological molecules (such as proteins or enzymes), some simplifications have to be made; otherwise, the computer requirements become extremely high (long calculations times, immense memory needs) or even exceed available computational resources. One of the recently developed techniques makes use of several such simple approximations where only a small (and the most relevant) part of the studied system is modeled with the QM method, while the remaining protein or enzyme is modeled with simple molecular mechanics (MM). These hybrid QM/MM methods are available in the so-called ONIOM scheme, which can be used to optimize the geometries of large chemical systems, such as proteins. However, the EPR parameters cannot be calculated with the ONIOM method, which is rather useful to obtain the proper geometry of the paramagnetic species.

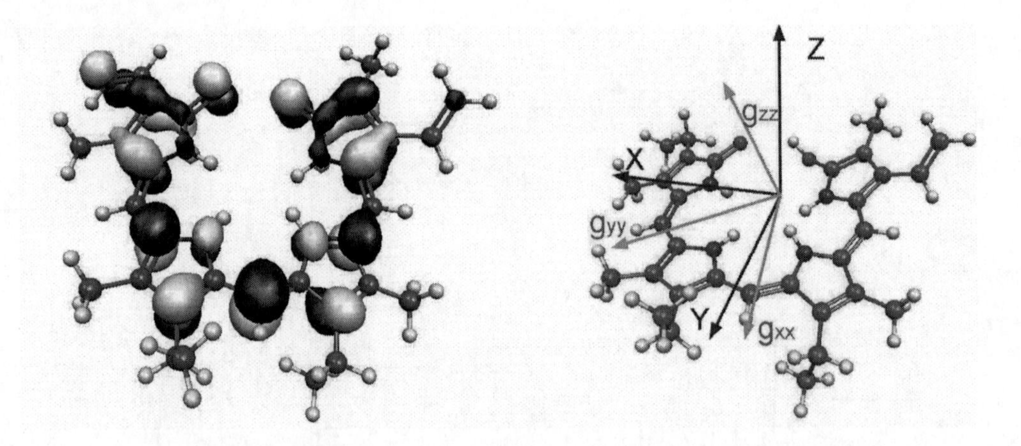

Figure 3.28. DFT-computed spin density (left) and the orientation of the g tensor in a billiverdin molecule carrying an unpaired electron.

Figure 3.28 represents the computed spin density and orientation of the g tensor for a billiverdin molecule, which is an important intermediate in biological electron transfer reactions. The biliverdin model was first optimized within a protein pocket in a model structure comporting more than 1000 atoms using the ONIOM method, and subsequently the EPR parameters were computed on such optimized geometry for the billiverdin fragment only.

Very recently calculations of EPR parameters in very large systems up to several thousands of atoms (e.g., a paramagnetic molecule docked in a surrounding protein) become possible with a new extension of DFT theory, the so-called orbital-free frozen-density embedding (FDE) formalism. The environment of an embedded subsystem is accounted for by means of the embedding potential depending explicitly on electron densities corresponding to the embedded subsystems (e.g., a docked molecule) and its environment (e.g., a protein).

3.12. EPR IN BIOLOGY AND IN BIOINORGANIC CHEMISTRY

Studies of biological molecules are a major goal in the application of modern EPR spectroscopy. As mentioned in the previous paragraphs, the paramagnetic species are not restricted to simple organic radicals; many common transition metals are also paramagnetic due to their electronic structure. Since the transition metal ions are essential in biology and especially in many enzymatic systems, where they catalyze specific biological reactions, EPR is *par excellence* the most suitable technique to obtain structural information around the paramagnetic metal centers. Most importantly, with the array of modern pulsed EPR techniques, one can specifically target the desired paramagnetic species and their immediate environment. On the other hand, to properly interpret spectroscopic data, use of theoretical and especially computational approaches, as seen in the previous paragraph, is essential in order to obtain insights into the electronic structure of the active biological sites. Quantum chemical calculations have proved to be of extreme utility in understanding complex mechanisms, in which metalloenzyme tandems are involved. In order to extract useful information from the spectroscopic data on complex biological species, many researchers conduct in parallel studies on the so-called bioinorganic models. These are usually synthetic molecules, which, although they reproduce the most important features of the models they are supposed to mimic, are much less complicated than the real systems. Such an approach has the advantage of providing the researcher with spectroscopic data that are much easier to interpret. In addition, collection of data from several similar models allows performing benchmark studies that can greatly enhance our understanding of the usually much more complicated spectral patterns arising from real biological molecules. Some of the biologically relevant systems that are the subject of intense research by both experimental EPR and theoretical methods are listed below:

a. Metal centers in enzymatic systems containing common transition-metal ions (Ni^{3+}, Mn^{2+}, Fe^{2+}, Fe^{3+}, Cu^{2+}, Co^{2+}).
b. Photosystems, in which metal centers (Mn, Fe) and more or less stable radicals (such as tyrosyl or quinol radicals) appear during the catalytic cycle of the reactions.
c. Organic ion radicals: very reactive species that are created along a course of redox reactions. They can be stabilized under certain conditions.
d. Porhphyrins and its derivatives, in which a light-induced triplet state is of importance in many biological centers.
e. Spin-labeled proteins or molecules.

It is impossible to cover here, even partially, the broad field of metalloprotein or metalloenzyme EPR. It is, however, useful to briefly discuss a few most representative examples of the transition metal–containing systems encountered in biology and studied by EPR and related techniques.

3.12.1. Copper

Copper is required for normal functioning of plants, animals, humans, and most microorganisms. It is incorporated into a variety of proteins that perform specific metabolic functions. Cu^{2+} is a d^9 ion (it carries nine 3d electrons in the valence shell), and it is characterized by an electronic spin $S = 1/2$ and nuclear spin of $I = 3/2$. (Note that there are two copper isotopes, ^{63}Cu and ^{65}Cu, with the same nuclear spin $I = 3/2$, but with slightly different g_N values.) This makes Cu^{2+} an ideal EPR target, which in addition exhibits a rather easy-to-interpret and characteristic pattern. The shape of the EPR powder spectrum of many biological samples containing copper depends on the close ligand environment surrounding the metal center. The geometry of the surrounding ligands modifies the shape of the *g* tensor and hence modifies its spectral signature. A typical copper spectrum exhibits an intense central peak centered around $g \approx 2$, and a small four-peak pattern centered around $g \gg 2$ (note that Cu hyperfine coupling is the spacing between these four transitions). The two characteristic *g* values are called parallel and perpendicular *g* ($g_{\|}$ and $g_{\perp}$). In Figure 3.29 is shown a characteristic EPR spectrum of a copper-containing protein, with highlighted $g_{\|}$, $g_{\perp}$ and hyperfine Cu coupling constant *A*.

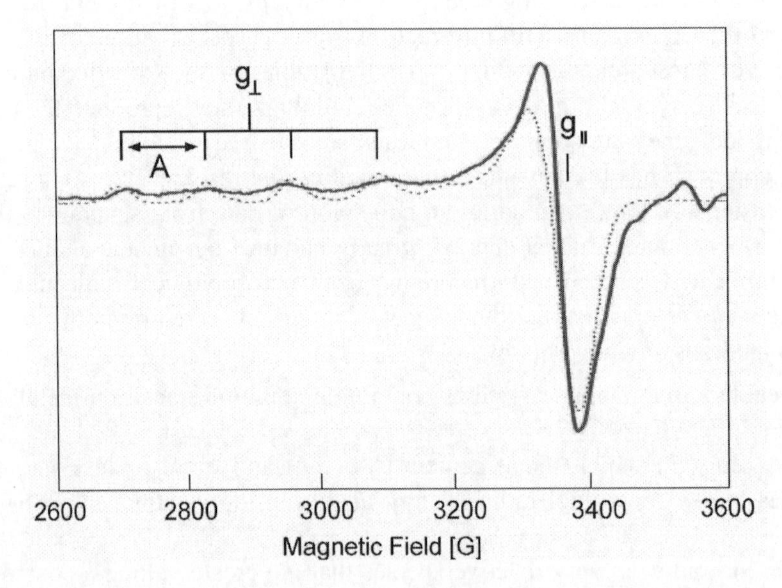

Figure 3.29. An EPR spectrum (powder pattern) of a copper-containing protein with characteristic *g* and *A* values. Spectrum adapted from [3].

3.12.2. Iron

Iron is important for many biological systems. Its presence is required in mammalian respiration, gene replication, and cell growth, and many of these activities involve electron transfer reactions. Iron is found in various oxidation states, the principal ones being ferrous Fe^{2+} and ferric Fe^{3+}, although some iron enzymes generate reactive intermediates of higher valence (Fe^{4+} or Fe^{5+}) during their catalytic cycle. Fe^{2+} is a d^6 ion (it possesses six 3d electrons in the valence shell), and it is characterized by two possible electronic spin states: high-spin $S = 2$ and low-spin $S = 0$. Fe^{3+} has one less electron (d^5), which results in another two states of electron spin: high-spin $S = 5/2$ and low spin $S = 1/2$ (note that the nuclear spin of the most naturally abundant (97.6%) iron isotopes, ^{54}Fe and ^{56}Fe, is zero). One of the important one-electron transfer reactions indispensable for DNA synthesis involves formation of an amino acid cysteinyl radical for conversion of ribonucleotides to deoxyribonucleotides. This reaction is catalyzed by the di-iron-containing ribonucleotide reductases. The ability of iron-containing species to accept and donate electrons is exposed in another class of compounds, the so-called iron–sulfur clusters found in many enzymes. In such clusters two or more iron ions are ligated by sulfur bridges. Iron–sulfur clusters participate in many complicated electron-transfer reactions and comport several different classes of compounds. In Figure 3.30 is shown an EPR spectrum of such a cluster (note that spectral patterns can significantly vary between different classes of FeS clusters) with characteristic values of g.

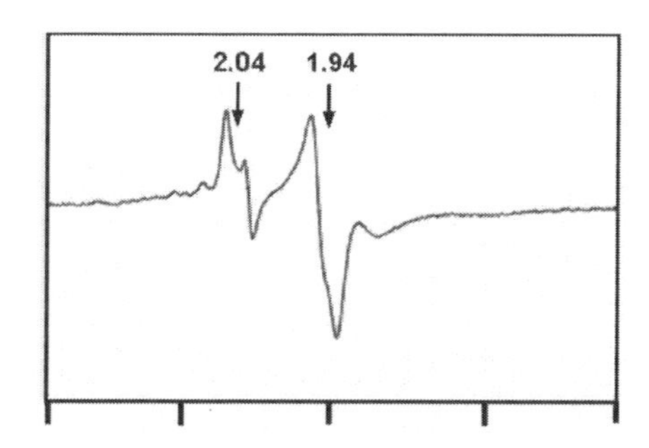

Figure 3.30. An EPR spectrum of the FeS cluster from *Rhodobacter capsulatus*. Adapted from [4].

Iron is also present in a number of porphyrin complexes known as hemes. One of the most well-known metalloproteins carrying a heme group is hemoglobin, used to transport oxygen in the human respiration chain. However, hemes are not only used for oxygen transport and storage, but also participate in a large number of key biological reactions. Some are simply oxidation reagents, while others can transfer oxygen atoms (oxidation) to substrates. Hemoproteins, which exemplify the numerous redox functions performed by hemes in biology, can be easily observed by EPR because of the presence of iron. As stated earlier, hemes can be observed in both high- and low-spin configurations, depending on the nature of the ligand binding to the iron center. Figure 3.31 shows two types of hemes in low- and high-spin states. They can be easily differentiated by their characteristic shifts in g values: Low-spin hemes exhibit g values close to 2 (in our example 1.93, 2.24, 2.38), while the high-spin species are characterized by the presence of very large g values that appear at low magnetic fields (here 7.44, 4.34, or 7.01, 4.80).

3.12.3. Manganese

Manganese is an essential trace element and appears to be present in a number of living organisms. Manganese-containing proteins and enzymes are found in all forms of life, and especially

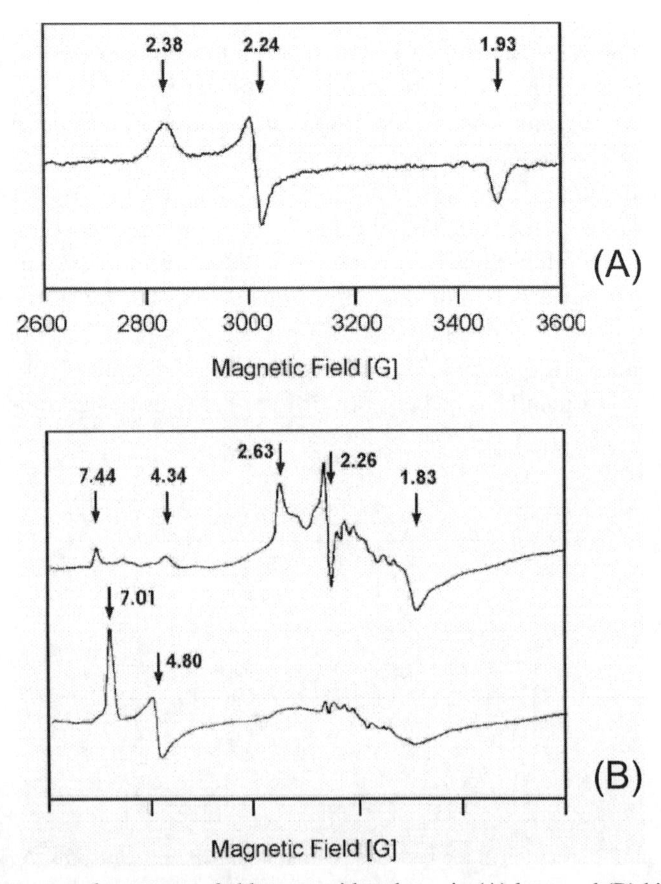

Figure 3.31. EPR spectra of two types of chloroperoxidase hems in (A) low- and (B) high-spin configurations. Adapted from [5] and [6], respectively.

in mitochondrial cells, where Mn ion is required for mitochondrial functions. In biological systems Mn can carry different oxidation states (Mn^{2+}, Mn^{3+}, Mn^{4+}), of which divalent Mn^{2+} is the most attractive from an EPR point of view. Mn^{2+} is a d^5 ion and in most of the biological systems exhibits a high-spin electronic state with $S = 5/2$. In addition, the ^{55}Mn nucleus has a nuclear spin of $I = 5/2$, giving rise to the characteristic six-line EPR pattern.

Enzymatic systems carrying manganese include, for example, superoxide dismutase (SOD), which catalyzes dismutation of superoxide into oxygen and hydrogen peroxide and is an important antioxidant in nearly all cells exposed to oxygen. In humans the excretion of nitrogenous waste from the catalysis of proteins is controlled by arginase, which hydrolyses arginine residues in the liver. An interesting example of an Mn-containing enzyme is oxalate decarboxylase, which catalyzes conversion of oxalate to formate and carbon dioxide. Figure 3.32 shows an EPR spectrum of this Mn-containing enzyme with characteristic a 6-line pattern arising from the manganese center. Manganese also plays a central role in the energy conversion that higher plants and algae complete via photosynthetic reactions, specifically water oxidation. This reaction is catalyzed by the membrane-bound Photosystem II (PS II) in its part called the Oxygen Evolving Complex (OEC). The central part of the OEC is an inorganic core including a mang-

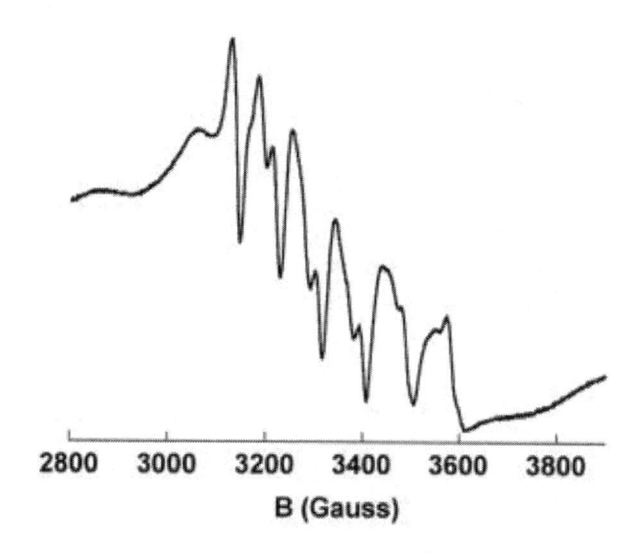

Figure 3.32. EPR spectrum showing typical signal from Mn^{2+} site observed in oxalate decarboxylase. Adapted from [7].

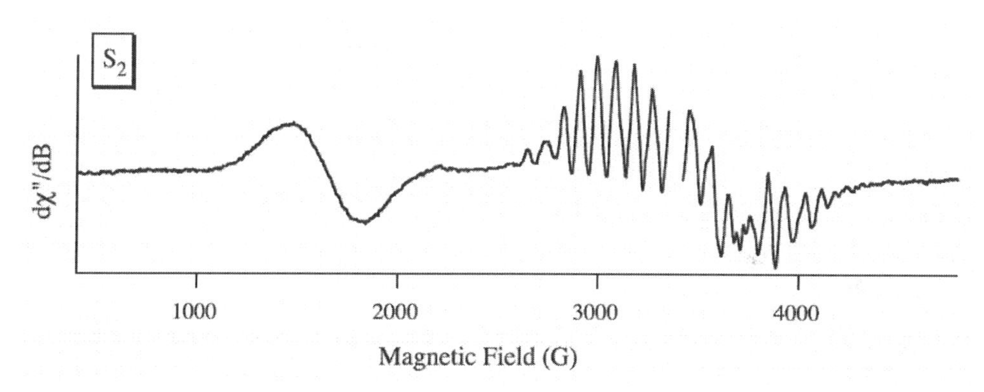

Figure 3.33. The EPR "multiline signal" arising from the Mn cluster in the oxygen evolving center of PSII. Adapted from [8].

nese cluster containing four Mn atoms, a Ca ion, and several cofactors. EPR spectroscopy of PS II has a long history and, combined with other spectroscopic techniques, has allowed for elucidation of the most important structural features of the manganese cluster and its surrounding a long time before the X-ray crystal structure of this complicated protein was available. Figure 3.33 shows the so-called "multiline signal" arising from the manganese cluster in the oxygen evolving center in Photosystem II. The multiline signal is characterized by a number of EPR lines, which are a consequence of the complicated couplings between the four manganese atoms of different oxidation states in the manganese cluster of PSII.

ACKNOWLEDGMENTS

The author expresses many thanks to Dr. Troy A. Stich and Dr. Stefan Stoll for useful suggestions, corrections, and discussion of the mansucript.

NOTE

1. Note that in an EPR experiment the detector is not physically placed on the X axis, as in an NMR probe, but the incident microwave from the sample is mixed with a reference microwave source. This allows to detect the actual phase of the microwave and hence the magnitude of the absorption in a given direction.

PROBLEMS

3.1. Draw an energy level scheme similar to that in Fig. 3.2 (disregard the nuclear Zeeman interaction) for a system with one unpaired electron, two equivalent nuclei of spin $I_1 = ½$, and one nucleus of spin $I_2 = 1$ with a hyperfine coupling four times greater for I_1 than for I_2. How many EPR lines should you obtain?

3.2. The spectral line corresponding to $g = 2.001$ is observed in an EPR spectrum at a field of 3467 G recorded in X-band (Microwave Freq. = 9.986 GHz). At what resonant field should this line be seen if the spectrum is recorded at high frequency on a W-band instrument (Microwave Freq. = 95.210 GHz)?

3.3. A phosphinyl radical (one unpaired electron on phosphorus atom R–P$^{\bullet}$H) gives a doublet of doublets ($I_P = 1/2$, $I_H = 1/2$, four lines) with a hyperfine coupling constant of 300 MHz for phosphorus P and another coupling constant of 30 MHz for proton H. If this compound is deuterated (H is replaced by a deuteron D), a new pattern is observed. The coupling constant remains the same for phosphorus but changes for the deuteron. Can you calculate a priori the coupling constant for D? Draw the corresponding spectra with H and with D.

3.4. The spectrum below was simulated for a mixture of CHD_2 and CH_2D radicals. Identify the lines belonging to each species and give the values for each of the hyperfine splittings. Compute the ratio a_H / a_D from the spectrum using a ruler. Compare it with the expected numerical value.

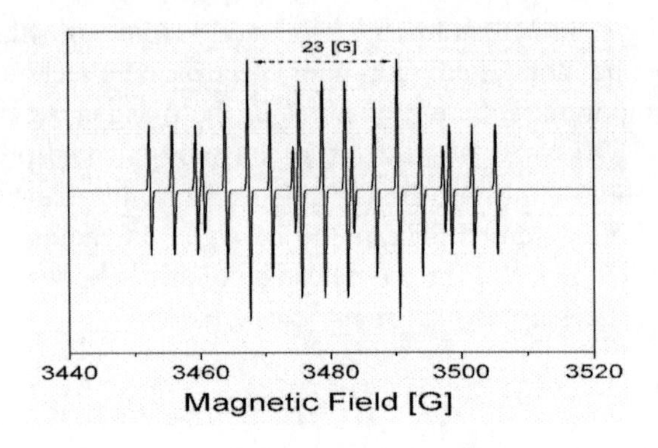

3.5. The spectrum below was simulated for a 1,3-butadiene anion radical. Draw schematically the molecule and identify the lines belonging to the equivalent protons. Draw a "stick diagram" for this spectrum.

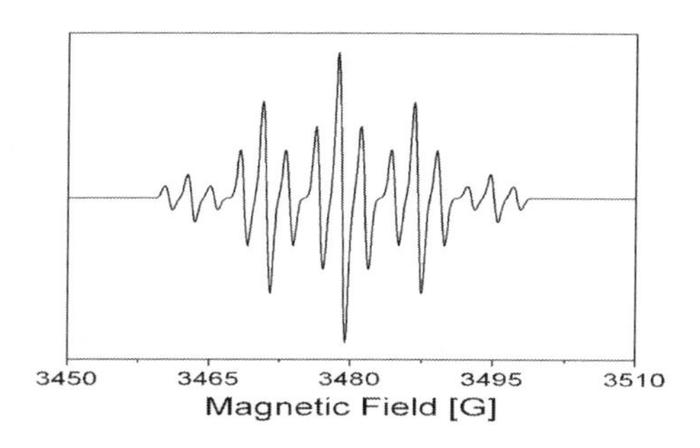

3.6. Calculate the spacing (in gauss) between two EPR transitions (lines at $g = 1.999$ and 2.001), measured at the X-, W-, and D-bands. Which one gives a better resolution, and thus higher accuracy?

FURTHER STUDY

Monographs

General EPR

1. Wertz JE, Bolton JR. 2007. *Electron paramagnetic resonance: elementary theory and practical applications*. New York: Chapman and Hall.
2. Atherton NM. 1993. *Principles of electron spin resonance*. Ellis Horwood Series in Physical Chemistry. Chichester: Ellis Horwood and PTR Prentice Hall.
3. Bersohn M, Baird JC. 1966. *An introduction to electron paramagnetic resonance*. New York: W.A. Benjamin.
4. Gordy W. 1980. *Theory and applications of electron spin resonance*. New York: John Wiley & Sons.
5. Poole CP. 1983. *Electron spin resonance*. Mineola, NY; Dover.
6. Schweiger A, Jeschke G. 2001. *Principles of pulse electron paramagnetic resonance*. Oxford: Oxford UP.
7. Eaton GR, Eaton SS, Salikhov KM. 1998. *Foundations of modern EPR*. Singapore: World Scientific.
8. An excellent compilation on the common EPR techniques at the introductory level is provided on the Bruker webpage: <http://www.bruker-biospin.com/whatisepr.html>.

EPR of Transition-Metal Ions

9. Abragam A, Bliney B. 1986. *Electron paramagnetic resonance of transition ions*. Oxford: Clarendon.
10. Mabbs FE, Collison D. 1992. *Electron paramagnetic resonance of transition metal compounds*. Studies in Inorganic Chemistry. Kidlington, Oxfordshire: Elsevier Science.
11. Pilbrow JR. 1990. *Transition-ion electron paramagnetic resonance*. Oxford: Clarendon.

REFERENCES

1. Hristova D. 2005. PhD Thesis, University of Basel, Switzerland.
2. Inanami O, Hashida S, Iizuka D, Horiuchi M, Hiraoka W, Shimoyama Y, Nakamura H, Inagaki F, Kuwabara M. 2005. Conformational change in full-length mouse prion: a site-directed spin-labeling study. *Biochem Biophys Res Commun* **335**:785–792.
3. Smith SR, Pala I, Benore-Parsons M. 2006. Riboflavin binding protein contains a type II copper binding site. *J Inorg Biochem* **100**:1730–1733.

4. Chevallet M, Dupuis A, Issartel J-P, Lunardi J, van Belzen R, Albracht SPJ. 2003. Two EPR-detectable [4Fe–4S] clusters, N2a and N2b, are bound to the NuoI (TYKY) subunit of NADH:ubiquinone oxidoreductase (Complex I) from Rhodobacter capsulatus. *Biochim Biophys Acta Bioenerg* **1557**:51–66.
5. Lefevre-Groboillot D, Frapart Y, Desbois A, Zimmermann JL, Boucher JL, Gorren AC, Mayer B, Stuehr DJ, Mansuy D. 2003. Two modes of binding of N-hydroxyguanidines to NO synthases: first evidence for the formation of iron–N-hydroxyguanidine complexes and key role of tetrahydrobiopterin in determining the binding mode. *Biochemistry* **42**:3858–3867.
6. Hollenberg PF, Hager LP, Blumberg WE, Peisach J. 1980. An electron paramagnetic resonance study of the high and low spin forms of chloroperoxidase. *J Biol Chem* **255**:4801–4807.
7. Chang CH, Svedruzic D, Ozarowski A, Walker L, Yeagle G, Britt RD, Angerhofer A, Richards NG. 2004. EPR spectroscopic characterization of the manganese center and a free radical in the oxalate decarboxylase reaction: identification of a tyrosyl radical during turnover. *J Biol Chem* **279**:52840–52849.
8. Britt RD, Campbell KA, Peloquin JM, Gilchrist ML, Aznar CP, Dicus MM, Robblee J, Messinger J. 2004. Recent pulsed EPR studies of the Photosystem II oxygen-evolving complex: implications as to water oxidation mechanisms. *Biochim Biophys Acta Bioenerg* **1655**:158–171.

4

THEORY AND APPLICATIONS OF BIOMOLECULAR NMR SPECTROSCOPY

James B. Ames
Department of Chemistry, University of California Davis

4.1. INTRODUCTION

Nuclear magnetic resonance (**NMR**) spectroscopy is an experimental technique that measures the interaction between electromagnetic radiation and atomic nuclei in atoms and molecules. Atoms bonded to form molecules exist in well-defined nuclear and electronic states (**stationary states**) that have discrete (**quantized**) energies. Energy in the form of electromagnetic radiation corresponding to the difference between the energies of the quantized nuclear states can promote the system from one quantum state to another, resulting in either **absorption** or **emission** of the radiation. Thus, if the system is going from a state i with energy E_i to a state j with energy E_j, the system changes in energy by an amount $\Delta E_{ij} = E_j - E_i$. In NMR spectroscopy, the difference in energy between the quantized nuclear states corresponds to the energy of a radiofrequency photon.

The energy of a photon can be described as a wave that has a **velocity** (speed and direction of wave propagation), **amplitude** (displacement of the wave perpendicular to velocity), **wavelength** (distance between wave crests), and **frequency** (number of crests that pass a fixed point per unit time). The relationship between the energy E and frequency ν of the radiation is given by **Planck's Law** ($E = h\nu$), where h is Planck's constant. The energy of a photon with a frequency of ν_{ij} will be absorbed or emitted by a quantized system going from state i to state j and is given by

$$\Delta E_{ij} = E_j - E_i = h\nu_{ij} .$$

Address correspondence to James B. Ames, PhD, Department of Chemistry, University of California Davis, One Shields Avenue, University of California Davis, Davis, CA 95616-8635 USA, 530 752 6358, 530 752 8995 (fax), <ames@chem.ucdavis.edu>.

T. Jue (ed.), *Biomedical Applications of Biophysics*,
Handbook of Modern Biophysics 3, DOI 10.1007/978-1-60327-233-9_4,
© Springer Science+Business Media, LLC 2010

The probability of absorbing a photon reaches a maximum when the frequency of the exciting radiation is exactly equal to ν_{ij}. This condition is called resonance, and for an NMR experiment the resonant frequency (ν_{ij}) is in the radiofrequency range in MHz. An NMR spectrum displays the probability of photon absorption (along the vertical axis) as a function of the radiation frequency (along the horizontal axis). What information can one get from an NMR spectrum? Each spectral peak position represents the spacing of quantized energy levels of individual atomic nuclei (called chemical shift) that can be interpreted in terms of models for molecular structure. In addition, the NMR spectral width of each peak is related to the rate of relaxation upon going from the excited to ground state and can be interpreted in terms of molecular dynamics.

4.2. NUCLEAR AND ELECTRONIC SPIN

Subatomic particles such as protons, electrons, and neutrons exhibit a property called **spin** that can be described in terms of its spinning motion for which the angular momentum is quantized. Electrons, protons, and neutrons exhibit two allowed spin states that are energetically equivalent (**degenerate**) in the absence of an external magnetic field, but become energetically nonequivalent (**nondegenerate**) in the presence of such a field, known as **Zeeman splitting**. Spin has a quantum number associated with it. Nuclear spin is designated by the quantum number I. The basic unit of nuclear spin, the spin on a proton, is $I = ½$. If the nucleus has $I > ½$, the nucleus is said to have a quadrupole moment. A quadrupolar spin exhibits not only Zeeman splitting (splitting of spin energy levels in a magnetic field), but also splitting of spin energy levels in the presence of an external electric field as well. We will concentrate on nuclei with $I = ½$, since these are most commonly encountered in biomolecular NMR experiments.

4.3. QUANTUM DESCRIPTION OF NUCLEAR SPIN

The spin quantum number I is the number of stationary spin states that the nucleus may occupy in any imposed magnetic field and is defined as follows: # of energy levels = $2I + 1$. Thus, for $I = ½$ nuclei (^{1}H, ^{13}C, ^{15}N, ^{19}F, ^{29}S, and ^{31}P) the nucleus can have two different spin states (i.e., $2(1/2) + 1 = 2$). For $I = 1$ nuclei (^{2}H and ^{14}N) the nucleus has three possible spin states, while $I = 3/2$ nuclei (^{11}B, ^{7}Li and ^{23}Na) can have four. In addition to the spin quantum number I, each individual spin state is designated a unique quantum number m, describing its orientation along the z-axis. Allowed values of m are given by

$$m = -I,\ -I+1,\ ...,\ I-1,\ I\,.$$

Therefore, the allowable spin orientations for $I = 1/2$ nuclei are defined by $m = -1/2$ or $+1/2$ (i.e., spin-up or spin-down), also designated as α ($m = +1/2$) and β ($m = -1/2$). For $I = 1$ nuclei, $m = -1, 0, 1$, and for $I = 3/2$ nuclei, $m = -3/2, -1/2, +1/2, +3/2$, and so on. All spin state orientations are degenerate in the absence of a magnetic field (or electric field for $I \geq ½$). However, the application of an external magnetic field (designated as the vector quantity $\vec{B}$) causes the various spin states to become nondegenerate, known as Zeeman splitting. The magnetic field $\vec{B}$ is typically reported in units of gauss (G) or tesla (T) where 1 T = 10,000 G.

The Zeeman splitting (that is, the difference in energies of the spin states) depends on the magnitude of $\vec{B}$, given by

$$\Delta E_{\alpha\beta} = \hbar\gamma B_0,$$

where B_0 is the length of $\vec{B}$, and γ is the **gyromagnetic ratio** of the nucleus, which is just a proportionality constant that describes the intrinsic receptivity for a particular type of nucleus. For example, γ is greatest for ^{1}H, explaining why ^{1}H NMR is much more sensitive than ^{13}C or ^{15}N NMR experiments. Figure 4.1 shows the extent of Zeeman splitting between two spin quantum states, $m = -1/2$ and $m = +1/2$, for a ^{1}H nucleus as a function of magnetic field intensity B.

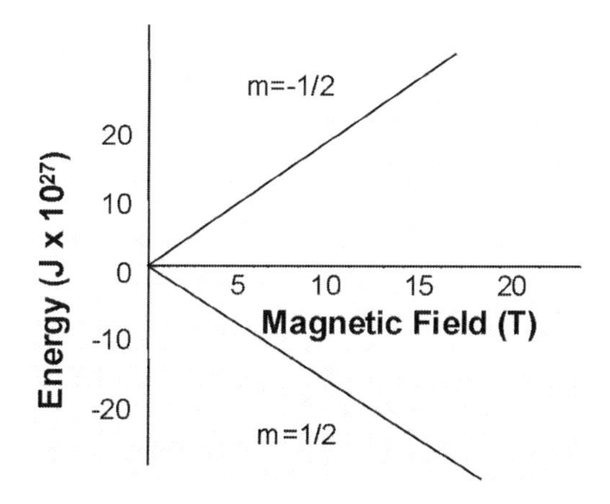

Figure 4.1. Energy splitting of nuclear spin levels for a ^{1}H nucleus as a function of magnetic field intensity. Splittings are calculated using a value of 1.54592×10^{-34} J/s for $\hbar$ and 26.7519×10^7 rad/s·T for γ.

Planck's Law says that the difference in energy is proportional to the frequency of transition $\alpha \rightarrow \beta$ ($\Delta E_{\alpha\beta} = \hbar\omega_{\alpha\beta} = h\nu_{\alpha\beta}$). In other words, $\omega_{\alpha\beta} = 2\pi\nu_{\alpha\beta} = \gamma B$, where $\nu_{\alpha\beta}$ is the spin transition frequency in Hz, and $\omega_{\alpha\beta}$ is the angular frequency in radians/sec, also called the **Larmor frequency**. The magnetic field of an NMR spectrometer is often referred to in terms of frequency units. Thus, a "500 MHz" NMR spectrometer has a magnetic field strength of ~11.74 T that will cause protons to resonate with a Larmor frequency of 500 MHz (5×10^9 Hz), ^{13}C will resonate at 125 MHz, and ^{15}N at 50 MHz.

4.4. SPIN-STATE POPULATIONS IN ENSEMBLES

The above quantum mechanical description was applied to a single and isolated spin. In NMR experiments we are dealing with huge numbers of spins (~10^{23}) that can interact with each other in concert (**coherent ensemble**). In an ensemble of coherent spins, the population of spins in the available quantized states at thermal equilibrium is determined by the relative energy of those states using the **Boltzmann equation**, with lower energy states being more populated. In UV-visible or infrared absorption spectroscopy, the energy difference between the electronic or vibrational states is large enough so that the lowest energy state (ground state) is nearly 100% populated at thermal equilibrium with negligible population of excited states at room temperature. In contrast, the splitting between nuclear spin states is energetically very small, even at high magnetic fields. For example, the splitting between the lower energy α-state ($m = +1/2$) and higher energy β-state ($m = -1/2$) for ^{1}H at 11.74 T is only 0.239 J/mol, which is much smaller than the available thermal energy: $E = RT$ ($RT = 2.48$ kJ/mol at 298 K), where R is the gas constant, equal to 8.314 J/mol-K, and T is temperature in degrees Kelvin. The relative population of the α and β states (N_α and N_β) at thermal equilibrium can be obtained from Boltzmann's equation:

$$\frac{N_\beta}{N_\alpha} = \exp\left(-\frac{\Delta E_{\alpha\beta}}{kT}\right),$$

where k is Boltzmann's constant (equal to R divided by Avogadro's number, $k = 1.38 \times 10^{-23}$ J/K), and T is absolute temperature. At 300 K, for protons at 11.74 T, Boltzmann's equation yields a ratio of N_β/N_α of 0.999992, corresponding to a very small population difference (i.e., ~1 parts in 10^5). In other words, $N_\alpha - N_\beta = N_\alpha - 0.999992 * N_\alpha = N_\alpha * 0.000008$. As the magnetic field intensity decreases, this difference becomes even smaller. It also becomes smaller as the gyromagnetic ratio decreases, so that other nuclei like ^{13}C and ^{15}N have even smaller population differences than ^{1}H at a given temperature and field strength. This very small population difference ($\Delta N_{\alpha\beta} = N_\alpha - N_\beta$) is proportional to the rate of photon absorption and explains in part the inherently low sensitivity of NMR.

An ensemble of nuclear spins can undergo net absorption or emission of radiation only as long as there is a population difference between the α and β states ($\Delta N_{\alpha\beta} = N_\alpha - N_\beta \neq 0$), and thus, although $\Delta N_{\alpha\beta}$ is small, it determines the intensity or signal strength of an NMR transition. The probability per unit time (P) of a single spin undergoing transition $\alpha \rightarrow \beta$ is nearly the same as that for $\beta \rightarrow \alpha$, because both transitions are nearly equally likely due to the small energy difference. The transition rate of $\alpha \rightarrow \beta$ for the ensemble is given by $P * N_\alpha$, and the transition rate of $\beta \rightarrow \alpha$ is given by $P * N_\beta$. The maximum rate of photon absorption is therefore given by

$$\text{rate} = P \Delta N_{\alpha\beta} \Delta E_{\alpha\beta}.$$

The population difference ($\Delta N_{\alpha\beta}$) eventually approaches zero if the sample is continuously irradiated. Once this happens, the transition is said to be **saturated** (transition rate = 0), and no further absorption of radiation can occur until a population difference is reestablished by spontaneous **relaxation** in which excited spins relax back to the ground state ($\beta \rightarrow \alpha$).

The similar rates for $\alpha \rightarrow \beta$ and the reverse transition $\beta \rightarrow \alpha$ cause a slow relaxation rate for reestablishing equilibrium after photon absorption. Hence, the long relaxation times and small equilibrium population differences $\Delta N_{\alpha\beta}$ both conspire to reduce the sensitivity of the NMR experiment. Nevertheless, these same factors are also the basis for the success of NMR as a spectroscopic technique, since long relaxation times give narrow linewidths. Even more importantly, nonequilibrium coherent spin ensembles can be readily prepared, are relatively stable, and can be operated on during the course of an NMR experiment. Relaxation in NMR takes place on a timescale (milliseconds to minutes) that is accessible to modern electronics, and one can perform multiple perturbations on a system of spins and observe the results of those perturbations. This is the basis for the huge number of multinuclear and multidimensional NMR experiments used for structural analysis of proteins and nucleic acids [1,2].

4.5. NUCLEAR SHIELDING AND CHEMICAL SHIFT

The magnetic environment of a nuclear spin is affected by a variety of influences, including valence shell orbital hybridization and bonding of the spin-active atom, charge and bond polarity, electronegativity of nearby atoms, and solvent effects. Changes in environment will cause small changes in the resonant frequency of the nuclear spin, known as **chemical shift**. This dependence of chemical shift upon the local electronic environment provides one of the most important sources of structural information from NMR.

If a nucleus is located in a region where the local induced field B_{loc} is in the same direction as the applied field B_0, the effective field detected by the nucleus is larger than the applied field, $B_{eff} = |B_0 + B_{loc}| > B_0$. The nucleus is said to be deshielded, and will resonate at a higher frequency than expected based only on the applied field. Conversely, a nucleus that is in a region in which $B_{eff} = |B_0 + B_{loc}| < B_0$ is shielded and resonates at a lower frequency than expected from the applied field, B_0. The extent of **shielding** or **deshielding** depends mostly on the strength of the applied field, since the induced local field (B_{loc}) caused by the surrounding electrons is field dependent. The proportionality between the chemical shift and the applied field is given by the **shielding constant** σ_i for a nucleus in a particular environment. We can express a relationship between the observed resonance frequency of nucleus i and the applied field as

$$\omega_i = \gamma B_0(1-\sigma_i).$$

Chemical shift is usually measured relative to a standard reference. Chemical shift is in units proportional to energy, such as frequency. One difficulty with using absolute frequency units is that small changes in the magnetic field will result in changes in the resonance frequencies of both the sample and the reference. However, if one takes the ratio of the differences in resonance frequencies between the signal of interest and the reference, the magnetic field term will cancel. This leaves δ, a dimensionless and magnetic field-independent measure of chemical shift given by

$$\delta_i = \frac{\omega_i - \omega_{ref}}{\omega_{ref}} = \frac{\nu_i - \nu_{ref}}{\nu_{ref}}.$$

The difference frequency ($\omega_i - \omega_{ref}$) is generally a million-fold smaller than the absolute reference frequency (ω_{ref}) (i.e., ~1 part per million (ppm)), and the ratio δ is reported as ppm relative to a standard.

NMR chemical shift is used to determine structural properties in proteins [1–5]. The chemical shifts of mainchain nuclei ($^1H\alpha$, $^{13}C\alpha$ and ^{13}CO) are characteristic of protein secondary structure [3]. In an α-helix, $\delta_{H\alpha}$ is ~ 1 ppm lower than what is in a random coil, whereas δ_{Ca} and δ_{CO} are both 5–10 ppm higher. Conversely, in a β-sheet, $\delta_{H\alpha}$ is greater than its corresponding random coil value and δ_{Ca} and δ_{CO} are both lower. A chemical shift index (CSI) has been devised to quantitatively calculate the secondary structure of proteins on the basis of these chemical shifts [3]. NMR chemical shifts are also used to calculate the pKa of His sidechains [6], the oxidation state of Cys sulfhydryl groups [7], and related folding properties [2,8].

4.6. NMR SCALAR COUPLING

If chemical shift were the only information from NMR, the technique would offer not much more than other types of spectroscopy, and, in fact, infrared spectroscopy would be even more informative concerning molecular structure and environment. The true value of NMR lies in the ability to detect coupling between nuclear spins and to delineate coupling pathways. Analysis of NMR scalar couplings provides atomic-level information about bonding patterns and stereochemistry in addition to the chemical environment from chemical shifts, making NMR the only technique aside from crystallography that can be used for atomic-level molecular structure determination.

There are two types of coupling observed in NMR. One involves transfer of magnetization through chemical bonds called J-coupling or scalar coupling, while the other involves magnetization transfer through space called dipolar coupling. Scalar coupling is propagated by interactions of nuclear spin with the spins of bonding electrons. Consider two nuclei (A and B) that are covalently bonded. Nucleus A can occupy either two spin states (α or β). Electrons that reside in bonding orbitals overlapping with nuclear spin A will be affected by the spin state of A, and the electron spin states will change slightly in energy in response to the spin of the nucleus. This perturbation of electronic spin states can be propagated to another nucleus (B) if nucleus B also overlaps the affected orbitals (i.e., A and B must be connected by covalent bonds less than 4 bonds apart). The result is a slight change in the resonance frequency of nucleus B depending on whether nucleus A is in the $m = +1/2$ or $m = -1/2$ state, and nuclei A and B are said to be J-coupled. Coupling is a two-way street, and if nucleus A affects B, nucleus B also affects A, and to the same extent. The difference between the resonance frequency of spin B when A is in the α state versus the β state is called the coupling constant and is usually designated as J_{AB}.

NMR scalar couplings are used to determine protein conformation using the Karplus relation [9,10]. In particular, the 3-bond J-coupling between amide proton and the adjacent alpha proton ($^3J_{NHa}$) in proteins is useful for determing the mainchain secondary structure. If $^3J_{NHa}$ of a particular amide proton is less than 6 Hz, then that amino acid residue is in an α-helix, whereas a $^3J_{NHa}$ value greater than 8 Hz indicates that residue resides in a β-sheet.

4.7. DIPOLAR COUPLING

Holding two magnets close together demonstrates that their magnetic dipoles interact. The direct through-space interaction between nuclear spins is called dipolar coupling, and although splittings due to dipolar coupling are not observed in NMR spectra of isotropic liquids, dipolar coupling is still important in a variety of phenomena, including nuclear spin relaxation and the nuclear Overhauser effect (NOE). In solids and oriented phases, splittings due to dipolar coupling are quite large (on the order of 100–1000 Hz).

The dipolar coupling interaction between two nuclear dipoles (μ_1 and μ_2) in a magnetic field B_0 is illustrated in Figure 4.2. The x and y components of the spin dipoles average to zero with time. The z components (μ_z) parallel to the applied field B_0 do not average out, and the force experienced by μ_2 due to the magnetic field of μ_1 depends on the internuclear distance and angle formed by the internuclear vector with respect to B_0. The observed dipolar splitting D_{AB} due to dipolar interactions between two nuclei A and B is summarized by

$$D_{AB} = \frac{m_0}{4\pi}\frac{m_1 m_2}{r^3}(3\cos^2\theta - 1),$$

where D_{AB} is the dipolar coupling constant between nucleus A and nucleus B, θ is the angle between B_0 and internuclear vector r_{AX}, and μ_0 is the permittivity of free space. In isotropic liquids, θ varies randomly (and rapidly) from 0 to π, and the average value of D_{AB} is 0. Hence, no dipolar splitting is observed in isotropic liquids. Note that when the term $3\cos^2\theta = 1$ ($\theta = 54.7°$)

the coupling also vanishes. Dipolar couplings are usually eliminated in solid-state NMR spectra by spinning the sample rapidly in the applied field around an axis at an angle $\theta = 54.7°$ with respect to the applied field in a technique known as magic angle spinning solid-state NMR.

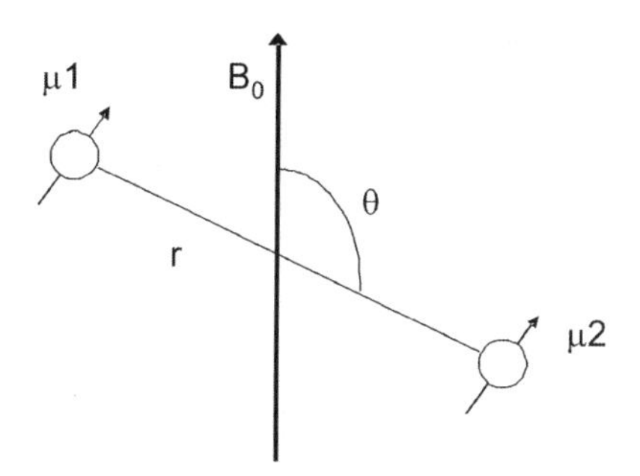

Figure 4.2. Dipolar coupling between two nuclear magnetic dipoles μ1 and μ2.

Dipolar coupling has two important consequences for solution NMR. First, nuclear spin relaxation is mediated by dipolar coupling (at least for $I = 1/2$ nuclei). As such, the extent of dipolar coupling will dictate the relaxation time and spectral width of an NMR transition and therefore determine a number of experimental variables such as repetition rate and detection sensitivity. The second consequence is the NOE, which is observed as a change in intensity in the signal of one nucleus when the signal of a nearby nucleus is perturbed. The NOE greatly increases the information content of NMR, since it identifies those nuclei that are close to each other in space. Note that since dipolar coupling is not mediated by bonding electrons, the two coupled spins do not have to be part of the same molecular in order for NOEs to be detected between them. This makes it possible to perform NMR experiments to investigate the structures of multimolecular complexes, solvent–solute interactions, and other interesting phenomena involving more than one molecule. The effective range of the NOE is less than 5 Å due to an r^{-6} distance dependence, where r is the internuclear distance. For protein structure analysis, one looks for NOEs between two protons in amino acids that are far apart in primary sequence, but are located close together in three-dimensional space. These long-range NOEs provide powerful distance constraints for calculating the three-dimensional structures of proteins using distance geometry and restrained molecular dynamics approaches [1,11].

4.8. NMR RELAXATION AND DYNAMICS

NMR provides information concerning molecular dynamics on a variety of timescales. Types of dynamical processes measured by NMR include conformational equilibria, molecular reorientation and rotational correlation times, chemical exchange processes, tautomerism, weak complex formation, and large-scale correlated motions in macromolecules. The range of timescales is quite large: from 10^{-10} s (motions that affect dipolar relaxation) to chemical exchange processes with time constants on the order of hours or days.

The most straightforward application of NMR for studying dynamics is the analysis of chemical shifts in a situation where chemical exchange is taking place. Consider an NMR-active nucleus that samples two distinct environments, A and B, each of which gives rise to resolved resonances. If exchange is slow, two different resonances will be observed, one at frequency ν_A and the other at ν_B, with relative intensities proportional to the fraction of time spent by the nucleus in each environment. As far as the NMR experiment is concerned, two different nuclei are being observed. In such a case, exchange is slow on the chemical shift timescale.

Now imagine that the rate of chemical exchange is speeded up. One consequence is that the lifetime of a spin state in a given environment is shortened, and so the NMR peaks get broader (**exchange broadening**), because the **Heisenberg Uncertainty Principle** says that the uncertainty in the energy levels (spectral width = full width at half height = $\Delta\nu_{12}$) is inversely proportional to the state lifetime (τ):

$$\Delta\nu_{12} = \frac{1}{\pi\tau}.$$

If the exchange rate increases even more, we reach a point where the rate of chemical exchange is similar to the difference in their spin frequencies, and the nuclei switch between sites at random time intervals that are, on average, comparable to the time for one period of a spin rotation. The result is an observed frequency that is distributed around a maximum at

$$\nu_{obs} = \chi_A \nu_A + (1 - \chi_A)\nu_B,$$

where χ_A is the fraction of exchanging nuclei occupying site A and $(1 - \chi_A)$ the fraction occupying site B.

Exchange rates increase with increasing temperature, and one can often move between the slow and fast exchange regimes by changing temperature. The temperature at which the two slow-exchange resonances at ν_A and ν_B collapse into a single resonance at ν_{obs} is called the coalescence temperature, and the exchange rate ($k_{c(AB)}$) at the coalescence temperature is given by

$$k_{c(AB)} = \frac{\pi \Delta\nu_{AB}}{\sqrt{2}}.$$

This equation allows NMR to measure the kinetics of exchange processes with activation energies in the range of 40–100 kJ/mol. Since $\Delta\nu_{AB}$ is field dependent, a range of rate constants at different coalescence temperatures can be measured for a given reaction using NMR spectrometers with different magnetic field strengths.

As the exchange rate increases such that $k_{AB} \gg \Delta\nu_{AB}$, the fast exchange regime is now reached. The spectral width ($\Delta\nu_{1/2}$ = full width at half height) is given by

$$\Delta\nu_{1/2} = \frac{1}{2\pi}(\nu_A - \nu_B)^2 k_{AB}^{-1}.$$

The spectral width in the fast exchange regime is inversely proportional to the exchange rate (k_{AB}), in contrast to being directly proportional to k_{AB} in the slow exchange regime. In the fast exchange regime, an increase in the exchange rate by raising the temperature leads to a much narrower spectral width. The transition between the slow and fast exchange regimes is illustrated in Figure 4.3. At slow exchange rates (1 s^{-1}), a nuclear spin at site A will remain there throughout the NMR experiment, and the spectral width will be determined by the excited state lifetime (lifetime broadening), giving rise to two sharp peaks. As the exchange rate increases (100 s^{-1}), exchange processes (A $\leftrightarrow$ B) become fast enough to compete with spin relaxation ($\beta \rightarrow \alpha$) for determining the observed excited state lifetime, giving rise to exchange broadening

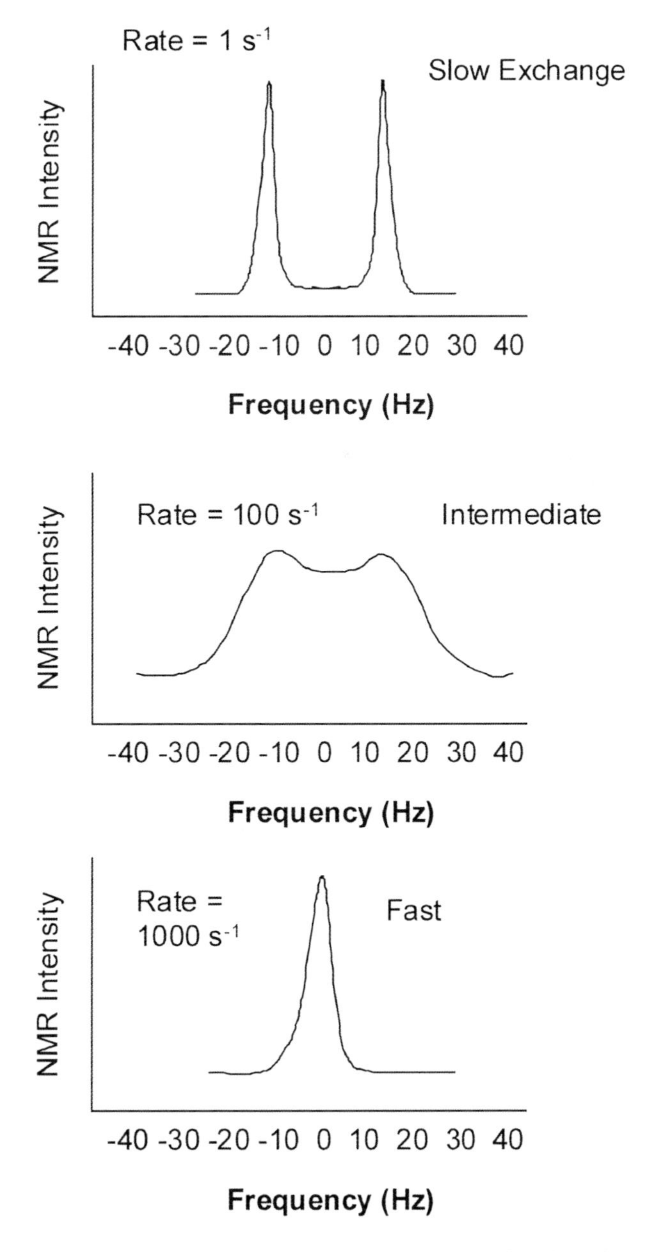

Figure 4.3. NMR spectra of spins from two equally populated sites as a function of exchange rate. The exchange rate is given in exchange events per second.

(two broad peaks). In the fast exchange regime (1000 s^{-1}), a single peak emerges at frequency ν_{obs}, which represents a weighted time average of the original two peaks from sites A and B, and the spectral width continues to sharpen as the exchange rate gets faster. In short, a comprehensive analysis of the spectral width as a function of temperature and/or magnetic field strength can be used to physically measure the exchange rate (k_{AB}).

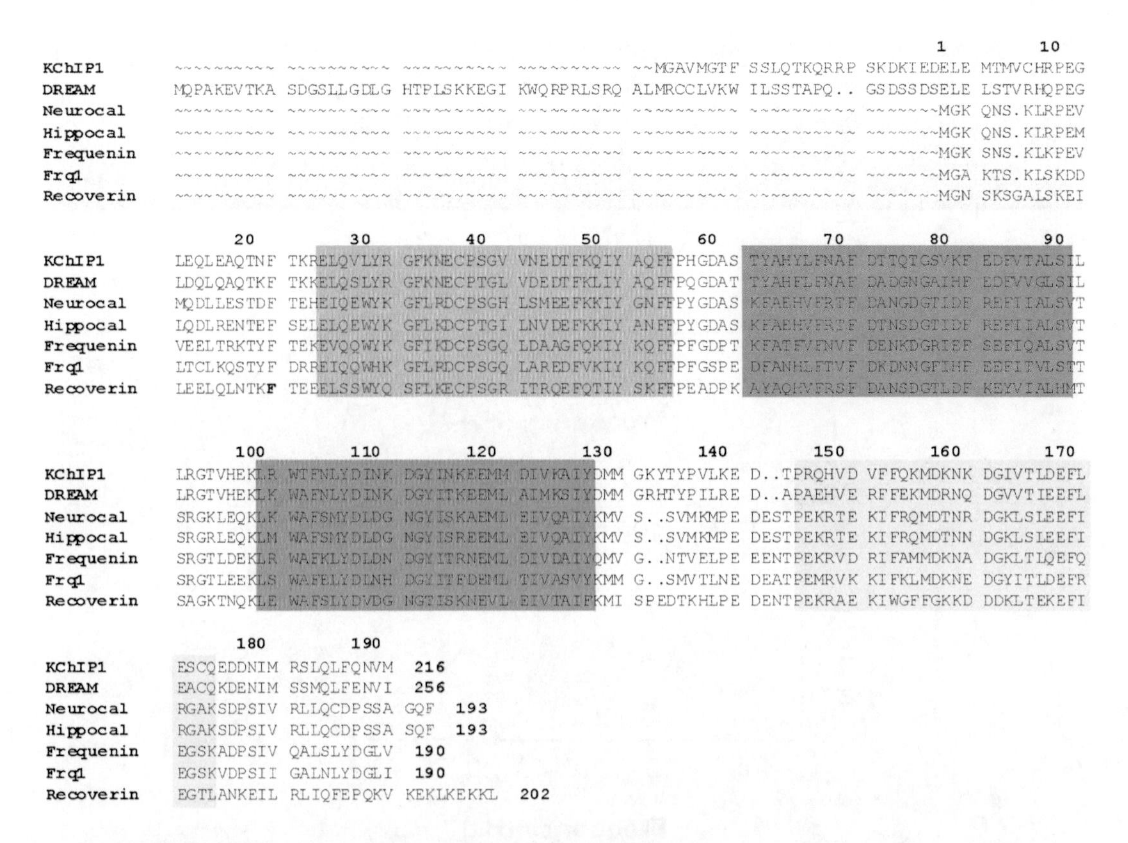

Figure 4.4. Amino-acid sequence alignment of human KChIP1 (Swiss-Prot accession no. Q9NZI2), mouse DREAM (Q9QXT8), bovine neurocalcin δ (P61602), human hippocalcin (Q5U068), drosophila frequenin (P37236), S. cerevisae Frq1 (Q06389), and bovine recoverin (P21457). The 29-residue EF-hand motifs are highlighted in color: EF-1 (green), EF-2 (red), EF-3 (cyan), EF-4 (yellow). Please visit http://extras.springer.com/ to view a high-resolution full-color version of this illustration.

4.9. NEURONAL CALCIUM SENSOR PROTEINS

Intracellular calcium (Ca^{2+}) regulates a variety of neuronal signal transduction processes in the brain and retina [12,13]. The effects of changes in neuronal Ca^{2+} are mediated primarily by an emerging class of neuronal calcium sensor (NCS) proteins [14–17] that belong to the EF-hand superfamily [18,19]. The human genome encodes 14 members of the NCS family [20]. The amino acid sequences of NCS proteins are highly conserved from yeast to humans (Fig. 4.4). Recoverin, the first NCS protein to be discovered, and the guanylate cyclase–activating proteins (GCAPs) are expressed exclusively in the retina, where they serve as Ca^{2+} sensors in vision [21–24[. Other NCS proteins are expressed in the brain and spinal cord: such as neurocalcin [25], frequenin (NCS-1) [26,27], K^+ channel interacting proteins (KChIPs) [28], DREAM/calsenilin [29,30], and hippocalcin [31,32]. Frequenin is also expressed outside the central nervous system [33], as well as in invertebrates including flies [26], worms [34], and yeast (Frq1) [35–37]. The common features of these proteins are an approximately 200-residue chain containing four EF-hand motifs, the sequence CPXG in the first EF-hand that markedly impairs its capacity to bind Ca^{2+}, and an amino-terminal myristoylation consensus sequence.

The structurally similar NCS proteins have remarkably different physiologic functions. Perhaps the best-characterized NCS protein is recoverin, which serves as a calcium sensor in retinal rod cells. Recoverin prolongs the lifetime of light-excited rhodopsin [38–40] by inhibiting rhodopsin kinase only at high Ca^{2+} levels [41–43]. Hence, recoverin makes receptor desensitization Ca^{2+} dependent, and the resulting shortened lifetime of rhodopsin at low Ca^{2+} levels may promote visual recovery and contribute to adaptation to background light. Recoverin was also identified as the antigen in cancer-associated retinopathy, an autoimmune disease of the retina caused by a primary tumor in another tissue [44,45]. Other NCS proteins in retinal rods include the GCAP proteins that activate retinal guanylate cyclase only at low Ca^{2+} levels and inhibit the cyclase at high Ca^{2+} [22,23,46]. GCAPs are important for regulating the recovery phase of visual excitation, and particular mutants are linked to various forms of retinal degeneration [47,48]. Yeast and mammalian frequenins bind and activate a particular PtdIns 4-OH kinase isoform (Pik1 gene in yeast) [33,35,49] required for vesicular trafficking in the late secretory pathway [50,51]. Mammalian frequenin (NCS-1) also regulates voltage-gated Ca^{2+} and K^{+} channels [52,53]. The KChIPs regulate the gating kinetics of voltage-gated, A-type K^{+} channels [28]. The DREAM/calsenilin/KChIP3 protein binds to specific DNA sequences in the prodynorphin and c-fos genes [29,54] and serves as a calcium sensor and transcriptional repressor for pain modulation [55,56]. Hence, the functions of the NCS proteins appear to be quite diverse and nonoverlapping.

Mass spectrometric analysis of retinal recoverin and some of the other NCS proteins revealed that they are myristoylated at the amino terminus [32,57,58]. Recoverin contains an N-terminal myristoyl (14:0) or related fatty acyl groups (12:0, 14:1, 14:2). Retinal recoverin and myristoylated recombinant recoverin, but not unmyristoylated recoverin, bind to membranes in a Ca^{2+}-dependent manner. Likewise, bovine neurocalcin and hippocalcin contain an N-terminal myristoyl group and both exhibit Ca^{2+}-induced membrane binding. These findings led to the proposal that NCS proteins possess a Ca^{2+}–myristoyl switch (Fig. 4.5). The covalently attached fatty acid is highly sequestered in recoverin in the calcium-free state. The binding of calcium to recoverin leads to extrusion of the fatty acid, making it available to interact with lipid bilayer membranes or other hydrophobic sites.

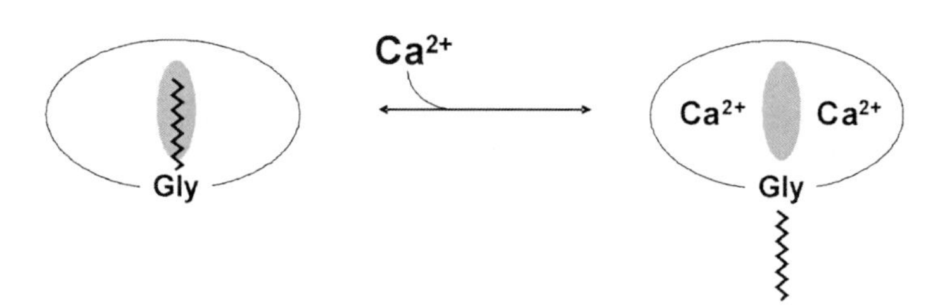

Figure 4.5. Schematic diagram of calcium–myristoyl switch in recoverin. The binding of two Ca^{2+} ions promotes extrusion of the myristoyl group and exposure of other hydrophobic residues (marked by the shaded oval).

4.10. NMR STRUCTURAL ANALYSIS OF CA^{2+}-MYRISTOYL SWITCH PROTEINS

In this section we review the NMR-derived structures of the Ca^{2+}–myristoyl switch protein recoverin, which serves as a Ca^{2+} sensor in vision. Ca^{2+}-induced conformational changes as well as structural interactions with membranes and/or downstream signaling proteins are examined to understand mechanisms of target recognition.

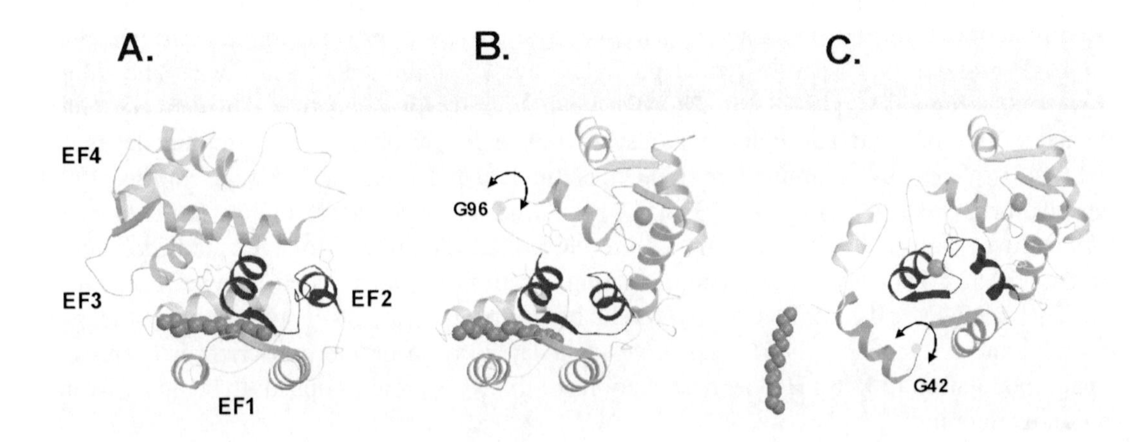

Figure 4.6. NMR-derived structures of myristoylated recoverin with 0 Ca^{2+} bound (A), 1 Ca^{2+} bound (B), and 2 Ca^{2+} bound (C). The first step of the mechanism involves binding of Ca^{2+} to EF-3, which causes minor structural changes within the EF-hand that sterically promote a 45° swiveling of the two domains, resulting in partial unclamping of the myristoyl group and dramatic rearrangement at the domain interface. The resulting altered interaction between EF-2 and EF-3 facilitates the binding of a second Ca^{2+} to the protein at EF-2 in the second step, which causes structural changes within the N-terminal domain that directly lead to ejection of the fatty acyl group. Please visit http://extras.springer.com/ to view a high-resolution full-color version of this illustration.

4.10.1. Nuclear Magnetic Resonance Spectroscopy

The structures of myristoylated recoverin in solution with 0, 1, and 2 Ca^{2+} bound were determined by nuclear magnetic resonance (NMR) spectroscopy as described previously [59–61] (Fig. 4.6). To elucidate the structure, resonances in the NMR spectra were first assigned to specific amino acid residues. Triple-resonance NMR experiments correlating ^{15}N, ^{13}C, and ^{1}H resonances (including HNCO, HNCACB, CBCACONNH, HBHACONNH, and CBCACOCAHA) were performed and analyzed to facilitate making the assignment of backbone resonances [62]. The backbone assignments served as the basis for assigning the sidechain resonances by analyzing ^{15}N- and ^{13}C-edited total correlation spectroscopy (TOCSY) and nuclear Overhauser effect spectroscopy (NOESY) experiments [63]. The complete sequence-specific assignments then served as the basis for analyzing ^{15}N- and ^{13}C-edited NOESY experiments to establish more than 3000 proton–proton distance relationships throughout the protein. In addition, dihedral angle information was deduced from analysis of J-couplings ($^{3}J_{NH\alpha}$) and chemical shift data [3]. Finally, the NMR-derived distances and dihedral angle information were used to calculate the three-dimensional structure using distance geometry and restrained molecular dynamics. The structure calculations were performed using the YASAP protocol [11] within the program X-PLOR [64], as described previously [63].

4.10.2. Calcium-Induced Conformational Changes

The NMR-derived structures of myristoylated recoverin in solution with 0, 1, and 2 Ca^{2+} bound are shown in Figure 4.6. In the Ca^{2+}-free state, the myristoyl group is sequestered in a deep hydrophobic cavity in the N-terminal domain. The cavity is formed by five α-helices. The two helices of EF-1 (residues 26–36 and 46–56), the exiting helix of EF-2 (residues 83–93), and entering helix of EF-3 (residues 100–109) lie perpendicular to the fatty acyl chain and form a box-like arrangement that surrounds the myristoyl group laterally. A long amphipathic α-helix near the N terminus (residues 4–16) packs closely against and runs antiparallel to the fatty acyl group, and serves as a lid on top of the four-helix box. The N-terminal residues Gly 2 and Asn 3 form a tight hairpin turn that connects the myristoyl group to the N-terminal helix. This turn positions the myristoyl group inside the hydrophobic cavity and gives the impression of a cocked trigger. The bond angle strain stored in the tight hairpin turn may help eject the myristoyl group from the pocket once Ca^{2+} binds to the protein.

The structure of myristoylated recoverin with one Ca^{2+} bound at EF-3 (half saturated recoverin, Fig. 4.6B) [60] represents a hybrid structure of the Ca^{2+}-free and Ca^{2+}-saturated states. The structure of the N-terminal domain (residues 2–92, green and red in Fig. 4.6) of half-saturated recoverin (Fig. 4.6B) resembles that of the Ca^{2+}-free state (Fig. 4.6A) and is very different from that of the Ca^{2+}-saturated form (Fig. 4.6C). Conversely, the structure of the C-terminal domain (residues 102–202, cyan and yellow in Fig. 4.6) of half-saturated recoverin more closely resembles that of the Ca^{2+}-saturated state. Most striking in the structure of half-saturated recoverin is that the myristoyl group is flanked by a long N-terminal helix (residues 5–17) and is sequestered in a hydrophobic cavity containing many aromatic residues from EF-1 and EF-2 (F23, W31, Y53, F56, F83, and Y86). An important structural change induced by Ca^{2+} binding at EF-3 is that the carbonyl end of the fatty acyl group in the half-saturated species is displaced far away from hydrophobic residues of EF-3 (W104 and L108; Fig. 4.6A,B) and becomes somewhat solvent exposed. By contrast, the myristoyl group of Ca^{2+}-free recoverin is highly sequestered by residues of EF-3 [59].

The structure of myristoylated recoverin with two Ca^{2+} bound shows the amino-terminal myristoyl group to be extruded [61] (Fig. 4.6C). The N-terminal eight residues are solvent exposed and highly flexible and thus serve as a mobile arm to position the myristoyl group outside the protein when Ca^{2+} is bound. The flexible arm is followed by a short α-helix (residues 9–17) that precedes the four EF-hand motifs, arranged in a tandem array, as was seen in the X-ray structure. Calcium ions are bound to EF-2 and EF-3. EF-3 has the classic "open conformation" similar to the Ca^{2+}-occupied EF-hands in calmodulin and troponin C. EF-2 is somewhat unusual and the helix-packing angle of Ca^{2+}-bound EF-2 (120°) in recoverin more closely resembles that of the Ca^{2+}-free EF-hands (in the "closed conformation") found in calmodulin and troponin C. The overall topology of Ca^{2+}-bound myristoylated recoverin is similar to the X-ray structure of unmyristoylated recoverin described above. The RMS deviation of the mainchain atoms in the EF-hand motifs is 1.5 Å in comparing Ca^{2+}-bound myristoylated recoverin to unmyristoylated recoverin. Hence, in Ca^{2+}-saturated recoverin, the N-terminal myristoyl group is solvent exposed and does not influence the interior protein structure.

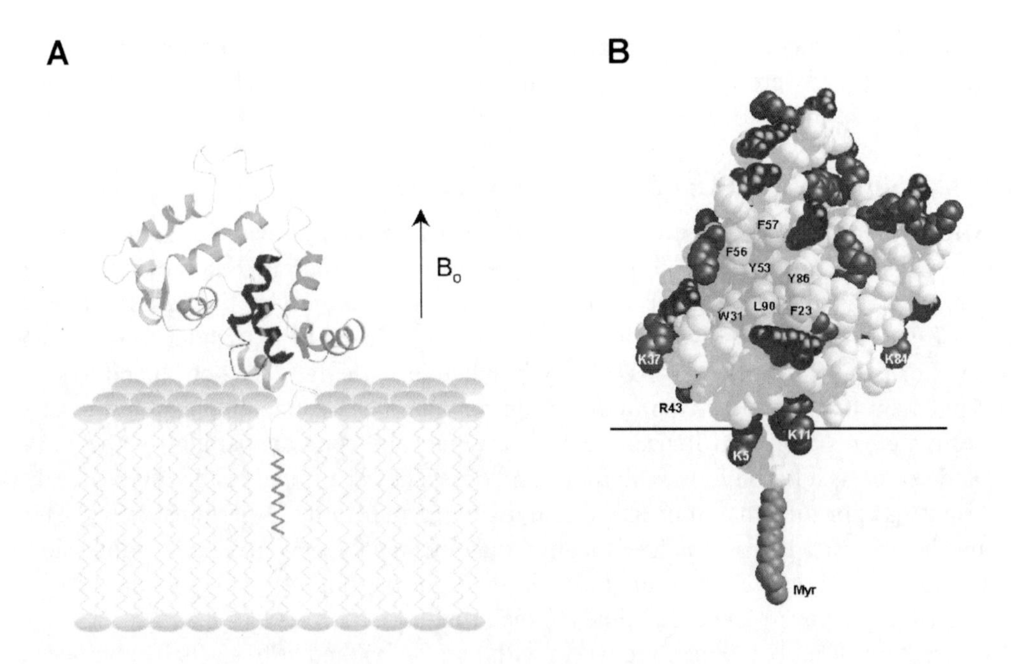

Figure 4.7. Mainchain structure (A) and space-filling representation (B) of myristoylated recoverin bound to oriented lipid bilayers determined by solid-state NMR [65]. Hydrophobic residues are yellow, bound Ca^{2+} ions are orange, and charged residues are red and blue. Please visit http://extras.springer.com/ to view a high-resolution full-color version of this illustration.

The Ca^{2+}-induced exposure of the myristoyl group (Figs. 4.5 and 4.6) enables recoverin to bind to membranes only at high Ca^{2+}. Recent solid-state NMR studies have determined the structure of Ca^{2+}-bound myristoylated recoverin bound to oriented lipid bilayer membranes (Fig. 4.7) [65]. Membrane-bound recoverin appears to retain approximately the same overall structure as it has in solution. The protein is positioned on the membrane surface such that its long molecular axis is oriented 45° with respect to the membrane normal. The N-terminal region of recoverin points toward the membrane surface, with close contacts formed by basic residues K5, K11, K22, K37, R43, and K84. This orientation of membrane-bound recoverin allows an exposed hydrophobic crevice (lined primarily by residues F23, W31, F35, I52, Y53, F56, Y86, and L90) near the membrane surface that may serve as a potential binding site for the target protein — rhodopsin kinase (Fig. 4.7B).

4.10.3. Structural Basis of Target Recognition

The N-terminal hydrophobic patch conserved in the structures of NCS proteins (Fig. 4.8) has been implicated previously in target recognition [66–68]. The recent X-ray structure of KChIP1 in a target complex with the Kv4.2 channel definitively demonstrates that the N-terminal hydrophobic patch of KChIP1 is indeed located at the intermolecular interface [69]. Hydrophobic residues L56, I77, Y78, F81, F82, F111, and L115 of KChIP1 form close intermolecular contacts with A223, W224, F227, A228, and A231 of Kv4.2. The corresponding residues in

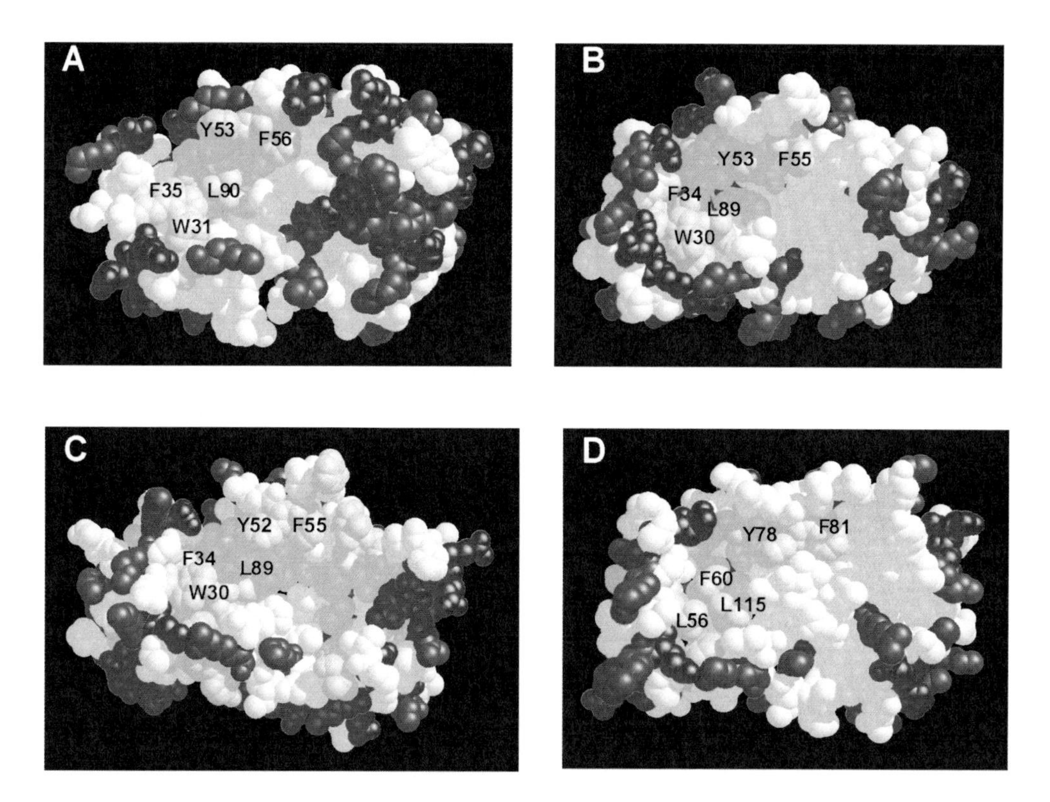

Figure 4.8. Space-filling representations of the Ca^{2+}-bound structures of recoverin (A), frequenin (B), neurocalcin (C), and KChIP1 (D). Exposed hydrophobic residues are yellow, neutral residues are white, and charge residues are red and blue. Please visit http://extras.springer.com/ to view a high-resolution full-color version of this illustration.

GCAP2 and recoverin are also important for target recognition [67,68]. The hydrophobic interfacial residues of Kv4.2 are located on the same face of an amphipathic helix (magenta) that interacts with a concave surface (or groove) of KChIP1 formed by helices of EF-1 (green) and EF-2 (red) (Fig. 4.8A).

A similar type of hydrophobic target helix interaction was seen previously in the structures of calmodulin bound to the M13 segment of myosin light-chain kinase (Fig. 4.9B) [70,71] and calcineurin B-subunit interacting with the catalytic A-subunit (Fig. 4.9C) [72]. This common structural framework appears to be a recurring theme for target recognition by EF-hand proteins generally. What distinguishes the different target interactions are the precise contour properties of the target-binding site in each case formed by particular residues from EF-1 and EF-2. For example, in calmodulin, a number of methionine residues (M36, M51, and M72) help form deep holes beneath the surface of the binding site that are recognized by aromatic rings of Trp residues strategically positioned in the target helix. By contrast, KChIP1 contains a number of aromatic residues (F53, Y78, F81, and F111), forming an exposed flat surface that interacts with Ala methyl groups on the opposing face of the target helix. The hydrophobic residues of KChIP1 at the target interface are highly conserved in all NCS proteins, suggesting a common structural motif for target recognition. However, a few important interfacial residues of KChIP1 are not conserved (e.g., V55, L56, L118, and L119), suggesting that residues at these positions in other NCS proteins may be important for conferring target specificity. Indeed, these residues in yeast Frq1 correspond to Q29, W30, T92, and S93 that have been implicated to interact specifically with phosphatidylinositol 4-kinase isoform, Pik1 in our preliminary structural analysis (unpublished results, also see [36]).

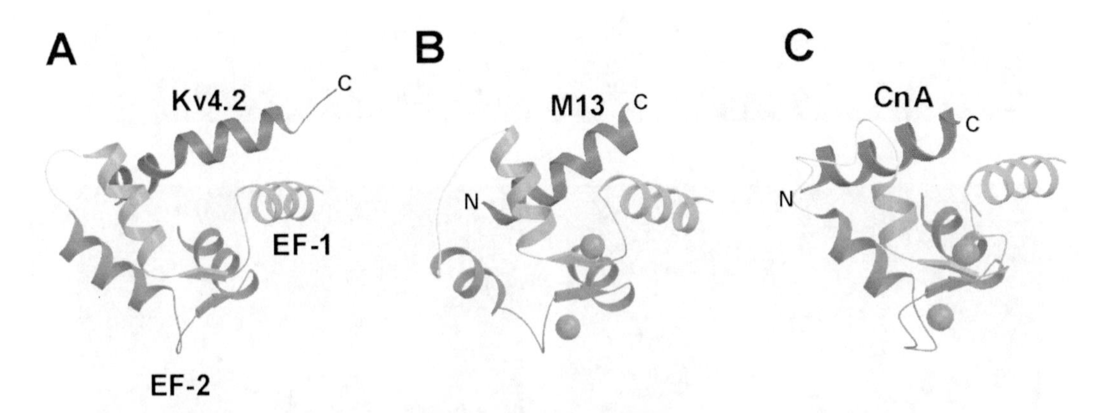

Figure 4.9. Ribbon diagrams illustrating intermolecular interactions for KChIP1 bound to Kv4.2 [69] (A), calmodulin bound to M13 peptide [70] (B), and calcineurin B bound to calcineurin A [72] (C). In each case, a target helix (magenta) is inserted in a groove formed by the helices of EF-1 (green) and EF-2 (red). The intermolecular interactions are mostly hydrophobic, as described in the text. Please visit http://extras.springer.com/ to view a high-resolution full-color version of this illustration.

4.11. SUMMARY

In summary, we have used NMR spectroscopy to solve the molecular structures of NCS proteins and examined structural determinants important for target recognition. The overall main-chain structures and the presence of an exposed hydrophobic patch in the N-terminal domain are highly conserved in all NCS proteins. Differences in their surface charge density and protein dimerization properties may help to explain target specificity and functional diversity of NCS proteins. In the future, atomic resolution structures of additional NCS proteins bound to their respective target proteins are needed to improve our understanding of how this structurally conserved family of proteins can uniquely recognize their diverse biological targets.

PROBLEMS

4.1. Using Boltzmann's equation, calculate the equilibrium ratios of populations in the two spin states of ^{1}H, ^{15}N, and ^{29}S in a 5.87-T field (250 MHz 1H) at 298 K. How does the sign of the gyromagnetic ratio (positive for ^{1}H and negative for ^{15}N and ^{29}Si) affect your calculations?

4.2. Consider the following system: proton A is coupled to proton B, which exchanges with site C to which A is not coupled. All chemical shifts are distinct, with an equilibrium constant $K_{BC} = 1$ for site exchange. Sketch the spectrum you expect for each of the following situations:

a. Slow exchange on the chemical shift and J-coupling timescale.
b. Fast exchange on the J-coupling scale, but slow on the chemical shift timescale.
c. Fast exchange on both timescales.

4.3. Spin populations can be thought of in a very chemical way in terms of "concentrations" of spins (e.g., [α] or [β]). The usual diagram for an $I = ½$ system using a chemist formalism is:

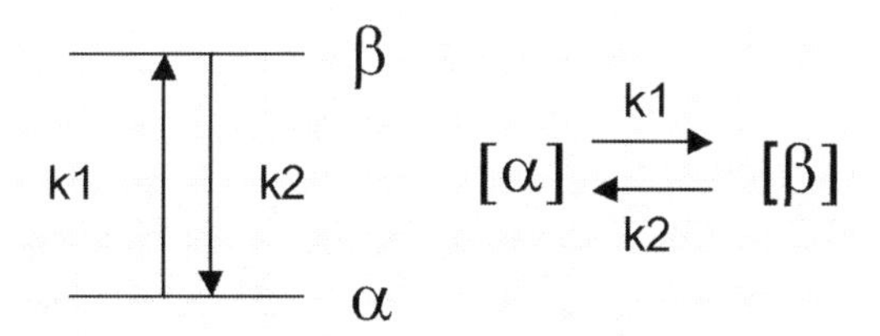

where $k1$ and $k2$ are rate constants for interconversion of spin populations with concentrations given by [α] and [β]. Calculate ΔG^0 for this reaction for ^1H in an 11.74-T magnetic field. Also, calculate the equilibrium constant for this reaction ($K_{eq} = [\alpha]_{eq}/[\beta]_{eq}$). Note that $\Delta S = 0$ since there is no disorder change associated with a single spin flip.

FURTHER STUDY

Cavanagh J, Fairbrother WJ, Palmer AG, Skelton NJ. 1996. *Protein NMR spectroscopy: principles and practice*. New York: Academic Press.

Faust B. 1995. NMR of whole body fluids. *Educ Chem* 32:22.

Guntert P. 2008. Automated structure determination from NMR spectra. *Eur Biophys J* 37:1031–1035.

Hore PJ. 1995. *Nuclear magnetic resonance*. New York: Oxford UP.

Mittermaier A, Kay LE. 2006. New tools provide insights in nmr studies of protein dynamics. *Science* 312:224–226.

REFERENCES

1. Clore GM, Gronenborn AM. 1997. NMR structures of proteins and protein complexes beyond 20,000 M(r). *Nat Struct Biol* **4**(1):849–853.
2. Wuthrich K. 1986. *NMR of proteins and nucleic acids*. New York: John Wiley and Sons.
3. Wishart DS, Sykes BD, Richards FM. 1992. The chemical shift index: a fast and simple method for the assignment of protein secondary structure through NMR spectroscopy. *Biochemistry* **31**(24):1647–1651.
4. Berjanskii MV, Wishart DS. 2008. Application of the random coil index to studying protein flexibility. *J Biomol NMR* **40**(1):31–48.
5. Cornilescu G, Delaglio F, Bax A. 1999. Protein backbone angle restraints from searching a database for chemical shift and sequence homology. *J Biomol NMR* **13**(3):289–302.
6. Schubert M, Poon DK, Wicki J, Tarling CA, Kwan EM, Nielsen JE, Withers SG, McIntosh LP. 2007. Probing electrostatic interactions along the reaction pathway of a glycoside hydrolase: histidine characterization by NMR spectroscopy. *Biochemistry* **46**(25):7383–7395.
7. Sharma D, Rajarathnam K. 2000. 13C NMR chemical shifts can predict disulfide bond formation. *J Biomol NMR* **18**(2):165–171.
8. Loh SN, Kay MS, Baldwin RL. 1995. Structure and stability of a second molten globule intermediate in the apomyoglobin folding pathway. *Proc Natl Acad Sci USA* **92**(12):5446–5450.

9. Pagano K, Fogolari F, Corazza A, Vigilino P, Esposito G. 2007. Estimation of 3JHN-Halpha and 3JHalpha-Hbeta coupling constants from heteronuclear TOCSY spectra. *J Biomol NMR* **39**(3):213–222.
10. Anglister J, Grzesiek S, Wang AC, Ren H, Klee CB, Bax A. 1994. ^{1}H, ^{13}C, ^{15}N nuclear magnetic resonance backbone assignments and secondary structure of human calcineurin B. *Biochemistry* **33**(12):3540–3547.
11. Nilges M, Gronenborn AM, Brunger AT, Clore GM. 1988. Determination of three-dimensional structures of proteins by simulated annealing with interproton distance restraints: application to crambin, potato carboxypeptidase inhibitor and barley serine proteinase inhibitor 2. *Protein Eng* **2**(1):27–38.
12. Augustine GJ, Santamaria F, Tanaka K. 2003. Local calcium signaling in neurons. *Neuron* **40**(2):331–346.
13. Berridge MJ, Lipp P, Bootman MD. 2000. The versatility and universality of calcium signalling. *Nat Rev Mol Cell Biol* **1**(1):11–21.
14. Burgoyne RD, O'Callaghan DW, Hasdemir B, Haynes LP, Tepikin AV. 2004. Neuronal Ca2+-sensor proteins: multitalented regulators of neuronal function. *Trends Neurosci* **27**(4):203–209.
15. Ames JB, Tanaka T, Stryer L, Ikura M. 1996. Portrait of a myristoyl switch protein. *Curr Opin Struct Biol* **6**(4):432–438.
16. Braunewell KH, Gundelfinger ED. 1999. Intracellular neuronal calcium sensor proteins: a family of EF-hand calcium-binding proteins in search of a function. *Cell Tissue Res* **295**(1):1–12.
17. Burgoyne RD, Weiss JL. 2001. The neuronal calcium sensor family of Ca2+-binding proteins. *Biochem J* 353:1–12.
18. Ikura M. 1996. Calcium binding and conformational response in EF-hand proteins. *Trends Biochem Sci* **21**:14–17.
19. Moncrief ND, Kretsinger, RH, Goodman, M. 1990. Evolution of EF-hand calcium-modulated proteins. *J Mol Evol* **30**:522–562.
20. Weiss JL, Burgoyne RD. 2002. Neuronal calcium sensor proteins. In: *Handbook of cell signaling*, pp. 79–82. Ed R Bradshaw. San Diego: Academic Press.
21. Dizhoor AM, Ray S, Kumar S, Niemi G, Spencer M, Rrolley D, Walsh KA, Philipov PP, Hurley JB, Stryer L. 1991. Recoverin: a calcium sensitive activator of retinal rod guanylate cyclase. *Science* **251**:915–918.
22. Dizhoor AM, Lowe DG, Olsevskaya EV, Laura RP, Hurley JB. 1994. The human photoreceptor membrane guanylyl cyclase, RetGC, is present in outer segments and is regulated by calcium and a soluble activator. *Neuron* **12**(6):1345–1352.
23. Palczewski K, Subbaraya I, Gorczyca WA, Helekar BS, Ruiz CC, Ohguro H, Huang J, Zhao X, Crabb JW, Johnson RS. 1994. Molecular cloning and characterization of retinal photoreceptor guanylyl cyclase-activating protein. *Neuron* **13**(2):395–404.
24. Palczewski K, Polans AS, Baehr W, Ames JB. 2000. Ca(2+)-binding proteins in the retina: structure, function, and the etiology of human visual diseases. *Bioessays* **22**:337–350.
25. Hidaka H, Okazaki K. 1993. Neurocalcin family: a novel calcium-binding protein abundant in bovine central nervous system. *Neurosci Res* **16**(2):73–77.
26. Pongs O, Lindemeier J, Zhu XR, Theil T, Engelkamp D, Krah-Jentgens I, Lambrecht HG, Koch KW, Schwemer J, Rivosecchi R, Mallart A, Galceran J, Canal I, Barbas JA, Ferrus A. 1993. Frequenin—a novel calcium-binding protein that modulates synaptic efficacy. *Neuron* **11**:15–28.
27. McFerran BW, Graham ME, Burgoyne RD. 1998. Neuronal Ca2+ sensor 1, the mammalian homologue of frequenin, is expressed in chromaffin and PC12 cells and regulates neurosecretion from dense-core granules. *J Biol Chem* **273**(35):22768–22772.
28. An WF, Bowlby MR, Betty M, Cao J, Ling HP, Mendoza G, Hinson JW, Mattsson KI, Strassle BW, Trimmer JS, Rhodes KJ. 2000. Modulation of A-type potassium channels by a family of calcium sensors. *Nature* **403**(6769):553–556.
29. Carrion AM, Link WA, Ledo F, Mellstrom B, Naranjo JR. 1999. DREAM is a Ca2+-regulated transcriptional repressor. *Nature* **398**(6722):80–84.
30. Buxbaum JD, Choi EK, Luo Y, Lilliehook C, Crowley AC, Merriam DE, Wasco W. 1998. Calsenilin: a calcium-binding protein that interacts with the presenilins and regulates the levels of a presenilin fragment. *Nat Med* **4**(10):1177–1181.
31. Kobayashi M, Takamatsu K, Saitoh S, Miura M, Noguchi T. 1992. Molecular cloning of hippocalcin, a novel calcium-binding protein of the recoverin family exclusively expressed in hippocampus [published erratum appears in *Biochem Biophys Res Commun* 1993 Oct **29**;196(2):1017]. *Biochem Biophys Res Commun* **189**(1):511–517.

32. Kobayashi M, Takamatsu K, Saitoh S, Noguchi T. 1993. Myristoylation of hippocalcin is linked to its calcium-dependent membrane association properties. *J Biol Chem* **268**(25):18898–18904.
33. Kapp Y, Melnikov S, Shefler A, Jeromin A, Sagi R. 2003. NCS-1 and phosphatidylinositol 4-kinase regulate IgE receptor-triggered exocytosis in cultured mast cells. *J Immunol* **171**:5320–5327.
34. Gomez M, De Castro E, Guarin E, Sasakura H, Kuhara A, Mori I, Bartfai T, Bargmann CI, Nef P. 2001. Ca2+ signaling via the neuronal calcium sensor-1 regulates associative learning and memory in C. elegans. *Neuron* **30**(1):241–248.
35. Hendricks KB, Wang BQ, Schnieders EA, Thorner J. 1999. Yeast homologue of neuronal frequenin is a regulator of phosphatidylinositol-4-OH kinase. *Nature Cell Biol* **1**:234–241.
36. Huttner IG, Strahl T, Osawa M, King DS, Ames JB, Thorner J. 2003. Molecular interactions of yeast frequenin with Pik1. *J Biol Chem* **278**(7):4862–4874.
37. Hamasaki-Katagiri N, Molchanova T, Takeda K, Ames JB. 2004. Fission yeast homolog of neuronal calcium sensor-1 (Ncs1p) regulates sporulation and confers calcium tolerance. *J Biol Chem* **279**(13):12744–12754.
38. Kawamura S. 1993. Rhodopsin phosphorylation as a mechanism of cyclic GMP phosphodiesterase regulation by S-modulin. *Nature* **62**(6423):855–857.
39. Erickson MA, Lagnado L, Zozulya S, Neubert TA, Stryer L, Baylor DA. 1998. The effect of recombinant recoverin on the photoresponse of truncated rod photoreceptors. *Proc Natl Acad Sci USA* **95**(11):6474–6479.
40. Makino CL, Dodd RL, Chen J, Burns ME, Roca A, Simon MI, Baylor DA. 2004. Recoverin regulates light-dependent phosphodiesterase activity in retinal rods. *J Gen Physiol* **123**(6):729–741.
41. Calvert PD, Klenchin VA, Bownds MD. 1995. Rhodopsin kinase inhibition by recoverin. Function of recoverin myristoylation. *J Biol Chem* **270**(41):24127–24129.
42. Klenchin VA, Calvert PD, Bownds MD. 1995. Inhibition of rhodopsin kinase by recoverin: further evidence for a negative feedback system in phototransduction. *J Biol Chem* **270**(27):16147–16152.
43. Chen CK, Inglese J, Lefkowitz RJ, Hurley JB. 1995. Ca(2+)-dependent interaction of recoverin with rhodopsin kinase. *J Biol Chem* **270**(30):18060–18066.
44. Polans AS, Buczylko J, Crabb J, Palczewski K. 1991. A photoreceptor calcium binding protein is recognized by autoantibodies obtained from patients with cancer-associated retinopathy. *J Cell Biol* **112**:981–989.
45. Subramanian L, Polans AS. 2004. Cancer-related diseases of the eye: the role of calcium and calcium-binding proteins. *Biochem Biophys Res Commun* **322**(4):1153–1165.
46. Palczewski K, Sokal I, Baehr W. 2004. Guanylate cyclase-activating proteins: structure, function, and diversity. *Biochem Biophys Res Commun* **322**(4):1123–1130.
47. Semple-Rowland SL, Gorczyca WA, Buczylko J, Helekar BS, Ruiz CC, Subbaraya I, Palczewski K, Baehr W. 1996. Expression of GCAP1 and GCAP2 in the retinal degeneration (rd) mutant chicken retina. *FEBS Lett* **385**(1):47–52.
48. Sokal I, Li N, Surgucheva I, Warren MJ, Payne AM, Bhattacharya SS, Baehr W, Palczewski K. 1998. GCAP1 (Y99C) mutant is constitutively active in autosomal dominant cone dystrophy. *Mol Cell* **2**(1):129–133.
49. Strahl T, Grafelmann B, Dannenberg J, Thorner J, Pongs O. 2003. Conservation of regulatory function in calcium-binding proteins: human frequenin (neuronal calcium sensor-1) associates productively with yeast phosphatidylinositol 4-kinase isoform, Pik1. *J Biol Chem* **278**(49):49589–49599.
50. Hama H, Schnieders EA, Thorner J, Takemoto JY, DeWald DB. 1999. Direct involvement of phosphatidylinositol 4-phosphate in secretion in the yeast Saccharomyces cerevisiae. *J Biol Chem* **274**:34294–34300.
51. Walch-Solimena C, Novick P. 1999. The yeast phosphatidylinositol-4-OH kinase Pik1 regulates secretion at the Golgi. *Nature Cell Biol* **1**:523–525.
52. Nakamura TY, Pountney DJ, Ozaita A, Nandi S, Ueda S, Rudy B, Coetzee WA. 2001. A role for frequenin, a Ca2+-binding protein, as a regulator of Kv4 K+-currents. *Proc Natl Acad Sci USA* **98**(22):12808–12813.
53. Weiss JL, Archer DA, Burgoyne RD. 2000. Neuronal Ca2+ sensor-1/frequenin functions in an autocrine pathway regulating Ca2+ channels in bovine adrenal chromaffin cells. *J Biol Chem* **275**(51):40082–40087.
54. Carrion AM, Mellstrom B, Naranjo JR. 1998. Protein kinase A-dependent derepression of the human prodynorphin gene via differential binding to an intragenic silencer element. *Mol Cell Biol* **18**(12):6921–6929.
55. Cheng HY, Pitcher GM, Laviolette SR, Whishaw IQ, Tong KI, Kockeritz LK, Wada T, Joza NA, Crackower M, Goncalves J, Sarosi I, Woodgett JR, Oliveira-dos-Santos AJ, Ikura M, van der Kooy D, Salter MW, Penninger JM. 2002. DREAM is a critical transcriptional repressor for pain modulation. *Cell* **108**(1):31–43.

56. Lilliehook C, Bozdagi O, Yao J, Gomez-Ramirez M, Zaidi NF, Wasco W, Gandy S, Santucci AC, Haroutunian V, Huntley GW, Buxbaum JD. 2003. Altered Abeta formation and long-term potentiation in a calsenilin knock-out. *J Neurosci* **23**(27):9097–9106.
57. Dizhoor AM, Ericsson LH, Johnson RS, Kumar S, Olshevskaya E, Zozulya S, Neubert TA, Stryer L, Hurley JB, Walsh KA. 1992. The NH2 terminus of retinal recoverin is acylated by a small family of fatty acids. *J Biol Chem* **267**(23):16033–16036.
58. Ladant D. 1995. Calcium and membrane binding properties of bovine neurocalcin expressed in *Escherichia coli. J Biol Chem* **270**:3179–3185.
59. Tanaka T, Ames JB, Harvey TS, Stryer L, Ikura M. 1995. Sequestration of the membrane-targeting myristoyl group of recoverin in the calcium-free state. *Nature* **376**(6539):444–447.
60. Ames JB, Hamasaki N, Molchanova T. 2002. Structure and calcium-binding studies of a recoverin mutant (E85Q) in an allosteric intermediate state. *Biochemistry* **41**(18):5776–5787.
61. Ames JB, Ishima R, Tanaka T, Gordon JI, Stryer L, Ikura M. 1997. Molecular mechanics of calcium-myristoyl switches. *Nature* **389**(6647):198–202.
62. Ames JB, Tanaka T, Stryer L, Ikura M. 1994. Secondary structure of myristoylated recoverin determined by three-dimensional heteronuclear NMR: implications for the calcium-myristoyl switch. *Biochemistry* **33**(35):10743–10753.
63. Tanaka T, Ames JB, Kainosho M, Stryer L, Ikura M. 1998. Differential isotype labeling strategy for determining the structure of myristoylated recoverin by NMR spectroscopy. *J Biomol NMR* **11**(2):135–152.
64. Brunger AT. 1992. X-PLOR, version 3.1: a system for x-ray crystallography and NMR. New Haven: Yale UP.
65. Valentine KG, Mesleh MF, Opella SJ, Ikura M, Ames JB. 2003. Structure, topology, and dynamics of myristoylated recoverin bound to phospholipid bilayers. *Biochemistry* **42**(21):6333–6340.
66. Krylov DM, Niemi GA, Dizhoor AM, Hurley JB. 1999. Mapping sites in guanylyl cyclase activating protein-1 required for regulation of photoreceptor membrane guanylyl cyclases. *J Biol Chem* **274**(16):10833–10839.
67. Tachibanaki S, Nanda K, Sasaki K, Ozaki K, Kawamura S. 2000. Amino acid residues of S-modulin responsible for interaction with rhodopsin kinase. *J Biol Chem* **275**:3313–3319.
68. Olshevskaya EV, Boikov S, Ermilov A, Krylov D, Hurley JB, Dizhoor AM. 1999. Mapping functional domains of the guanylate cyclase regulator protein, GCAP-2. *J Biol Chem* **274**(16):10823–10832.
69. Zhou W, Qian Y, Kunjilwar K, Pfaffinger PJ, Choe S. 2004. Structural insights into the functional interaction of KChIP1 with Shal-type K(+) channels. *Neuron* **41**(4):573–586.
70. Ikura M, Clore GM, Gronenborn AM, Zhu G, Klee CB, Bax A. 1992. Solution structure of a calmodulin-target peptide complex by multidimensional NMR. *Science* **256**(5057):632–638.
71. Meador WE, Means AR, Quiocho FA. 1992. Target enzyme recognition by calmodulin: 2.4 Å structure of a calmodulin–peptide complex. *Science* **257**(5074):1251–1255.
72. Griffith JP, Kim JL, Kim EE, Sintchak MD, Thomson JA, Fitzgibbon MJ, Fleming MA, Caron PR, Hsiao K, Navia MA. 1995. X-ray structure of calcineurin inhibited by the immunophilin-immunosuppressant FKBP12–FK506 complex. *Cell* **82**(3):507–522.

5

FRET and Its Biological Application as a Molecular Ruler

Jie Zheng
Department of Physiology and Membrane Biology, School of Medicine, University of California Davis

5.1. INTRODUCTION

Fluorescence resonance energy transfer (FRET) is a well-studied physical process in which two fluorescent molecules in close proximity interact with each other. During this interaction, energy is passed from one fluorophore (the donor) to the other fluorophore (the acceptor). This simple physical phenomenon can be found in many natural biological systems. For example, the luminescence of the green light-emitting Northwest Pacific jellyfish *Aequorea victoria* involves two proteins: aequorin and green fluorescent protein (GFP). The chemiluminescent aequorin emits a blue light by itself [1], but can pass its energy to GFP in the light organs of the jellyfish [2,3]. It is the fluorescence emission of GFP that yields the characteristic green light of the jellyfish.

The physical basis for FRET was first discussed by Theodor Förster in the 1940s [4] and can be summarized as follows. Light energy absorbed by the donor fluorophore brings an electron into one of the excited states (Fig. 5.1). Normally, this absorbed energy is released in the form of light (fluorescence emission) as the electron falls back to the ground state, S_0. (Other means of dissipating the absorbed energy include passing it to a nearby nonfluorescent molecule, electron transfer, heat, etc.). During the excited state of the donor, however, if the acceptor fluorophore is present within the electromagnetic field generated by the excited donor electron, the energy can also be passed to the acceptor, bringing its electron to the excited states. When the energy is eventually released from the acceptor, fluorescence emission is observed. Due to energy transfer through resonance coupling between the fluorophores, the donor fluorescence emission decreases; meanwhile, the acceptor fluorescence emission increases.

Address correspondence to Jie Zheng, Department of Physiology and Membrane Biology, School of Medicine, University of California Davis, One Shields Avenue, Davis, CA 95616 USA, 530 752-1241, <jzheng@ucdavis.edu>

T. Jue (ed.), *Biomedical Applications of Biophysics*,
Handbook of Modern Biophysics 3, DOI 10.1007/978-1-60327-233-9_5,
© Springer Science+Business Media, LLC 2010

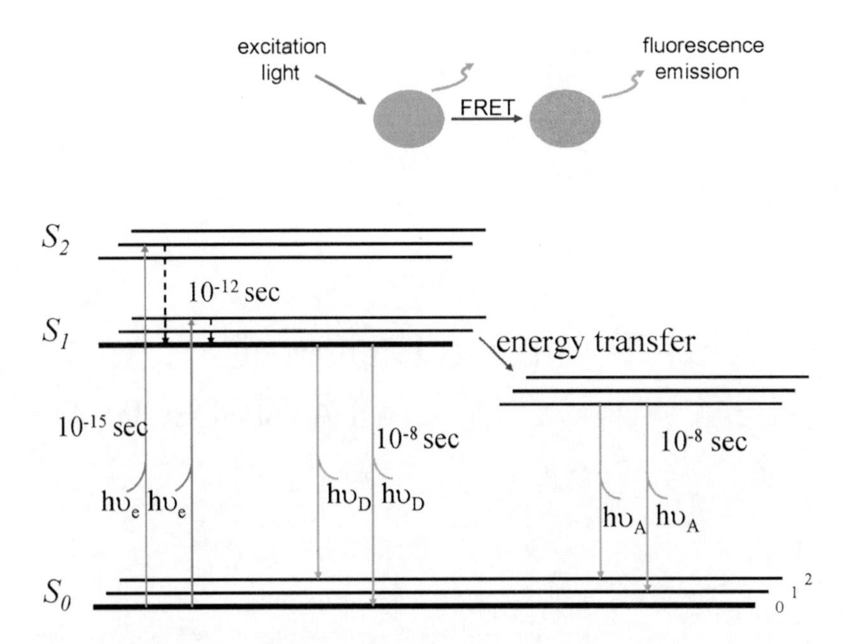

Figure 5.1. Jablonski diagram illustrating the energy coupling process. Arrows indicate energy transitions; S_0 is the ground state, S_1 and S_2 are excited states, each with a number of vibrational energy levels; $h\upsilon_e$ represents the photon energy absorbed by the donor, $h\upsilon_D$ and $h\upsilon_A$ are photon emission from the donor and acceptor, respectively; the rate of each transition process is shown in unit of second. On the top, a diagram showing a FRET pair and light path is shown in the same colors as the energy diagram. Note, however, that energy transfer is a non-radiative process with no photon involved. Please visit http://extras.springer.com/ to view a high-resolution full-color version of this illustration.

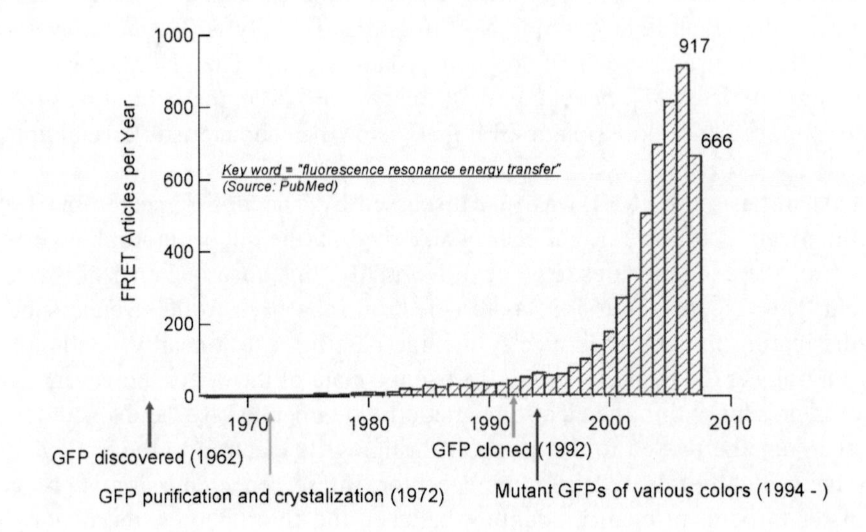

Figure 5.2. Number of FRET articles published each year. A PubMed search with the keyword "fluorescence resonance energy transfer" was used to construct this figure. The number of articles found in 2006 and 2007 (by mid-October) are shown. Also labeled are major events related to the fluorescent proteins. In light of the massive body of literature, examples used in this chapter will be mostly limited to ion channel studies. Please visit http://extras.springer.com/ to view a high-resolution full-color version of this illustration.

Applications of FRET methods in biomedical research began in the early 1970s (Fig. 5.2). These early applications were limited by means of conjugating fluorophores to biological samples [5]. However, with the rapid advance of molecular biology methods FRET has found increasing usage. Site-directed cysteine mutations into target proteins, for example, allow selective attachment of sulfhydrol-reactive organic fluorophores that are widely available. Nonetheless, an explosion of FRET applications in biomedical research occurred shortly after the cloning of GFP and the generation of many mutant fluorescent proteins of various colors, e.g., cyan fluorescent protein (CFP) and yellow fluorescent protein (YFP) [6]. GFP-based FRET provides researchers with a very powerful tool to detect molecular interactions in living cells and follow such interactions in a noninvasive manner during biological processes that can hardly be substituted by other methods. Research into better approaches in using these fluorescent proteins continuously yields new tools for biological studies [7,8].

5.2. DISTANCE DEPENDENCE OF FRET EFFICIENCY

The majority of FRET methods rely on measuring changes in the FRET efficiency — the efficiency of energy transfer between donor and acceptor. Resonance energy transfer can only occur within the electromagnetic field generated by the excited electron, which decays rapidly over distance. It is thus expected that FRET efficiency is strongly dependent on the distance between fluorophores. As described by Förster, the FRET efficiency, E, can be calculated as

$$E = \frac{R_0^6}{R^6 + R_0^6}, \tag{5.1}$$

in which R is the distance between the donor and the acceptor, and R_0, termed the Förster distance, is a characteristic distance for a particular fluorophore pair at which the FRET efficiency is equal to 50%. The relationship between FRET efficiency and distance was experimentally confirmed by Stryer and Haugland using fluorophores naphthyl and dansyl attached to the opposite ends of polyprolyl peptides with increasing numbers of proline [9]. For most useful FRET pairs, R_0 falls between 30 and 80 Å (1 Å = 0.1 nm). Pairs of fluorescent proteins, for example, normally have an R_0 value around 50–60 Å.

The Förster equation can be visualized graphically in Figure 5.3. For this example, a hypothetical FRET pair with an R_0 value of 50 Å is used. Two important features of FRET can be seen in the figure. First, FRET is extremely sensitive to distance. At a distance of around R_0 (point A), a change in the distance as small as 1 Å — a length smaller than that of a covalent bond — translates into a FRET efficiency change of 3%! Second, the distance sensitivity diminishes quickly when the FRET pair moves away from R_0. At shorter distances (point B), energy transfer approaches 100% and stays relatively constant. At longer distances (point C), the donor electrons release the energy mainly through fluorescence emission. Energy transfer becomes non-significant beyond about 100 Å (10 nm).

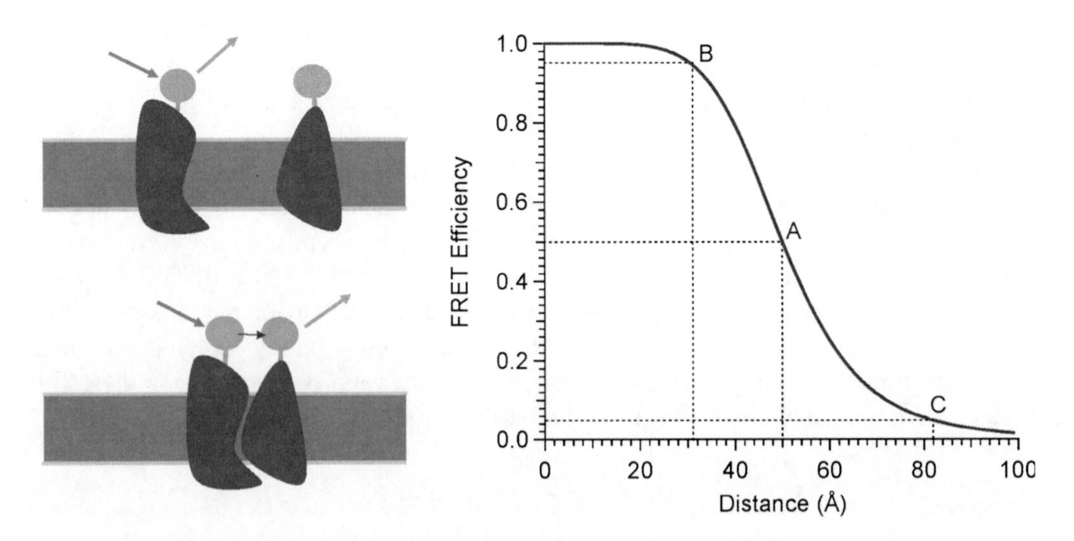

Figure 5.3. Distance dependence of the FRET efficiency. (Left) A schematic diagram illustrating two membrane proteins each labeled with a fluorophore. When they are far apart (top), there is no FRET between the fluorophores. The fluorophores are independent of each other. When the two proteins are interacting (bottom), the fluorophores are at close proximity. Energy transfer occurs. (Right) A FRET pair with an R_0 of 50 Å is used to construct this figure. Points A, B, and C represent 50, 95, and 5% FRET efficiency, respectively. Please visit http://extras.springer.com/ to view a high-resolution full-color version of this illustration.

5.3. SPECTRA OVERLAP BETWEEN FRET PAIRS

In order for energy transfer to occur, the two fluorophores must not only be in close proximity, but the amount of energy required to bring the acceptor electron to its excited states also has to match the amount of energy that is released by the donor electron. Reflecting this prerequisite, all FRET pairs exhibit spectra overlap. Specifically, the emission spectrum of the donor overlaps with the absorption spectrum of the acceptor. The extent of spectral overlap determines how well two fluorophores form a FRET pair. The relationship between the Förster distance and the spectra overlap can be described by the following equation:

$$R_0^{\,6} = 8.79 \times 10^{23} (\kappa^2 n^{-4} Q_D J(\lambda)), \tag{5.2}$$

in which R_0 is in Å, κ^2 is a unitless orientation factor (discussed below), n is the reflective index of the medium, Q_D is the quantum yield of the donor in the absence of acceptor, $J(\lambda)$ is the overlap integral between the donor emission spectrum and the acceptor absorption spectrum. Useful FRET pairs can often be identified by examining their spectral properties.

5.4. ORIENTATION FACTOR

Another prerequisite for energy transfer is that the FRET pair be oriented such that the electron dipole moments of the two fluorophores are at least partially aligned. Orientation of a FRET pair can greatly affect the energy transfer efficiency. Energy transfer is most efficient when the vectors for the two dipole moments are perfectly aligned. The efficiency drops when the two dipole vectors have an angle. When the two vectors are perpendicular to each other, the efficiency becomes zero — there is no FRET under this particular situation even when the two fluorophores are within FRET distances. The orientation factor, κ^2, can be calculated by the following equation:

$$\kappa^2 = (\cos\theta_T - 3\cos\theta_D \cos\theta_A)^2, \tag{5.3}$$

in which θ_T is the angle between the emission transition dipole of the donor and the absorption transition dipole of the acceptor. θ_D and θ_A are the angles between these dipoles and the vector joining the two fluorophores. The range for the κ^2 value is between 0 and 4. The relationship between FRET efficiency and the orientation factor of the fluorophores can be seen in Eq. (5.2). From the equation it is obvious that the orientation factor can significantly affect the FRET efficiency.

In practice, however, fluorophores are rarely fixed at a particular orientation. (One way to approximate the fixed fluorophore situation experimentally is to dissolve fluorophore molecules in high viscous media such as glycerol.) Movements of both the fluorophores and the molecules they are attached to keep changing the orientation of the transition dipoles. Such molecular tumbling during the excited state of the fluorophores allows a fluorophore pair to sample a wide range of possible orientations, making the effective κ^2 value a nonzero number somewhere between 0 and 4. A widely used assumption for freely mobile fluorophore pairs is that κ^2 is equal to 2/3.

5.5. ADVANTAGES OF FRET FOR BIOLOGICAL APPLICATIONS

From the discussion above, one can already see that fluorescence resonance energy transfer has a number of unique features that are highly suitable for applications in biomedical research. New approaches exploiting unique features of FRET are continuously being developed for live-cell measurements.

1. Optimal distance range: For most useful FRET pairs the characteristic distance R_0 is between 30 and 80 Å. As a result, the FRET efficiency drops to virtually zero when the distance is beyond 100 Å. This ensures that FRET reports only close proximity between molecules in this range. As a vast amount of biomolecules are of comparable size, the distance range of FRET makes it especially suitable to detect intermolecular interactions as well as conformational changes of individual proteins in biological systems. The resolution of optical methods in general is limited by the wavelength of the observation light, which is about 400–700 nm in the visible range. FRET provides exceptional sensitivity of angstrom distances that is 10^3–10^4 times higher than the optical limit!
2. Exquisite sensitivity: As the FRET efficiency has a sixth-power dependence on the distance between the pair of fluorophores, it can reflect very small changes in the distance between biomolecules to which they are attached. Biochemical reactions constantly move molecules around. Enzymatic activities are often carried out through protein conformational changes. These processes result in distance changes

between molecules or different positions of the same molecule. As FRET provides a way to detect distance changes at an angstrom resolution, it has been called a "molecular ruler" for biological research [10].

3. Biological samples are mostly transparent to FRET: Organic materials are generally transparent to the electromagnetic field. This allows FRET to "measure" the distance between two points in a biological system without the requirement of a direct physical linkage. For example, FRET can occur between two fluorophores on opposite sides of the cell membrane even though they are physically separated by the lipid bilayer. At the same time, cultured cells and thin tissue samples are also transparent to visible light, making fluorescence excitation and detection applicable to many biological preparations.
4. Predictability: Energy transfer is a simple, well-understood physical process. The outcome of an FRET experiment can often be quite reliably predicted if the underlying biological process is known. This has been convincingly demonstrated by the FRET measurements from dansyl-polyprolyl-naphthyl [9]. Internal control experiments can often be designed based on predictability. For example, in a study of conformational changes in the intracellular cyclic nucleotide-binding domain (CNBD) of CNG channels, Zagotta and colleagues measured FRET between a GFP molecule attached to CNBD and a dipicrylamine (DPA) inserted in the plasma membrane [11]. DPA is a small negatively charged fluorophore. Its location within the membrane can be controlled by transmembrane voltage — depolarization brings DPA to the inner leaflet of the membrane, and hyperpolarization brings it to the outer leaflet [12]. By shifting DPA within the cell membrane with changes in the transmembrane voltage, the authors measured FRET at two distances for each CNBD conformation — the FRET efficiency measured with the outer leaflet-bound DPA should be always lower because of the larger distance added by the membrane.
5. Noninvasive measurements: FRET is an optical measurement that can be applied to living cells under physiological conditions. There is often no need to isolate the molecules of interests (as opposed to many biochemical methods). The use of fluorescent proteins further enhances this advantageous feature of FRET measurements. Fluorophore labeling can be achieved at the DNA level, before protein synthesis and participation of biological functions. Genetically labeled fluorophores allow for studies of developing cells or those involving deep tissues or even whole animals.
6. Long-lasting measurements: Many FRET quantification methods are nondestructive and almost fully reversible. These methods can be used in living cells to follow biological processes that take a long time to complete. In transgenic animals, FRET can potentially report protein–protein interaction throughout their lifespan.

A number of methods have been developed to quantify FRET. As FRET leads to a decrease in donor emission and an increase in acceptor emission, one can quantify FRET by measuring the change in fluorescence intensity of either fluorophore. In addition to intensity measurements, FRET can be quantified by other attributes. For example, FRET will alter the mean lifetime of the excited state of the donor fluorophore as well as its photobleaching rate. It will also affect anisotropy as energy transfer between fluorophores reduces light polarization. Each of these approaches has its advantages and limitations, and is applicable to specific situations. Common contaminating factors in FRET measurements are discussed elsewhere [13]. In the following sections, several major and noticeable FRET quantification methods will be examined, followed by FRET applications in biological research. Because of the vast body of relevant literature, examples are given here with an emphasis on ion channel studies.

5.6. DONOR DEQUENCHING

Donor dequenching is a technique in which the acceptor is removed to eliminate FRET, leading to recovery of donor emission intensity. The amount of fluorescence intensity increase represents the fraction of energy that is transferred to the acceptor instead of being emitted as donor fluorescence emission. For this straightforward FRET quantification method, the FRET efficiency can be calculated as

$$E = 1 - \frac{F_{\mathrm{D}}}{F_{\mathrm{D}}'}, \tag{5.4}$$

in which F_{D} and F_{D}' are the donor fluorescence intensity before and after photobleaching of the acceptor, respectively. For this method, no sophisticated equipment is required. All that is needed is the means to excite the donor and the acceptor and measure their intensities. In addition, FRET efficiency is quantified by donor fluorescence intensities. This is advantageous because bleed-through emission from the acceptor is normally insignificant and the excitation light for the acceptor does not appreciably excite the donor. While measurement of the acceptor emission is susceptible to both bleed-through and crosstalk [13], it is nonetheless not an issue here, as the acceptor emission is monitored under direct excitation. Furthermore, an increase in fluorescence intensity is distinct from the commonly observed gradual decrease in fluorescence intensity that is associated with photobleaching. As one of the most commonly used FRET quantification methods, the donor dequenching approach has been utilized in many ion channel studies [14–18]. This is a destructive approach and cannot be repeated with the same sample.

5.7. ENHANCED ACCEPTOR EMISSION

FRET can be quantified by the increase in acceptor emission due to energy transfer, also known as sensitized emission. For this method, the FRET efficiency is calculated as

$$E = \frac{\varepsilon_{\mathrm{A}}}{\varepsilon_{\mathrm{D}}}\left(\frac{F_{\mathrm{A}}'}{F_{\mathrm{A}}} - 1\right), \tag{5.5}$$

in which F_{A} and F_{A}' are the acceptor intensity in the absence and in the presence of the donor, respectively, and ε_{D} and ε_{A} are the molar extinction coefficients for the donor and the acceptor. Unlike the donor dequenching approach, in which donor fluorescence intensity can be measured directly before and after FRET, the enhanced acceptor emission is often not directly measurable. This is because the donor fluorophore cannot always be easily removed from the experimental system. Further hindering measurements of the enhanced acceptor emission is the spectral overlap between the donor and acceptor fluorophores. As a result of these complications, F_A cannot always be directly measured. Many approaches have been developed to indirectly measure F_A [13,19,20]. In the following sections two effective recent approaches will be outlined.

5.8. THREE-CUBE FRET

In this fluorescence microscope-based FRET quantification method, F_A is estimated indirectly from the amount of acceptor molecules in the experimental system and the efficiency of the recording system to excite the acceptor. The latter is estimated from a sample of pure acceptor molecules. In order to estimate F_A and F_A', the donor fluorescence emission in the acceptor range has to be removed. This contaminating bleed-through fluorescence is estimated from a sample of pure donor molecules. The name of the three-cube (or 3^3) method comes from the fact that three filter cubes are required to carry out all measurements [21]. The principle underlying this approach has been outlined previously by Clegg [22]. Examples of application of this method can be found in studies of the association between the voltage-gated Ca^{2+} channel and calmodulin [21,23,24], between acid-sensing ion channel (ASIC) subunits [25], and between N-methyl-d-aspartate (NMDA) receptor subunits [26].

5.9. SPECTRA FRET

Realizing the high reliability of identifying multiple fluorescence emission components with spectroscopic measurements, a spectroscopy-based FRET quantification method termed "spectra FRET" was recently developed [13,27,28]. Like the three-cube FRET approach, spectra FRET is based on the framework outlined by Clegg [22], in which the FRET efficiency is estimated as the enhanced emission of the acceptor. Two filter cubes without an emission filter are used in conjunction with a spectrograph and a CCD camera. The emission spectra from donor-only samples, acceptor-only samples, and donor-acceptor samples are collected and compared, from which the FRET efficiency is calculated according to Eq. (5.5). As the spectral properties of each fluorophore are unique, spectra FRET allows reliable separation of the fluorescence emissions of the donor and acceptor, and identification of possible contaminating fluorescence sources. In addition, as FRET efficiency is expected to be independent of the wavelength at which it is quantified, spectra FRET provides a convenient means to quantify FRET at a wide wavelength range to ensure accuracy. Furthermore, spectra FRET can be selectively conducted at distinct cellular locations. Examples of spectra FRET applications can be found in studies of cyclic nucleotide-gate channels [29–32], CLC-0 chloride channels [33], and TRP channels [34].

5.10. FLUORESCENCE INTENSITY RATIO BETWEEN DONOR AND ACCEPTOR

In many applications researchers are concerned with whether FRET is present or whether it changes without the need to quantify the exact amount of FRET. In these cases, a simple and sensitive FRET measurement is the ratio between the donor and the acceptor fluorescence emissions:

$$\text{FRET ratio} = \frac{F_D}{F_A}. \tag{5.6}$$

The high sensitivity of this FRET index comes from the fact that FRET changes both F_D and F_A but in opposite directions. As this approach does not attempt to estimate FRET efficiency, it is not necessary to clearly separate F_D and F_A. For example, one can simply use the donor excitation light combined with a beam splitter to simultaneously measure the fluorescence emissions at the donor and the acceptor ranges. This type of recording is advantageous for time course studies as no change in recording conditions is needed. For this reason, it has been widely used in biological studies [35–42].

5.11. LIFETIME MEASUREMENTS

The lifetime of the excited state of a fluorophore is determined by the sum of the rates for all transitions through which the absorbed photon energy can be released. Under normal conditions, the dominant pathway is fluorescence emission. The lifetime of the fluorophore excited state, which is approximately equal to the inverse of the rate of fluorescence emission, is in the order of 10^{-8} s. As FRET adds another pathway for the donor to release the acquired photon energy (see Fig. 5.1), the mean lifetime of the donor excitation state is shortened. The mean lifetime of the donor fluorophore with and without an acceptor, τ_{DA} and τ_D, respectively, can be directly measured from the fluorescence decay time course. The FRET efficiency can be calculated as

$$E = 1 - \frac{\tau_{DA}}{\tau_D}. \tag{5.7}$$

Ramadass and colleagues used lifetime measurements to study the interaction of the TRPV4 channel and the microfilament in living cells [43]. To do so they labeled TRPV4 with CFP and actin with YFP. In HaCaT keratinocyte cells transfected with TRPV4-CFP, the decay of CFP fluorescence can be approximated as a double-exponential process with time constants of 1.0 and 3.6 ns. When TRPV4-CFP and actin-YFP were co-expressed, the two time constants were estimated to be 0.8 and 3.3 ns, respectively, suggesting that TRPV4 interacted with actin to bring CFP and YFP into the FRET distance. FRET quantification based on donor lifetime measurements can also be found in numerous other studies [44–46].

Lifetime measurements have a number of advantages. (1) It is often possible to measure donor emission at a wavelength range where there is little contamination from acceptor emission. (2) All fluorescence measurements are conducted under identical conditions. (3) Lifetime measurement is independent of the fluorescence intensity, and thus can be easily compared across samples. (4) Because of the intensity independence, lifetime measurement is also less sensitive to photobleaching, which is inevitable in repetitive measurements.

5.12. FRET QUANTIFICATION THROUGH PHOTOBLEACHING RATE MEASUREMENTS

Photobleaching is a process in which the excited fluorophore undergoes photodestruction and becomes nonfluorescent. As discussed in the previous section, FRET reduces the excited state lifetime of the donor fluorophore. As a result, the chance of photobleaching is reduced. Effectively, the presence of an acceptor protects the donor from being photobleached; this protective effect is directly proportional to FRET efficiency:

$$E = 1 - \frac{\tau_{DA}}{\tau_D}, \tag{5.8}$$

in which τ_D and τ_{DA} are the time constants of donor photobleaching in the absence and presence of the acceptor, respectively. This unique phenomenon provides a means to indirectly detect the presence of FRET. One can compare the rate of photobleaching of the donor fluorophore in the absence of the acceptor to that in the presence of the acceptor. If the rate is lower when the acceptor is present, it indicates that FRET occurs. This irreversible approach is very simple to carry out — only one excitation light source and one emission detection are needed.

Isacoff and colleagues utilized photobleaching to detect FRET between fluorescein (donor) and tetramethylrhodamine (acceptor) that are attached to the S4 transmembrane domain of voltage-gated Shaker potassium channels [47]. Photobleaching was induced by repetitive light pulsing. The remaining donor fluorescence intensity was monitored. When only fluorescein was attached to S4 photobleaching was 60% complete in about 3.5 seconds. Attaching fluorescein and tetramethylrhodamine to diagonal S4's at a 1:1 stoichiometry lengthened the photobleaching time to about 4.5 seconds. Attaching fluorescein and tetramethylrhodamine to all the four S4's of a channel at a 1:3 stoichiometry further lengthened the photobleaching time to almost 10 seconds. FRET efficiency at the 1:3 stoichiometry is expected to be much higher than at 1:1 stoichiometry because (1) the three tetramethylrhodamine molecules surrounding the fluorescein molecule provide multiple pathways for energy transfer, and (2) the distance between fluorophores attached to neighboring S4's should be $\sqrt{2}$ smaller than the diagonal distance, hence a much higher FRET efficiency between these fluorophores.

5.13. PROTEIN COMPLEX FORMATION

A vast amount of biological functions are carried out by protein complexes. Formation of these complexes enhances the specificity and speed of biological processes. In photoreceptor neurons (rod and cone cells of the retina), the phototransduction cascade involves activation of a chain of enzymatic proteins including rhodopsin, G-protein, phosphodiesterase, and CNG channels. All of these proteins are found in the disc membrane in the outer segment of the rod and cone cells. Similarly, in the olfactory system signaling proteins are located in the membrane of tiny cilia on the top of olfactory neurons. Numerous biochemical and molecular methods — e.g., coimmunoprecipitation, disulfide linkage, and colocolization — have been utilized to study protein–protein interactions. FRET adds a new approach with its unique niches. It is a noninvasive measurement that can be carried out in real time and under physiological conditions. It also offers one of the most sensitive measurements because FRET reports only proximity within about 10 nm. With the development of multiple GFP mutants that form nice FRET pairs, FRET has emerged as a powerful tool for detecting protein–protein interaction.

FRET has been used to demonstrate associations between the voltage-gated Ca^{2+} channel Ca(v)3.3 and putative channel proteins CatSper1 and CatSper2 in sperm [48], Kv1.3 and Kv1.5 in macrophage [49], P2X and nicotinic channels [50], TRPC3 and TRPC4 [51], Ca^{2+} channel primary subunit and auxiliary subunit [52], TRPC4 and TRPC5 [53], CD3 of T cells and Kv1.3 [54], ENaC and CFTR [55], etc.

5.14. SUBUNIT STOICHIOMETRY OF PROTEIN COMPLEX

FRET not only serves as a convenient way to detect the subunit composition of a protein complex, in many cases it can also be used to determine the exact subunit stoichiometry. For example, if positive FRET signals can be detected when the donor and acceptor are attached to the same type of subunit, this indicates that the subunit type exists at least in two copies in the complex. CNG channels in the rod photoreceptor neurons exist as tetramers that are composed of CNGA1 and CNGB1 subunits. One can predict that if the subunit stoichiometry between CNGA1 and CNGB1 is 2:2 then co-expression of CNGA1-CFP, CNGA1-YFP, and CNGB1 should yield FRET. Similarly, co-expression of CNGA1, CNGB1-CFP, and CNGB1-YFP should also yield FRET. If the subunit stoichiometry is instead 3:1, only the first co-expression but not the second will yield FRET. Conversely, if the subunit stoichiometry is 1:3, only the second but not the first co-expression should yield FRET. Using this simple logic, it was determined that rod CNG channels are composed of three CNGA1 and a single CNGB1 subunit [30]. The subunit stoichiometries of several other channels have been determined this way. These include the olfactory CNG channel that is composed of two CNGA2, one CNGA4, and one CNGB1b [32], temperature-sensitive TRP channels in which TRPV1, TRPV2, TRPV3, and TRPV4 subunits can co-assemble randomly without fixed stoichiometry [34], an acid-sensing ion channel (ASIC) that is composed of either two 1a and two 1b, or two 1a and two 2b, or two 2a and two 2b subunits [25], the NMDA receptor that is composed of two NR1 and two NR2 subunits in a dimer-of-dimers arrangement [26], and the epithelial Na^+ channel that is composed of at least two alpha, two beta, and two gamma subunits [56]. In these studies, the existence of FRET between different subunits served as positive confirmation that all subunits are co-assembled into the channel complex. Subunit compositions have also been studied by FRET in heteromeric TRPC channels [57], TRPV channels [58], and ASIC/ENaC channels [59].

The limitation of this FRET approach is that, while it is easy to distinguish between a single copy of a subunit and two copies, it is hard, if possible, to use FRET to distinguish between two copies and higher orders. The epithelial Na^+ channel mentioned above, for example, has been proposed to be composed of three copies each of the alpha, beta, and gamma subunits [60,61]. Other fluorescence-based approaches have been tested to determine higher orders of subunits co-assembly [32,62,63].

5.15. BINDING OF LIGANDS OR MODULATORY MOLECULES

Molecular interactions in biological systems are often not permanent. Dynamic interactions rely on an intact cellular environment and physiological conditions. Their sensitivity to cellular environment and transient nature often make dynamic interactions much harder to detect. Binding of calmodulin, a Ca^{2+}-dependent regulatory protein is highly dynamic. Calmodulin has four Ca^{2+} binding sites. The affinities of these binding sites for Ca^{2+} are different, allowing calmodulin to respond to local Ca^{2+} events with great sensitivity. Association of calmodulin with many proteins links these Ca^{2+} events to the activity level of these proteins. For example, binding of calmodulin to voltage-gated Ca^{2+} channels serves as a negative feedback signal to regulate channel activity.

To directly follow calmodulin binding to voltage-gated Ca^{2+} channels, Yue and colleagues labeled the channel and calmodulin with CFP and YFP. FRET signals were measured in cultured cells co-expressing these fluorescent constructs with the three-cube method [21]. It was found that even before calmodulin is activated by Ca^{2+} ions it is already attached to the L-type, P/Q-type, and R-type Ca^{2+} channels. Activation of calmodulin causes it to alter the way it binds to the channel, through which it exerts modulatory effects on channel activities [23,24]. Interaction between calmodulin and Ca^{2+} channels has also been recorded using a net FRET approach [64]

The activity of many ion channels is modulated by calmodulin. For example, calmodulin downregulation of CNG channel activities plays a critical role in visual and olfactory adaptation. The interaction between calmodulin and CNG channels has been studied by FRET in isolated membrane patches [29]. Fluorescent calmodulin carrying an Alexa488 fluorophore was directly applied to the intracellular side of membrane patches that contained rod CNG channels carrying a CFP. FRET due to binding of the fluorescent calmodulin to the channel caused a reduction in CFP fluorescence, which was reversible upon washing off of calmodulin. By comparing the time course of FRET changes to the time course of current inhibition, it was concluded that binding of calmodulin was the rate-limiting step in this process. Binding of calmodulin to CNG channels disrupts a specific N–C termini interaction [65]. This process can be directly observed by FRET when the interacting domains are labeled by CFP and YFP. In this case, nonfluorescent calmodulin was applied to the intracellular side of CNG channels in cell-free membrane patches [31]. A decrease in FRET was observed during the time course of current inhibition, suggesting that calmodulin physically separated the two interacting domains.

Additional studies of ion channel modulation by calmodulin can be found in the literature [66]. In addition, binding of many other regulatory molecules can be detected with FRET. Examples include relative movements between the N and C termini of Kir6.2 induced by ATP-binding [67], association of microtubule plus-end tracking protein EB1 with voltage-gated Kv1 potassium channels, blockade of L-type Ca^{2+} channels by a fluorescent diltiazem analogue [68], G-protein modulation of Ca^{2+} channel α and β subunits interaction [69], etc.

5.16. CONFORMATIONAL REARRANGEMENTS

FRET measurements can be used to monitor real-time protein conformational changes. Indeed, as FRET has exquisite sensitivities to distance changes as small as 1 Å, it offers one of the most sensitive means to follow protein structural changes during protein function under physiological conditions. Examples can be found in elegant applications of FRET to the study of voltage sensor movements during cation channel activation. For this purpose, Bezanilla, Selvin, and colleagues used lanthanide-based resonance energy transfer (LRET), in which a lanthanide donor (which has an extremely long-lived excited state on the millisecond timescale) transfers energy to a conventional organic fluorescent acceptor [70]. One advantage of lanthanides for FRET methods is that their extremely long lifetimes allow clean separation of donor and acceptor emission components. In addition, emission from lanthanides is not polarized, thus reducing the influence of the κ^2 factor. Labeling of lanthanides such as terbium can be achieved through a chelator that also enhances the extinction coefficient of lanthanides [71]. Using chelated terbium coupled to fluorescein, these researchers were able to record small, angstrom-scale relative movements between S4 segments of Shaker potassium channels. A similar approach using LRET has been applied to prokaryotic Na^+ channel NaChBac and K^+ channel KvAP [72,73]. The Isacoff group approached the question of Shaker channel S4 movements with a conventional FRET pair, fluorescein–rhodamine [47]. FRET was measured through the slowing of donor photobleaching as discussed in Section 5.12.

Vertical movements of S4 relative to the plasma membrane have also been measured in the Shaker K^+ channels with FRET. The Bezanilla group labeled S4 positions with sulphorhodamine or 7-fluorobenz-2-oxa-1,3-diazole-4-sulphonamide (ABD), which formed a FRET pair with DPA (described above in §5.5) that was dissolved in the plasma membrane [74]. The Selvin group labeled S4 with chelated terbium (donor) and the channel pore with fluorescein-attached scorpion toxin [75]. In both cases, only very small FRET changes were detected, indicating that S4 does not move substantially in the vertical direction during voltage sensing.

Like other enzymatic activities, activation and inactivation of ion channels often involves complex conformational changes. Applications of FRET measurements in monitoring these conformational changes can be found in studies of voltage-gated $Ca_v1.2$ channels [76], voltage-gated Kv2.1 channels [77], ryanodine receptors [78], L-type Ca^{2+} channels [79], P2X channels [80], gramicidin channels [81,82], etc.

5.17. INTRACELLULAR EVENT INDICATORS

When a donor and an acceptor are linked together by a linker that can respond to changes in the intracellular environment by changing its conformation, a FRET-based molecular sensor is constructed. Many such fusion constructs are designed based on mutant GFPs. Earlier versions link blue fluorescent protein (BFP) and GFP together with a peptide that is protease-sensitive [83,84]. Proteolysis activities in cells expressing these sensors will cut the peptide link; as a result, the FRET signal drops. Later versions used CFP and YFP to avoid problems associated with BFP such as dimness and lack of photostability [85,86]. Many FRET-based intracellular event indicators are based on a similar concept. For example, Tsien and colleagues designed an intracellular Ca^{2+} indicator by linking CFP and YFP with calmodulin (a Ca^{2+}-activated modulatory protein) and M13 (which contains a calmodulin-binding sequence) [87]. Dubbed "cameleon," this construct changes its fluorescence when an increase in intracellular Ca^{2+} concentration leads to occupation of the Ca^{2+} sites in calmodulin and binding of calmodulin to M13. The color change is due to increased FRET as a result of changes in probably both the distance and orientation associated with the conformational change in the calmodulin-M13 linker. The Ca^{2+} sensitivity range of cameleons can be tuned by mutations in the Ca^{2+}-binding sites of calmodulin. Similarly linking CFP and YFP together with type II protein kinase A makes a cAMP sensor [88]; switching the linker to a cGMP-dependent protein kinase makes it a cGMP sensor [89,90]. Based on this versatile design, many intracellular event indicators have been developed in the past years to monitor intracellular activities of Ras and Rap1 [91], Ran [92], tyrosine kinases [93,94, serine/threonine kinases [95], etc.

ACKNOWLEDGMENTS

The work in my laboratory is supported by funding from the American Heart Associations and National Institutes of Health, and is conducted partially in a facility constructed with support from Research Facilities Improvement Program Grant C06-RR-12088-01 from the National Center for Research Resources. I also thank David P. Jenkins for review of the manuscript.

PROBLEMS

The figure below (please visit http://extras.springer.com/ to view a high-resolution full-color version) illustrates structural changes associated with the binding of cGMP to the ligand-binding domain of a cyclic nucleotide-gated (CNG) channel: the α–helix shown in red moves to the left, becoming closer to the rest of the ligand-binding domain. The distance between the base of the ligand-binding domain and three positions (1, 2, and 3) along the α–helix in the unliganded state (left) are 35, 25, and 16 Å, respectively; they become 22, 19, and 14 Å, respectively, when cGMP is bound (right). A FRET pair with R_0 = 40 Å is chosen to monitor the structural change with FRET. If the donor (D) is attached to the base of the ligand-binding domain as shown, where should the acceptor be attached to? How much FRET change do you expect to see?

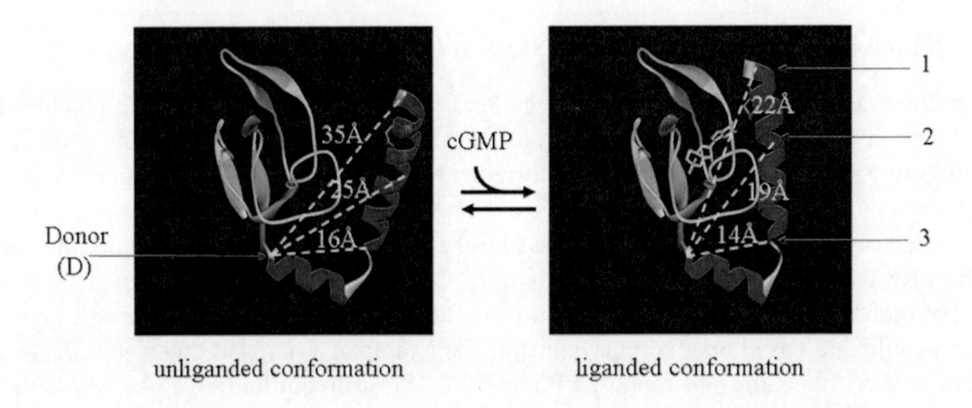

Please visit http://extras.springer.com/ to view a high-resolution full-color version of this illustration.

FURTHER STUDY

Lakowicz J. 2006. *Principles of fluorescence spectroscopy*, 3rd ed, New York: Springer.
Tsien RY. 1998. *The green fluorescence protein, Annu Rev Biochem* **67**:509–544.
Clegg RM. 1992. Fluorescence resonance energy transfer and nucleic acids, *Methods Enzymol* **211**:353–388.
Selvin PR. 1995. *Fluorescence resonance energy transfer, Methods Enzymol* **246**:300–334.

REFERENCES

1. Shimomura O. 1995. A short story of aequorin. *Biol Bull* **189**(1):1–5.
2. Shimomura O, Johnson FH, Saiga Y. 1962. Extraction, purification and properties of aequorin, a bioluminescent protein from the luminous hydromedusan, *Aequorea. J Cell Comp Physiol* **59**:223–239.
3. Morin JG, Hastings JW. 1971. Energy transfer in a bioluminescent system. *J Cell Physiol* **77**(3):313–318.
4. Foster T. 1948. Intermolecular energy migration and fluorescence. *Ann Phys* **2**:55–75. [English translation in *Biological Physics: Key Papers in Physics*. Ed EV Mielczarek, E Greenbaum, RS Knox. 1993. New York: American Institute of Physics.]
5. Lakowicz J. 2006. *Principles of fluorescence spectroscopy*, 3rd ed. New York: Plenum.
6. Tsien RY. 1998. The green fluorescent protein. *Annu Rev Biochem* **67**:509–544.
7. Giraldez T, Hughes TE, Sigworth FJ. 2005. Generation of functional fluorescent BK channels by random insertion of GFP variants. *J Gen Physiol* **126**(5):429–438.
8. Guerrero G, Siegel MS, Roska B, Loots E, Isacoff EY. 2002. Tuning FlaSh: redesign of the dynamics voltage range, and color of the genetically encoded optical sensor of membrane potential. *Biophys J* **83**(6):3607–3618.
9. Stryer L, Haugland RP. 1967. Energy transfer: a spectroscopic ruler. *Proc Natl Acad Sci USA* **58**(2):719–726.
10. Stryer L. 1978. Fluorescence energy transfer as a spectroscopic ruler. *Annu Rev Biochem* 47:819-46.
11. Taraska, J.W. and W.N. Zagotta. 2007. Structural dynamics in the gating ring of cyclic nucleotide-gated ion channels. *Nat Struct Mol Biol*, **14**(9):854–860.
12. Chanda B, Blunck R, Faria LC, Schweizer FE, Mody I, Bezanilla F. 2005. A hybrid approach to measuring electrical activity in genetically specified neurons. *Nat Neurosci* **8**(11):1619–1626.
13. Takanishi CL, Bykova E, Cheng W, Zheng J. 2006. GFP-based FRET analysis in live cells. *Brain Res* **1091**(2):132–139.
14. Amiri H, Schultz G, Schaefer M. 2003. FRET-based analysis of TRPC subunit stoichiometry. *Cell Calcium* **33**(5–6):463–470.
15. Riven I, Iwanir S, Reuveny E. 2006. GIRK channel activation involves a local rearrangement of a preformed G protein channel complex. *Neuron* **51**(5):561–573.
16. Riven I, Kalmanzon E, Segev L, Reuveny E. 2003. Conformational rearrangements associated with the gating of the G protein-coupled potassium channel revealed by FRET microscopy. *Neuron* **38**(2):225–235.
17. Leuranguer V, Papadopoulos S, Beam KG. 2006. Organization of calcium channel beta1a subunits in triad junctions in skeletal muscle. *J Biol Chem* **281**(6):3521–3527.
18. Corry B, Rigby P, Liu ZW, Martinac B. 2005. Conformational changes involved in MscL channel gating measured using FRET spectroscopy. *Biophys J* **89**(6):L49–L51.
19. Gordon GW, Berry G, Liang XH, Levine B, Herman B. 1998. Quantitative fluorescence resonance energy transfer measurements using fluorescence microscopy. *Biophys J* **74**(5):2702–2713.
20. Xia Z, Liu Y. 2001. Reliable and global measurement of fluorescence resonance energy transfer using fluorescence microscopes. *Biophys J* **81**(4):2395–23402.
21. Erickson MG, Alseikhan BA, Peterson BZ, Yue DT. 2001. Preassociation of calmodulin with voltage-gated Ca(2+) channels revealed by FRET in single living cells. *Neuron* **31**(6):973–985.
22. Clegg RM. 1992. Fluorescence resonance energy transfer and nucleic acids. *Methods Enzymol* **211**:353–388.
23. Erickson MG, Liang H, Mori MX, Yue DT. 2003. FRET two-hybrid mapping reveals function and location of L-type Ca^{2+} channel CaM preassociation. *Neuron* **39**(1):97–107.
24. Mori MX, Erickson MG, Yue DT. 2004. Functional stoichiometry and local enrichment of calmodulin interacting with Ca^{2+} channels. *Science* **304**(5669):432–435.
25. Gao Y, Liu SS, Qiu S, Cheng W, Zheng J, Luo JH. 2007. Fluorescence resonance energy transfer analysis of subunit assembly of the ASIC channel. *Biochem Biophys Res Commun* **359**(1):143–150.
26. Qiu S, Hua YL, Yang F, Chen YZ, Luo JH. 2005. Subunit assembly of N-methyl-d-aspartate receptors analyzed by fluorescence resonance energy transfer. *J Biol Chem* **280**(26):24923–24930.
27. Zheng J. 2006. Spectroscopy-based quantitative fluorescence resonance energy transfer analysis. *Methods Mol Biol* **337**:65–78.
28. Zheng J, Zagotta WN. 2003. Patch-clamp fluorometry recording of conformational rearrangements of ion channels. *Sci STKE* **2003**(176):PL7.
29. Trudeau MC, Zagotta WN. 2004. Dynamics of Ca^{2+}-calmodulin-dependent inhibition of rod cyclic nucleotide-gated channels measured by patch-clamp fluorometry. *J Gen Physiol* **124**(3):211–223.

30. Zheng J, Trudeau MC, Zagotta WN. 2002. Rod cyclic nucleotide-gated channels have a stoichiometry of three CNGA1 subunits and one CNGB1 subunit. *Neuron* **36**(5):891–896.
31. Zheng J, Varnum MD, Zagotta WN. 2003. Disruption of an intersubunit interaction underlies Ca^{2+}-calmodulin modulation of cyclic nucleotide-gated channels. *J Neurosci* **23**(22):8167–8175.
32. Zheng J, Zagotta WN. 2004. Stoichiometry and assembly of olfactory cyclic nucleotide-gated channels. *Neuron* **42**(3):411–421.
33. Bykova EA, Zhang XD, Chen TY, Zheng J. 2006. Large movement in the C terminus of CLC-0 chloride channel during slow gating. *Nat Struct Mol Biol* **13**(12):1115–1119.
34. Cheng W, Yang F, Takanishi CL, Zheng J. 2007. Thermosensitive TRPV channel subunits coassemble into heteromeric channels with intermediate conductance and gating properties. *J Gen Physiol* **129**(3):191–207.
35. Kerr R, Lev-Ram V, Baird G, Vincent P, Tsien RY, Schafer WR. 2000. Optical imaging of calcium transients in neurons and pharyngeal muscle of *C. elegans*. *Neuron* **26**(3):583–594.
36. Adams SR, Harootunian AT, Buechler YJ, Taylor SS, Tsien RY. 1991. Fluorescence ratio imaging of cyclic AMP in single cells. *Nature* **349**(6311):694–697.
37. Koshimizu TA, Kretschmannova K, He ML, Ueno S, Tanoue A, Yanagihara N, Stojilkovic SS, Tsujimoto G. 2006. Carboxyl-terminal splicing enhances physical interactions between the cytoplasmic tails of purinergic P2X receptors. *Mol Pharmacol* **69**(5):1588–1598.
38. Lippiat JD, Albinson SL, Ashcroft FM. 2002. Interaction of the cytosolic domains of the Kir6.2 subunit of the K(ATP) channel is modulated by sulfonylureas. *Diabetes* **3**(51 Suppl):S377–S380.
39. George CH, Jundi H, Thomas NL, Scoote M, Walters N, Williams AJ, Lai FA. 2004. Ryanodine receptor regulation by intramolecular interaction between cytoplasmic and transmembrane domains. *Mol Biol Cell* **15**(6): 2627–2638.
40. Hein P, Frank M, Hoffmann C, Lohse MJ, Bunemann M. 2005. Dynamics of receptor/G protein coupling in living cells. *EMBO J* **24**(23):4106–4114.
41. Pond BB, Berglund K, Kuner T, Feng G, Augustine GJ, Schwartz-Bloom RD. 2006. The chloride transporter Na(+)-K(+)-Cl– cotransporter isoform-1 contributes to intracellular chloride increases after in vitro ischemia. *J Neurosci* **26**(5):1396–1406.
42. Hernandez VH, Bortolozzi M, Pertegato V, Beltramello M, Giarin M, Zaccolo M, Pantano S, Mammano F. 2007. Unitary permeability of gap junction channels to second messengers measured by FRET microscopy. *Nat Methods* **4**(4):353–358.
43. Ramadass R, Becker D, Jendrach M, Bereiter-Hahn J. 2007. Spectrally and spatially resolved fluorescence lifetime imaging in living cells: TRPV4-microfilament interactions. *Arch Biochem Biophys* **463**(1):27–36.
44. Biskup C, Zimmer T, Benndorf K. 2004. FRET between cardiac Na+ channel subunits measured with a confocal microscope and a streak camera. *Nat Biotechnol* **22**(2):220–224.
45. Biskup C, Kelbauskas L, Zimmer T, Benndorf K, Bergmann A, Becker W, Ruppersberg JP. Stockklausner C, Klocker N. 2004. Interaction of PSD-95 with potassium channels visualized by fluorescence lifetime-based resonance energy transfer imaging. *J Biomed Opt* **9**(4):753–759.
46. Zelazny E, Borst JW, Muylaert M, Batoko H, Hemminga MA, Chaumont F. 2007. FRET imaging in living maize cells reveals that plasma membrane aquaporins interact to regulate their subcellular localization. *Proc Natl Acad Sci USA* **104**(30):12359–12364.
47. Glauner KS, Mannuzzu LM, Gandhi CS, Isacoff EY. 1999. Spectroscopic mapping of voltage sensor movement in the Shaker potassium channel. *Nature* **402**(6763):813–817.
48. Zhang D, Chen J, Saraf A, Cassar S, Han P, Rogers JC, Brioni JD, Sullivan JP, Gopalakrishnan M. 2006. Association of Catsper1 or -2 with Ca(v)3.3 leads to suppression of T-type calcium channel activity. *J Biol Chem* **281**(31):22332–22341.
49. Vicente R, Escalada A, Villalonga N, Texido L, Roura-Ferrer M, Martin-Satue M, Lopez-Iglesias C, Soler C, Solsona C, Tamkun MM, Felipe A. 2006. Association of Kv1.5 and Kv1.3 contributes to the major voltage-dependent K+ channel in macrophages. *J Biol Chem* **281**(49):37675–37685.
50. Khakh BS, Fisher JA, Nashmi R, Bowser DN, Lester HA. 2005. An angstrom scale interaction between plasma membrane ATP-gated P2X2 and alpha4beta2 nicotinic channels measured with fluorescence resonance energy transfer and total internal reflection fluorescence microscopy. *J Neurosci* **25**(29):6911–6920.
51. Poteser M, Graziani A, Rosker C, Eder P, Derler I, Kahr H, Zhu MX, Romanin C, Groschner K. 2006. TRPC3 and TRPC4 associate to form a redox-sensitive cation channel: evidence for expression of native TRPC3-TRPC4 heteromeric channels in endothelial cells. *J Biol Chem* **281**(19):13588–13595.

52. Takahashi SX, Miriyala J, Tay LH, Yue DT, Colecraft HM. 2005. A CaVbeta SH3/guanylate kinase domain interaction regulates multiple properties of voltage-gated Ca^{2+} channels. *J Gen Physiol* **126**(4):365–377.
53. Schindl R, Frischauf I, Kahr H, Fritsch R, Krenn M, Derndl A, Vales E, Muik M, Derler I, Groschner K, Romanin C. 2008. The first ankyrin-like repeat is the minimum indispensable key structure for functional assembly of homo- and heteromeric TRPC4/TRPC5 channels. *Cell Calcium* **43**(3):260–269.
54. Panyi G, Bagdany M, Bodnar A, Vamosi G, Szentesi G, Jenei A, Matyus L, Varga S, Waldmann TA, Gaspar R, Damjanovich S. 2003. Colocalization and nonrandom distribution of Kv1.3 potassium channels and CD3 molecules in the plasma membrane of human T lymphocytes. *Proc Natl Acad Sci USA* **100**(5):2592–2597.
55. Berdiev BK, Cormet-Boyaka E, Tousson A, Qadri YJ, Oosterveld-Hut HM, Hong JS, Gonzales PA, Fuller CM, Sorscher EJ, Lukacs GL, Benos DJ. 2007. Molecular proximity of CFTR and ENaC assessed by fluorescence resonance energy transfer. *J Biol Chem* **282**(50):26481–36488.
56. Staruschenko A, Medina JL, Patel P, Shapiro MS, Booth RE, Stockand JD. 2004. Fluorescence resonance energy transfer analysis of subunit stoichiometry of the epithelial Na+ channel. *J Biol Chem* **279**(26):27729–27734.
57. Hofmann T, Schaefer M, Schultz G, Gudermann T. 2002. Subunit composition of mammalian transient receptor potential channels in living cells. *Proc Natl Acad Sci USA* **99**(11):7461–7466.
58. Hellwig N, Albrecht N, Harteneck C, Schultz G, Schaefer M. 2005. Homo- and heteromeric assembly of TRPV channel subunits. *J Cell Sci* **118**(Pt 5):917–928.
59. Meltzer RH, Kapoor N, Qadri YJ, Anderson SJ, Fuller CM, Benos DJ. 2007. Heteromeric assembly of acid-sensitive ion channel and epithelial sodium channel subunits. *J Biol Chem* **282**(35):25548–25559.
60. Snyder PM, Cheng C, Prince LS, Rogers JC, Welsh MJ. 1998. Electrophysiological and biochemical evidence that DEG/ENaC cation channels are composed of nine subunits. *J Biol Chem* **273**(2):681–684.
61. Eskandari S, Snyder PM, Kreman M, Zampighi GA, Welsh MJ, Wright EM. 1999. Number of subunits comprising the epithelial sodium channel. *J Biol Chem* **274**(38):27281–27286.
62. Staruschenko A, Adams E, Booth RE, Stockand JD. 2005. Epithelial Na+ channel subunit stoichiometry. *Biophys J* **88**(6):3966–3975.
63. Ulbrich MH, Isacoff EY. 2007. Subunit counting in membrane-bound proteins. *Nat Methods* **4**(4):319–321.
64. Singh A, Hamedinger D, Hoda JC, Gebhart M, Koschak A, Romanin C, Striessnig J. 2006. C-terminal modulator controls Ca^{2+}-dependent gating of Ca(v)1.4 L-type Ca^{2+} channels. *Nat Neurosci* **9**(9):1108–1116.
65. Varnum MD, Zagotta WN. 1997. Interdomain interactions underlying activation of cyclic nucleotide-gated channels. *Science* **278**(5335):110–113.
66. Maximciuc AA, Putkey JA, Shamoo Y, Mackenzie KR. 2006. Complex of calmodulin with a ryanodine receptor target reveals a novel, flexible binding mode. *Structure* **14**(10):1547–1556.
67. Tsuboi T, Lippiat JD, Ashcroft FM, Rutter GA. 2004. ATP-dependent interaction of the cytosolic domains of the inwardly rectifying K+ channel Kir6.2 revealed by fluorescence resonance energy transfer. *Proc Natl Acad Sci USA* **101**(1):76–81.
68. Brauns T, Prinz H, Kimball SD, Haugland RP, Striessnig J, Glossmann H. 1997. L-type calcium channels: binding domains for dihydropyridines and benzothiazepines are located in close proximity to each other. *Biochemistry* **36**(12):3625–3631.
69. Hummer A, Delzeith O, Gomez SR, Moreno RL, Mark MD, Herlitze S. 2003. Competitive and synergistic interactions of G protein beta(2) and Ca(2+) channel beta(1b) subunits with Ca(v)2.1 channels, revealed by mammalian two-hybrid and fluorescence resonance energy transfer measurements. *J Biol Chem* **278**(49):49386–49400.
70. Cha A, Snyder GE, Selvin PR, Bezanilla F. 1999. Atomic scale movement of the voltage-sensing region in a potassium channel measured via spectroscopy. *Nature* **402**(6763):809–813.
71. Selvin PR. 1995. Fluorescence resonance energy transfer. *Methods Enzymol* **246**:300–334.
72. Richardson J, Blunck R, Ge P, Selvin PR, Bezanilla F, Papazian DM, Correa AM. 2006. Distance measurements reveal a common topology of prokaryotic voltage-gated ion channels in the lipid bilayer. *Proc Natl Acad Sci USA* **103**(43):15865–15370.
73. Sandtner W, Bezanilla F, Correa AM. 2007. In vivo measurement of intramolecular distances using genetically encoded reporters. *Biophys J* **93**(9):L45–L47.
74. Chanda B, Asamoah OK, Blunck R, Roux B, Bezanilla F. 2005. Gating charge displacement in voltage-gated ion channels involves limited transmembrane movement. *Nature* **436**(7052):852–856.

75. Posson DJ, Ge P, Miller C, Bezanilla F, Selvin PR. 2005. Small vertical movement of a K+ channel voltage sensor measured with luminescence energy transfer. *Nature* **436**(7052):848–851.
76. Kobrinsky E, Schwartz E, Abernethy DR, Soldatov NM. 2003. Voltage-gated mobility of the Ca^{2+} channel cytoplasmic tails and its regulatory role. *J Biol Chem* **278**(7):5021–5028.
77. Kobrinsky E, Stevens L, Kazmi Y, Wray D, Soldatov NM. 2006. Molecular rearrangements of the Kv2.1 potassium channel termini associated with voltage gating. *J Biol Chem* **281**(28):19233–19240.
78. George CH, Jundi H, Walters N, Thomas NL, West RR, Lai FA. 2006. Arrhythmogenic mutation-linked defects in ryanodine receptor autoregulation reveal a novel mechanism of Ca^{2+} release channel dysfunction. *Circ Res* **98**(1):88–97.
79. Kobrinsky E, Kepplinger KJ, Yu A, Harry JB, Kahr H, Romanin C, Abernethy DR, Soldatov NM. 2004. Voltage-gated rearrangements associated with differential beta-subunit modulation of the L-type Ca(2+) channel inactivation. *Biophys J* **87**(2):844–857.
80. Fisher JA, Girdler G, Khakh BS. 2004. Time-resolved measurement of state-specific P2X2 ion channel cytosolic gating motions. *J Neurosci* **24**(46):10475–10487.
81. Harms GS, Orr G, Montal M, Thrall BD, Colson SD, Lu HP. 2003. Probing conformational changes of gramicidin ion channels by single-molecule patch-clamp fluorescence microscopy. *Biophys J* **85**(3):1826–1838.
82. Borisenko V, Lougheed T, Hesse J, Fureder-Kitzmuller E, Fertig N, Behrends JC, Woolley GA, Schutz GJ. 2003. Simultaneous optical and electrical recording of single gramicidin channels. *Biophys J* **84**(1):612–622.
83. Heim R, Tsien RY. 1996. Engineering green fluorescent protein for improved brightness, longer wavelengths and fluorescence resonance energy transfer. *Curr Biol* **6**(2):178–182.
84. Mitra RD, Silva CM, Youvan DC. 1996. Fluorescence resonance energy transfer between blue-emitting and red-shifted excitation derivatives of the green fluorescent protein. *Gene* **173**(1 Spec No):13–17.
85. Mahajan NP, Harrison-Shostak DC, Michaux J, Herman B. 1999. Novel mutant green fluorescent protein protease substrates reveal the activation of specific caspases during apoptosis. *Chem Biol* **6**(6):401–409.
86. Luo KQ, Yu VC, Pu Y, Chang DC. 2001. Application of the fluorescence resonance energy transfer method for studying the dynamics of caspase-3 activation during UV-induced apoptosis in living HeLa cells. *Biochem Biophys Res Commun* **283**(5):1054–1060.
87. Miyawaki A, Llopis J, Heim R, McCaffery JM, Adams JA, Ikura M, Tsien RY. 1997. Fluorescent indicators for Ca^{2+} based on green fluorescent proteins and calmodulin. *Nature* **388**(6645):882–887.
88. Warrier S, Ramamurthy G, Eckert RL, Nikolaev VO, Lohse MJ, Harvey RD. 2007. cAMP microdomains and L-type Ca^{2+} channel regulation in guinea-pig ventricular myocytes. *J Physiol* **580**(Pt.3):765–776.
89. Honda A, Adams SR, Sawyer CL, Lev-Ram V, Tsien RY, Dostmann WR. 2001. Spatiotemporal dynamics of guanosine 3',5'-cyclic monophosphate revealed by a genetically encoded, fluorescent indicator. *Proc Natl Acad Sci USA* **98**(5):2437–2442.
90. Sato M, Hida N, Ozawa T, Umezawa Y. 2000. Fluorescent indicators for cyclic GMP based on cyclic GMP-dependent protein kinase Ialpha and green fluorescent proteins. *Anal Chem* **72**(24):5918–5924.
91. Mochizuki N, Yamashita S, Kurokawa K, Ohba Y, Nagai T, Miyawaki A, Matsuda M. 2001. Spatio-temporal images of growth-factor-induced activation of Ras and Rap1. *Nature* **411**(6841):1065–1068.
92. Kalab P, Weis K, Heald R. 2002. Visualization of a Ran-GTP gradient in interphase and mitotic Xenopus egg extracts. *Science* **295**(5564):2452–2456.
93. Kurokawa K, Mochizuki N, Ohba Y, Mizuno H, Miyawaki A, Matsuda M. 2001. A pair of fluorescent resonance energy transfer-based probes for tyrosine phosphorylation of the CrkII adaptor protein in vivo. *J Biol Chem* **276**(33):31305–31310.
94. Ting AY, Kain KH, Klemke RL, Tsien RY. 2001. Genetically encoded fluorescent reporters of protein tyrosine kinase activities in living cells. *Proc Natl Acad Sci USA* **98**(26):15003–15008.
95. Zhang J, Ma Y, Taylor SS, Tsien RY. 2001. Genetically encoded reporters of protein kinase A activity reveal impact of substrate tethering. *Proc Natl Acad Sci USA* **98**(26):14997–15002.

6

INTRODUCTION TO MODERN TECHNIQUES IN MASS SPECTROMETRY

Caroline S. Chu and Carlito B. Lebrilla
Department of Chemistry, University of California Davis

6.1. INTRODUCTION

Mass spectrometry has emerged as an invaluable technique with a wide array of applications ranging from clinical to biodefense. With the development of different ionization techniques and mass analyzers, even challenging samples can be analyzed, thereby making mass spectrometry an important analytical tool in the field of biophysics. Mass spectrometry is the only technique that offers the combination of high sensitivity (attomole) with structural information. While other analytical techniques may provide higher sensitivity, these techniques do not provide structural information. Conversely, other techniques may provide more complete structures but have significantly less sensitivity. The different ionization techniques allow for the examination of analytes ranging from small metabolites to large macromolecular assemblies. In this chapter the major components are described rather than the possible applications, which would require volumes. With the major concepts in hand, the student is encouraged to read specific reviews regarding the kinds of applications of interest to the researcher.

A mass spectrometer separates ions based on the mass-to-charge ratio (m/z). There are three major components of a mass spectrometer: (1) an ion source generates gas-phase ions from the sample of interest, (2) a mass analyzer separates ions based upon their mass-to-charge (m/z), and (3) a detector monitors the ion current and converts it to a signal that gets stored by a data system. A general workflow for a mass spectrometer is illustrated in Figure 6.1.

Address correspondence to Carlito B. Lebrilla, Department of Chemistry, University of California Davis, One Shields Avenue, Davis, California 95616, USA, 530 752-6364, 530 752-8995 (fax), <cblebrilla@ucdavis.edu>.

T. Jue (ed.), *Biomedical Applications of Biophysics*,
Handbook of Modern Biophysics 3, DOI 10.1007/978-1-60327-233-9_6,
© Springer Science+Business Media, LLC 2010

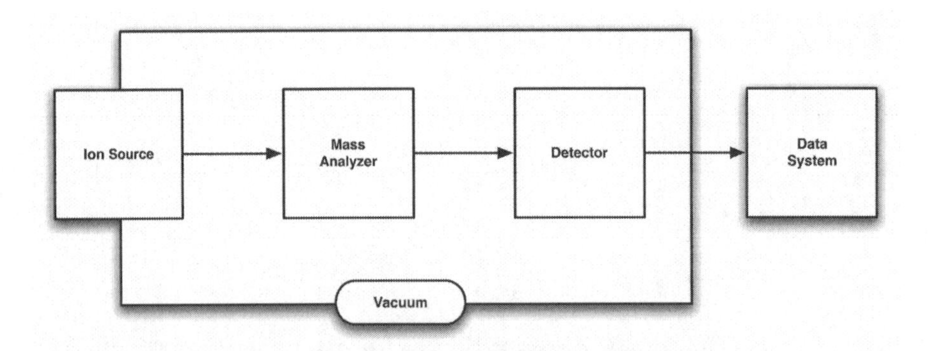

Figure 6.1. Basic components of a mass spectrometer.

6.2. IONIZATION TECHNIQUES (ION SOURCES)

There are a variety of ion sources used to generate ions for mass spectral analysis. Molecules that are volatile, such as those separated by gas chromatography, are best ionized in the gas phase by bombarding them with electrons. For this purpose electron-impact ionization (EI) and its derivative chemical ionization (CI) are optimal methods. Large volatile or thermally labile compounds are ionized by desorbing them directly from solid or liquid samples. Matrix-assisted laser desorption/ionization (MALDI) is used for solid samples while electrospray ionization (ESI) is used for liquid samples. In 2002 these two latter techniques were recognized with the award of the Nobel Prize in chemistry to John Fenn for ESI and Koichi Tanaka for MALDI.

6.3. ELECTRON IMPACT

In electron-impact ionization (electron impact, EI) ions are generated by directing an electron beam of energetic electrons into the molecule of interest. EI was first introduced by Dempster in 1921 for analysis of the isotopes of lithium and magnesium [1]. In the simplest form, an electron beam is produced by electrically heating a filament to a temperature that causes electrons to be emitted. The emitted electrons are directed through the source towards an anode (trap) located on the opposing end of the source. A small magnet is often used to guide the electron motion into a spiral path, thus increasing the path length and increasing the potential for electron–molecule interactions. The electron energy is typically set to 70 eV, for positive ionization mode, where the molecule will absorb about 14 eV as internal energy sufficient to eject an electron [2]. This condition results in extensive fragmentation. For this reason, EI is sometimes termed a "hard" ionization method, while CI, MALDI, and ESI are termed "soft" ionization because they produce more molecular ion and less fragmentation than EI. Lowering the electron energy can minimize fragmentation but ionization efficiency is also reduced [2,3].

6.4. CHEMICAL IONIZATION

Chemical ionization (CI) is essentially electron impact with a reagent gas resulting in ionization of the much more abundant reagent gas that reacts through ion–molecule reactions with the analyte. CI was introduced by Munson and Field in 1966 [4]. The ionization source setup for CI is analogous to EI; however, a reagent gas is introduced in the ionization source. In EI the source pressure is typically less than 10^{-6} torr, while in CI the source pressure is significantly higher at 0.1–2.0 torr. During this time the analyzer region is still maintained at 10^{-6} to 10^{-8} torr. Methane is a common reagent gas employed in CI, but other types of reagent gas such as ammonia and isobutane are also used depending on the application. Ions produced with CI are closed-shell, non-radical ions. Moreover, unlike EI, CI yields less fragmentation; therefore, intact molecular ions are readily obtained.

6.5. ELECTROSPRAY IONIZATION

In electrospray ionization (ESI) gaseous ions are generated directly from a liquid solution. ESI was initially introduced by Malcolm Dole at the International Symposium on Macromolecular Chemistry in Tokyo in 1966 and later published in 1968 [5]. In ESI the analyte is dissolved in a mixture of water and an organic solvent, such as acetonitrile or methanol. Oftentimes formic acid or acetic acid is added to the sample mixture to facilitate protonation of the analyte for the positive ion mode, while an ammonia solution is added to the sample to facilitate deprotonation in the negative ion mode. The major processes involved during ion formation are: (1) production of charged droplets at the electrospray capillary tip, (2) shrinkage of the charged droplets by solvent evaporation and repeated droplet disintegration resulting in small highly charged droplets, and (3) formation of gas-phase ions from the small and highly charged droplets (Fig. 6.2).

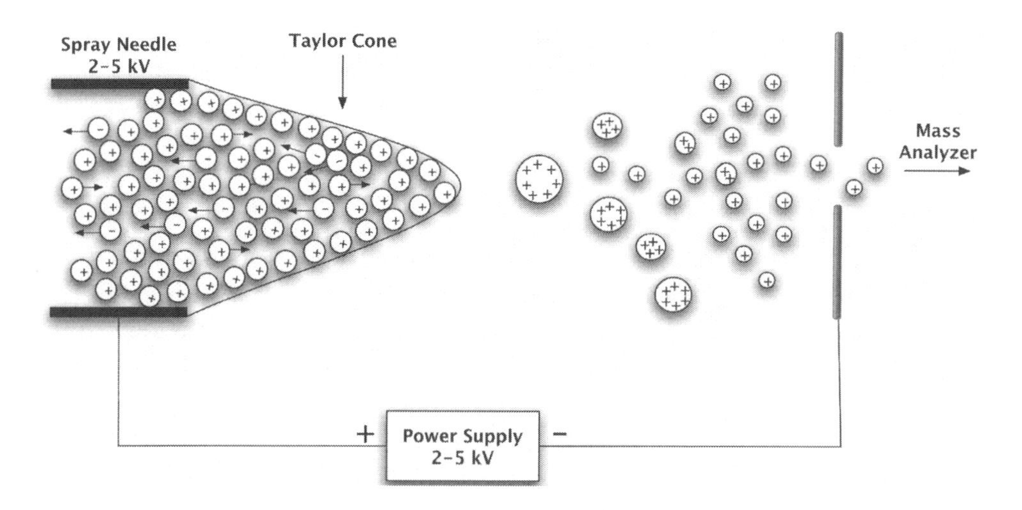

Figure 6.2. Basic schematic of the electrospray ionization process.

To produce the charged droplets at the electrospray capillary tip a voltage (2–5kV) is applied to the metal capillary located at a distance from the counter electrode. The applied voltage causes separation of the positive and negative ions, sometimes producing electrochemical conditions in the solution. In positive ionization mode positive ions will migrate to the surface of the liquid at the capillary tip while the negative ions retard back into the capillary. Accumulation of positive ions at the surface and the pull of the electric field cause the solution to distort at the tip, forming a Taylor cone. At the tip of the Taylor cone the fluid breaks down as positively charged droplets overcome the surface tension of the liquid and are emitted into the atmosphere. If the applied electric field is sufficiently high, a resulting fine jet emerges from the cone tip, breaking up into smaller droplets. The electric field, E_c, around the capillary tip is high ($E_c \approx 10^6$ V/m) and can be determined as follows:

$$E_c = \frac{2V_c}{r_c \ell n\left(\frac{4d}{r_c}\right)}, \tag{6.1}$$

where V_c is the applied potential, r_c is the capillary outer radius, and d is the distance from the capillary tip to the counter electrode. The electric field, E_c, is directly related to the applied potential, V_c, but is inversely related to the capillary outer radius, r_c. The electric field at the capillary tip, E_0, is given by

$$E_0 = \sqrt{\left(\frac{2\gamma\cos\theta}{\varepsilon_0 r_c}\right)}, \tag{6.2}$$

where γ is the surface tension of the liquid, θ is the half angle of the Taylor cone (49.3°), ε_0 is the permittivity of vacuum (also known as the permittivity of free space), and r_c is the radius of the capillary. It is important to note that the surface tension of the liquid, γ, is related to the electric field, E_0. With increasing surface tension (such as with pure water), a higher electric field is needed to produce an electric corona discharge.

The charged droplets then undergo shrinkage through solvent evaporation facilitated by the thermal energy in the form of heated gas. The radii of the charged droplets continue to decrease at constant charge q until the radii approach the Rayleigh limit, where the coulombic repulsion between the charges overcomes the surface tension. The Rayleigh limit, q_{RY}, can be determined as follows:

$$q_{RY} = 8\pi\sqrt{(\varepsilon_0 \gamma R^3)} \tag{6.3}$$

where ε_0 is the permittivity of vacuum, γ is the surface tension of the liquid, and R is the radius of the droplet. Equation (6.3) illustrates the condition where the electrostatic repulsion equals the surface tension of the liquid, γ, which keeps the droplet intact. When the droplets approach the Rayleigh limit the droplets undergo coulombic fission, where smaller (offspring) droplets are developed carrying less charge.

Once the smaller charged droplets are formed the formation of gas-phase ions occurs by means of two possible pathways — the charge residue mechanism (CRM) and the ion evaporation mechanism (IEM). The charge residue mechanism, proposed by Dole and coworkers in 1968 [6], describes the formation of gas-phase ions undergoing coulombic fission at the Rayleigh limit until all the solvent molecules are evaporated, leaving the charge analyte behind. The ion evaporation mechanism, first proposed by Iribarne and Thomson in 1976 [7], describes the event when the droplets reach a certain radius ($R \approx$ 10–20 nm) through solvent evaporation and coulombic fission, resulting in the ion getting directly emitted from the droplet into the gas phase.

An advantage of ESI is the ability to ionize large biological and thermally labile molecules while minimizing fragmentation. The method is highly suitable for proteins, but also other applications such as oligonucleotides, oligosaccharides, and glycopeptides. However, ESI is sensitive to high salt concentrations, requiring thorough desalting prior to electrospray ionization.

6.6. MATRIX-ASSISTED LASER DESORPTION/IONIZATION

Matrix-assisted laser desorption/ionization involves the production of ions by laser irradiation of a crystallized mixture consisting of analyte and matrix (Fig. 6.3). MALDI was first introduce in 1988 by Hillenkamp [8] and Tanaka [9]. MALDI has been widely used for the analysis of peptides, proteins, glycopeptides, oligosaccharides, and oligonucleotides. Important factors involved in MALDI ionization include the laser wavelength and the matrix. The laser wavelength is an important parameter used to excite the matrix molecules. A laser operating at a wavelength of 337 nm from a nitrogen laser is commonly used. However, 3rd and 4th harmonics of the Nd:YAG laser, different excimer lasers, infrared (IR) lasers (Er:YAG), and carbon dioxide lasers are also utilized.

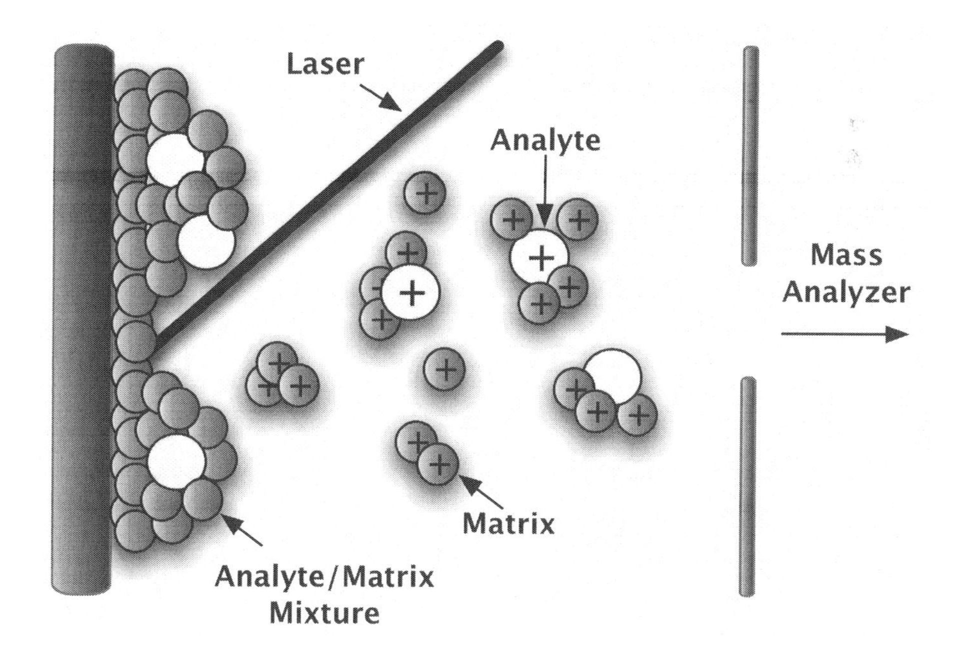

Figure 6.3. Schematic of the MALDI process.

Table 6.1. Common Matrices Used in MALDI

Matrix	Applications
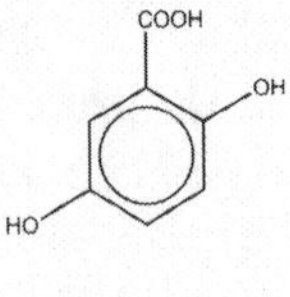 COOH, OH, HO 2,5 Dihydroxybenzoic acid, DHB	Carbohydrates, peptides, proteins, and synthetic polymers
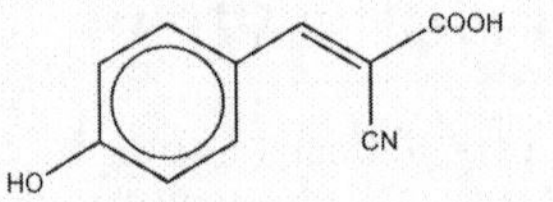 COOH, CN, HO 4-Hydroxy-α-cyanocinnamic acid, 4HCCA	Fragmentation, fullerenes, peptides, proteins
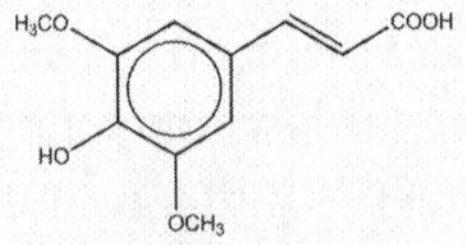 H_3CO, COOH, HO, OCH_3 Trans-3,5-dimethoxy-4hydroxy cinnamic acid, Sinapic acid or sinapinic acid, SA	Peptides, proteins
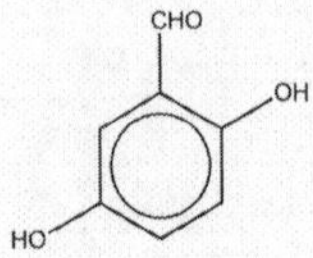 CHO, OH, HO 2,5 Dihydroxyacetophenone, DHAP	Carbohydrates, peptides, proteins
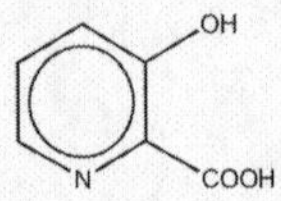 OH, N, COOH 3-Hydroxy-picolinic acid	Nucleic acids

Adapted from *MALDI MS. A practical guide to instrumentation, methods and applications, 2007* [10].

The matrix is the key component for MALDI ionization since it serves as a medium absorbing the energy from the laser and facilitating ionization of the analyte. Matrices can be classified as being "hot" or "cool" depending on the degree of fragmentation produced during the ionization process. A "hot" matrix yields a large degree of fragment ions, while a "cool" matrix yields ions with less internal energy. Table 6.1 describes some commonly used matrices and their applications. There are no defined guidelines for selecting a matrix, but some considerations must be made when choosing a matrix. The matrix should have a high absorptivity for the laser radiation wavelength. The matrix should be capable of forming a fine crystalline solid during codeposition with the analyte. The matrix should not compete with the analyte for the selected cations. In general, the matrix will depend heavily on the sample and the instrumentation used, although a few such as those in Table 6.1 are more commonly used.

The ionization processes in MALDI cannot yet be explained by a single mechanism. Several have been proposed in the MALDI plume and secondary ion formation (produced by ion/molecule reactions). Zenobi and Knochenmuss [11] provide a detailed discussion to the mechanisms involved in MALDI. In general, positive ions can be generated by cationization $[M+X]^+$ (where X = H, Li, Na, K, Cs, etc.), and negative ions can be generated by deprotonation $[M-H]^-$. Most of the ions observed are singly charged; however, multiply charged ions are sometimes observed for large proteins. An advantage of MALDI over ESI is its ability to tolerate relatively high concentrations of salts and buffers.

6.7. FOURIER TRANSFORM ION CYCLOTRON RESONANCE

Fourier transform ion cyclotron resonance mass spectrometry (FT-ICR MS) was first introduced in 1974 by Comisarow and Marshall [12,13], inspired by earlier developments by Lawrence and Livingston [14] in 1932 on ion cyclotron resonance (ICR). Modern instruments have the analyzer (or ICR) cell located within the bore of a high-field magnet. The ICR cell is illustrated in Figure 6.4, where a radiofrequency (RF), a sinusoidal wave, is applied to a pair of excitation plates to excite the ions that are detected by orthogonal plates, and, lastly, voltages are applied to trapping plates and inner plates (not shown), located at the ends of the ICR cell, to trap the ions within the ICR cell.

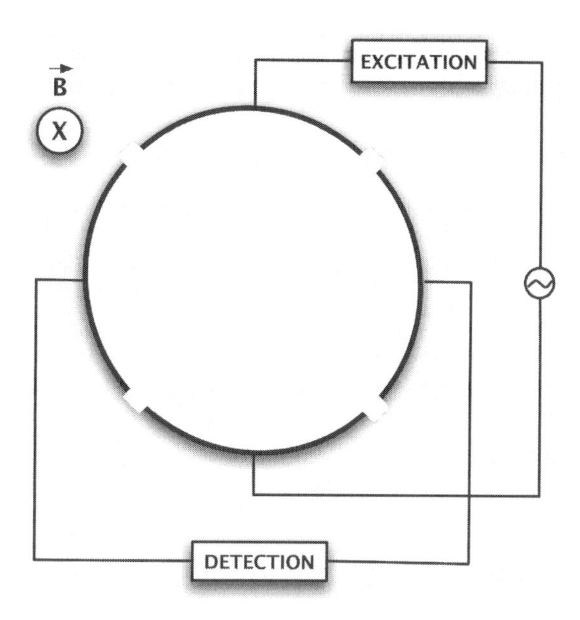

Figure 6.4. Schematic of the ICR cell.

The ions traveling in the presence of a magnetic field will precess at a frequency characteristic to its *m/q* value. Three types of motion contribute to an ion's motion in the ICR cell: (1) cyclotron motion where the ions precess around magnetic field lines, (2) trapping motion where the ions move back an forth between the trapping plates, and (3) magnetron motion where the ions move in a slow circular drift as the center of the cyclotron orbit follows the path of constant electrostatic potential in the ICR cell [15,16]. The most important motion is the cyclotron motion. The cyclotron frequency can be calculated by first using the Lorentz force. The charged particle, with a velocity component, will experience a force, the Lorentz force, as defined by the equation

$$\vec{F} = q\vec{E} + q\left(\vec{v} \times \vec{B}\right), \tag{6.4}$$

where $\vec{F}$ is the force, q is the charge on the charged particle, $\vec{E}$ is the electric field, $\vec{v}$ is the velocity of the charged particle, and $\vec{B}$ is the magnetic field. Considering the primary force being exerted on the charged particle to be from the magnetic field and in the absence of the electric field causes $\vec{E} = 0$, the Lorentz force equation can be simplified to

$$\vec{F} = q\left(\vec{v} \times \vec{B}\right). \tag{6.5}$$

As a result, the magnetic force is given by the cross product $\left(\vec{v} \times \vec{B}\right)$, which is equal to zero when the motion of the ion is moving parallel to the magnetic field and does not experience a magnetic field while the ion's velocity remains unchanged. The Lorentz force is greatest when the ion is moving perpendicular to the magnetic field ($\vec{v} \perp \vec{B}$), and the magnetic force on the ion is then

$$F = qvB. \tag{6.6}$$

The ions with velocity perpendicular to the magnetic field enter the ICR cell, experience a Lorentz force, and tend to precess in a circulating trajectory, called the cyclotron motion, represented in Figure 6.5.

The centrifugal force, F_1, is the outward directed (pseudo) force based upon the ion's mass, m, and the ion's angular acceleration, a. Substituting the ion's velocity, $\vec{v}$, and the circular path of radius, r, for angular acceleration, a, by definition of Newton's second law, the centrifugal force, F_1, becomes

$$F_1 = ma = \frac{mv^2}{r}. \tag{6.7}$$

While under the influence of the magnetic field, a stable circular trajectory of radius, r, is achieved when the outward directed centrifugal force, F_1, is balanced by the inward-directed Lorentz force, F (Eq. (6.8)):

$$F = F_1, \tag{6.8}$$

$$qvB = \frac{mv^2}{r}. \tag{6.9}$$

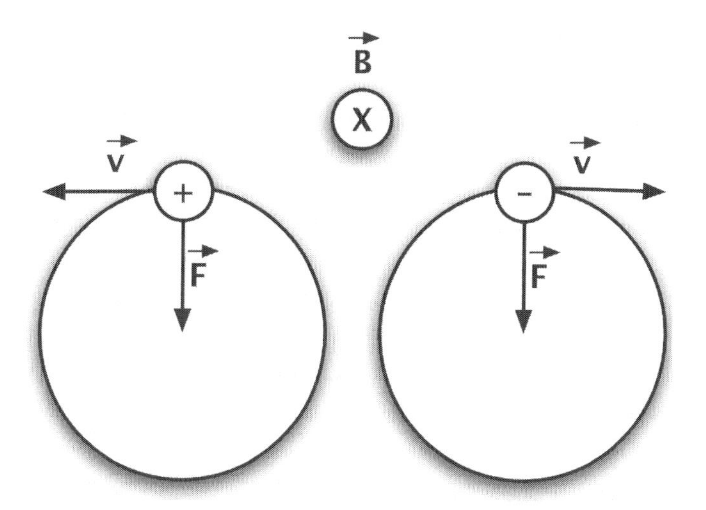

Figure 6.5. Ion cyclotron motion is the result of the balance between inward-acting Lorentz magnetic force, *F*, created by the magnetic field (perpendicular to the plane of the paper) for positively charged (left) and negatively charged (right) ions.

By solving for v/r, one can determine an ion's cyclotron frequency, ω (rad·sec^{-1}).

$$\omega = \frac{v}{r} = \frac{qB}{m}. \tag{6.10}$$

Oftentimes, the cyclotron frequency, ω, is defined in terms of "cycles per second," Hertz (Hz or sec^{-1}); therefore, incorporating the circumference of a circle, 2π radians, yields the cyclotron frequency, ω_c, in units of Hz:

$$\omega_c = \frac{\omega}{2\pi} = \frac{qB}{2\pi m}. \tag{6.11}$$

This equation indicates that the mass of the ion, m, is inversely related to the cyclotron frequency, ω_c, where the lighter ions with have a higher cyclotron frequency and the heavier ions will have a lower cyclotron frequency. By rearranging the equation, the ion's mass-to-charge ratio, m/q, can be calculated:

$$\frac{m}{q} = \frac{B}{2\pi\omega_c}, \tag{6.12}$$

$$q = ze, \tag{6.13}$$

$$\frac{m}{q} = \frac{m}{z} = \frac{eB}{2\pi\omega_c}. \tag{6.14}$$

The ion's mass-to-charge can further be simplified in terms of charge state, z, by using the definition of total charge, q, where q is the product of the net number of electronic charges, z, and the charge per electron, e (coulombs, C) (Eqs. (6.13) and (6.14)). As a result, the cyclotron frequency, ω_c, for any ion can be determined based upon the ion's mass-to-charge, m/z.

FT-ICR MS has the capabilities of high mass accuracies (part per million, ppm level) and high mass resolution (10^6, full width at half height or FWHH). These inherent capabilities of FT-ICR have been used for many biological applications, from clinical biomarker discovery [17–20], quantitative analysis of oligosaccharides in human milk [21,22], glycoproteomics [20, 23–25], proteomics [26–28], and even petroleomics [29,30]. Park and Lebrilla [31] thoroughly discuss the application of FT-ICR for the analysis of oligosaccharides.

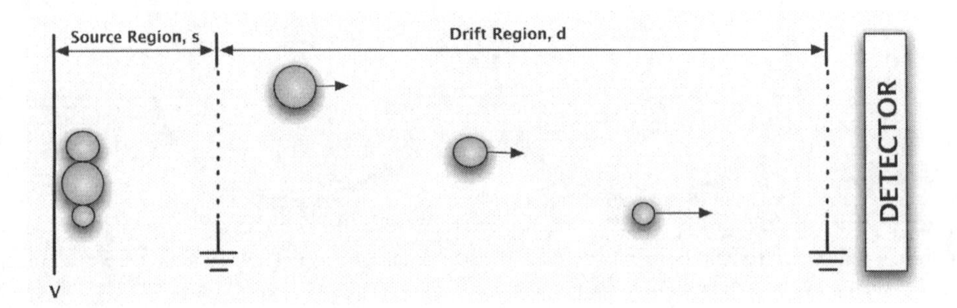

Figure 6.6. Schematic of a simplified time-of-flight mass analyzer.

6.8. TIME OF FLIGHT

The time-of-flight (TOF) mass analyzer was first introduced by Stephens in 1946 [32]. As illustrated in Figure 6.6, the basic components of a TOF mass analyzer include a short source-extraction region (*s*), a drift region (*d*), and a detector. The ions are accelerated in the presence of an electric field into a field-free drift region where the ions travel across this region at velocities inversely proportional to the square root of their masses. With this relationship lighter ions will reach the detector prior to the heavier ions.

The energy of the electric field, E, that the ions experience is described in the following equation:

$$E = qV\,, \tag{6.15}$$

where q is the ion's charge and V is the applied voltage. The kinetic energy, E_k, of the ion accelerating through the source-extraction region is defined as half the product of the ion's mass, m, and the square of the ion's velocity, v:

$$KE = E_k = \frac{mv^2}{2}\,. \tag{6.16}$$

All ions traveling through the source-extraction region will have the same kinetic energy, E_k, and experience the same electric field, E:

$$E = E_k\,, \tag{6.17}$$

$$qV = \frac{mv^2}{2}\,. \tag{6.18}$$

As the ions traverse the drift region, the ions velocities will be defined as

$$v = \sqrt{\frac{2qV}{m}}. \tag{6.19}$$

The velocities of the different ions are inversely proportional to the square root of the ion masses; thus, the flight time for each ion to travel across the drift region will differ accordingly. Substituting Equation (1.19) with the definition of velocity, v, where d is the distance traveled in the drift region and t is the time, the equation can then be solved for m/z for any ion as

$$v = \frac{d}{t}, \tag{6.20}$$

$$v = \frac{d}{t} = \sqrt{\frac{2zeV}{m}}, \tag{6.21}$$

$$\frac{m}{z} = \frac{2eVt^2}{d^2}, \tag{6.22}$$

By rearranging Equation (6.22), the flight time for an ion can be determined from their m/z:

$$t = d\sqrt{\frac{m}{2ezV}}. \tag{6.23}$$

The mass range on the TOF is theoretically unlimited; however, the resolution for a simple TOF is often poor.

A major modification to the TOF was introduced by Mamyrin et al. in 1973 with the addition of the reflectron to improve resolution [33]. In a reflectron TOF analysis of an ion mirror (composed of a series of decelerating electrodes) is used to reflect ions primarily to increase the path length. The reflectron also focuses ions of the same m/z by decreasing the spread in translational energies. The ions with the largest kinetic energy will penetrate deeper into the electric field, while slower ions will penetrate less. The ions then exit the reflectron with less spatial and energy spread, thereby significantly increasing resolution (Fig. 6.7).

TOF can be coupled to essentially any type of ionization method. However, it was MALDI that increased its popularity. The pulsed nature of the MALDI method made it ideal for TOF analysis. More recently TOF instruments have also been coupled to liquid chromatography (LC/MS) through ESI. The TOF analyzer serves as the MS detector of the compounds separated by high-performance liquid chromatography. An exciting development with this method is employment of microfluidic chip devices developed by Yin et al., allowing for nano-LC/MS [34,35] separation on a nanoflow column integrated into a microchip device. The microfluidic chip is a laminated polyimide device composed of two columns. An enrichment column traps and concentrates the sample, and an analytical column separates the analytes. This technology has been applied to proteomics and glycomics analysis of biological samples, as well as analysis of posttranslational modifications (PTMs) such as phosphorylation [36] and oligosaccharides [37–40].

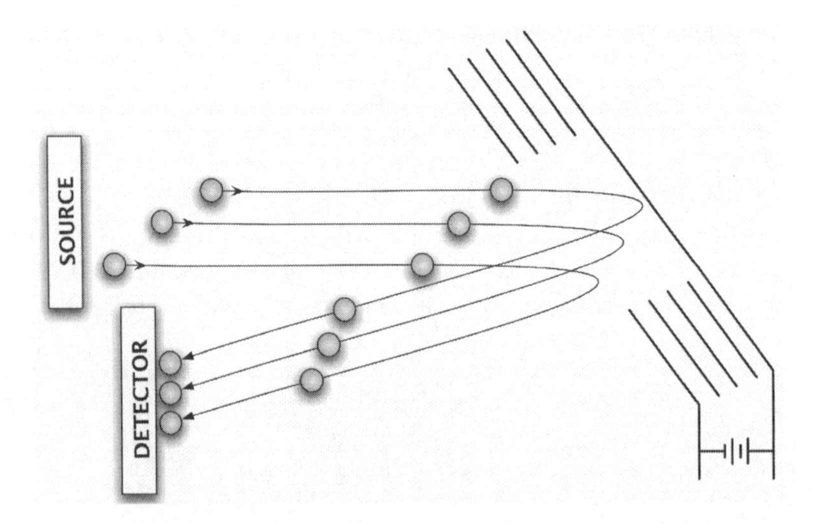

Figure 6.7. Schematic of a reflectron time-of-flight mass analyzer.

6.9. TANDEM MASS SPECTROMETRY

Tandem mass spectrometry is an important technique to obtain additional structural information for a particular sample. The ion of interest is isolated in the gas phase and induced to fragment through a single tandem event (MS) or multiple tandem events (MS/MS or MS^n, where n is the number of events). Similar to a puzzle, the fragmented "product" ions are then pieced together to determine the structure of the selected "precursor" ion. There are several approaches to achieve fragmentation that include collision-induced dissociation (CID), infrared multiphoton dissociation (IRMPD), electron-capture dissociation (ECD), and electron-transfer dissociation (ETD).

6.10. COLLISION-INDUCED DISSOCIATION

Collision-induced dissociation (CID), also referred to as collision-activated dissociation (CAD), involves transport of the ion to a collision region where the ion acquires translational energy and is collided with a gas (nitrogen, argon, or helium). The translational or kinetic energy is converted to vibrational energy causing fragmentation.

CID has been applied for the analysis of biomolecules such as peptides and carbohydrates [41–44] (Fig. 6.8). CID of peptides results in cleavage of the amide bonds resulting in b- and y-type fragment ions, depending on whether the charge carrier remains with the N or O terminus (Fig. 6.9). CID is widely used for proteomic analysis of proteins, where the proteins are digested with an enzyme, commonly trypsin, to generate peptides. The tryptic peptides are then fragmented with CID to determine the peptide sequence. Tryptic digestion is ideal for tandem

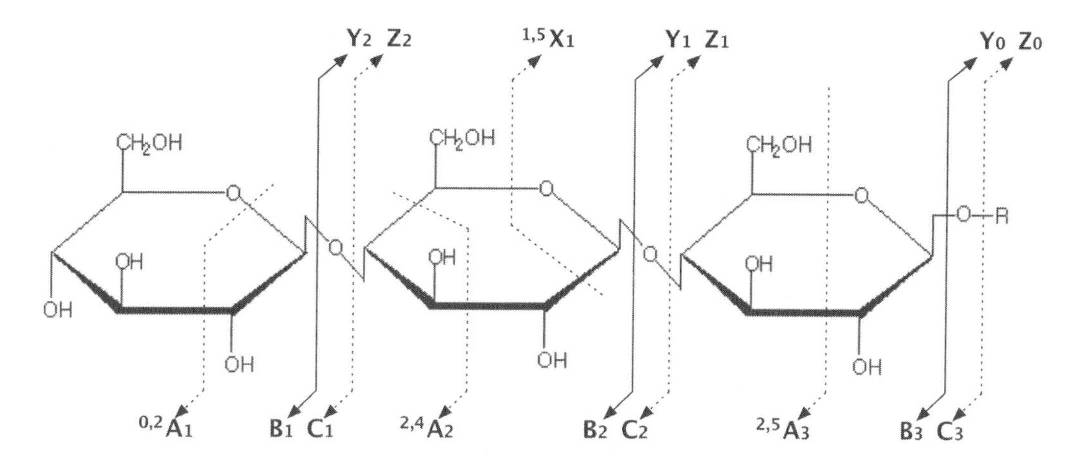

Figure 6.8. Fragmentation nomenclature of oligosaccharides [41].

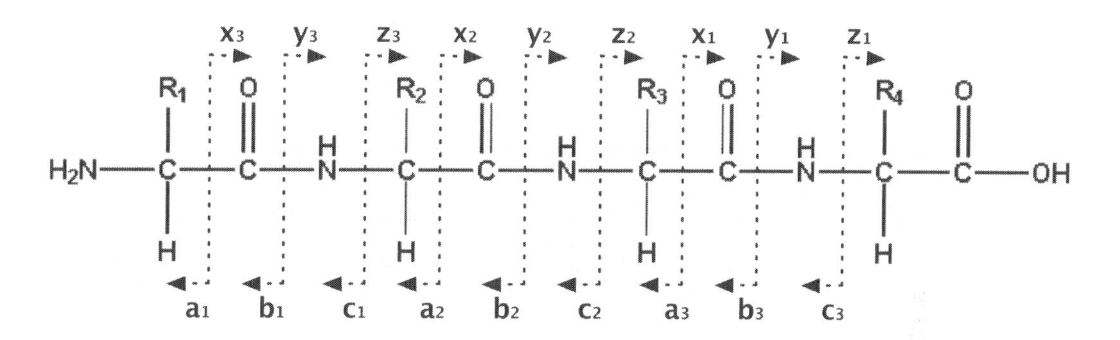

Figure 6.9. Fragmentation nomenclature of peptides [46].

MS, as it generally limits the size of peptides to 20 amino acids or less. CID is the most common tandem MS technique. It offers robustness and yields structural information on nearly all classes of analytes. In some cases numerous tandem MS events can be coupled (MS^n).

6.11. INFRARED MULTIPHOTON DISSOCIATION

Infrared multiphoton dissociation (IRMPD) involves the isolation of a precursor ion and irradiation of the precursor ion with IR photons to yield fragment ions. Similar to CID, IRMPD is a low-energy process. Unlike CID, where the energy is deposited only in the precursor ion, in IRMPD the energy is not only deposited into the precursor ion but also into the fragment ions, thereby yielding more fragmentation information without additional MS/MS events.

The energetics of IRMPD fragmentation follows the lowest dissociation threshold, whereas CID may access higher-energy fragmentation pathways. IRMPD has been utilized for the analysis of (glyco)peptides [23,24], oligonucleotides, and oligosaccharides [45–49]. In addition, IRMPD has been applied in combination with ECD [50–54] and CID [27,28].

6.12. ELECTRON CAPTURE DISSOCIATION

Electron-capture dissociation (ECD) was first developed by Zubarev et al. [55] in 1998 to fragment multiply charged peptides and even proteins. ECD involves the capture of free electrons by polypeptide polycations, leading to charge neutralization and resulting in an excited radical species that immediately undergoes bond cleavage:

$$MH_2^{2+} + e^- \rightarrow MH^{+\bullet} + H^\bullet . \tag{6.28}$$

Applications of ECD include peptides, polymers, oligonucleotides, and oligosaccharides. The advantages of ECD are analysis of PTMs, extensive sequence coverage of the polypeptide backbone, and fragmentation of disulfide bonds [56–58]. In ECD, fragmentation of peptides occurs at the amine bond, yielding c- and z-type ions, in comparison to CID and IRMPD, which yield amide cleavages leading to b- and y-type ions (Fig. 6.9). However, ECD has been limited mainly to FT ICR mass spectrometry. ECD has recently been applied with a radiofrequency ion trap [59].

6.13. ELECTRON TRANSFER DISSOCIATION

Electron-transfer dissociation (ETD), first introduced in 2004 by Coon [60] and Syka et al. [61], is analogous to ECD, where an anionic species is used instead of electrons for fragmentation of peptides or intact proteins. In ETD a multiprotonated peptide is allowed to interact with electron-rich anions, causing an electron transfer from anion to peptide. This results in a radical cation, which causes the peptide to fragment in a manner similar to ECD. The reaction can be conceptualized below, where $A^{-\bullet}$ is the electron-rich anion reacting with a multiply charged cation:

$$MH_2^{2+} + A^{-\bullet} \rightarrow MH^{+\bullet} + AH . \tag{6.29}$$

The reaction between the anion and the protonated peptide cations in ETD yield c- and z-type fragment ions. ETD has been applied in proteomic studies [62], including PTMs, such as phosphorylation [61,63,64]. A recent discussion of the applications of ETD for shotgun proteomics and top–down proteomics has been published [65].

6.14. SUMMARY

Mass spectrometry has the ability to characterize a sample rapidly, accurately, and with high sensitivity. Over the past 20 years mass spectrometry has been integrated into the workflow of many scientific applications. Further improvements are anticipated over the next 10 years offering even higher sensitivity and better resolution with lower cost. Moreover, further integration with other platforms such as high-performance liquid chromatography, gas chromatography, capillary electrophoresis, and ion mobility are expected to enhance the capabilities of the methods even further. Mass spectrometry has already had a tremendous impact in modern biological science but will continue to be well integrated in general studies of biological systems.

ACKNOWLEDGMENTS

The authors gratefully acknowledge Milady R. Niñonuevo and Cora Monce for their technical assistance.

PROBLEMS

1. Find the radius of gyration of a low-energy ion of *m/z* 1000 in a 9.4 T magnetic field. Let $v - 100$ m/s and $\omega = 6.7542 \times 10^5\ s^{-1}$.
2. Find the translational energy for an ion of *m/z* 1257 when it has been accelerated to a radius of gyration of 1 cm.

FURTHER STUDY

Cotter RJ. 1997. *Time-of-flight mass spectrometry*. Washington, DC: American Chemical Society.

Ekman R, Silberring J, Westman-Brinkmalm A, Kraj A. 2009. *Mass spectrometry instrumentation, interpretation, and applications*. Hoboken, NJ: John Wiley & Sons.

Hillenkamp F, Peter-Katalinic J, eds. 2007. *MALDI MS: A practical guide to instrumentation, methods and applications*. Weinheim: Wiley-VCH.

McIver RT, McIver JR. 2006. *Fourier transform mass spectrometry: principles and applications*. Lake Forest, IL: IonSpec.

REFERENCES

1. Dempster AJ. 1921. Positive ray analysis of lithium and magnesium. *Phys Rev* **18**:415–422.
2. Watson JT, Sparkman OD. 2007. *Introduction to mass spectrometry*. Chichester: John Wiley & Sons.
3. Ekman R, Silberring J, Westman-Brinkmalm A, Kraj A. 2009. *Mass spectrometry instrumentation, interpretation, and applications*. Hoboken, NJ: John Wiley & Sons.
4. Munson MSB, Field FH. 1966. Chemical ionization mass spectrometry, I: general introduction. *J Am Chem Soc* **88**:2621–2630.
5. Fenn JB. 2003. Electrospray wings for molecular elephants (Nobel lecture). *Angew Chem, Int Ed* **42**:3871–3894.
6. Dole ML, Mack LL, Hines RL, Mobley RC, Ferguson LD, Alice MB. 1968. Molecular beams of macroions. *J Chem Phys* **49**:2240–2249.
7. Iribarne JV, Thomson BA. 1976. On the evaporation of small ions from charged droplets. *J Chem Phys* **64**:2287.
8. Karas M, Hillenkamp F. 1988. Laser desorption ionization of proteins with molecular masses exceeding 10,000 daltons. *Anal Chem* **60**:2299–2301.
9. Tanaka K, Waki H, Ido Y, Akita S, Yoshida Y, Yoshida T. 1988. Protein and Polymer Analyses up to *m/z* 100000 by laser ionization time-of-flight mass spectrometry. *Rapid Commun Mass Spectrom* **2**:151–153.
10. Hillenkamp F, Peter-Katalinic J. 2007. MALDI MS: *a practical guide to instrumentation, methods and applications*. Weinheim: Wiley-VCH Verlag.
11. Zenobi R, Knochenmuss R. 1998. Ion formation in MALDI mass spectrometry. *Mass Spectrom Rev* **17**:337–366.

12. Comisarow MB, Marshall AG. 1974. Fourier transform ion cyclotron resonance spectroscopy. *Chem Phys Lett* **25**:282–283.
13. Comisarow MB, Marshall AG. 1996. The early development of Fourier transform ion cyclotron resonance (FT-ICR) spectroscopy. *J Mass Spectrom* **31**:581–585.
14. Lawrence EO, Livingston MS. 1932. The production of high speed light ions without the use of high voltages. *Phys Rev* **40**:19–35.
15. Marshall AG, Grosshans PB. 1991. Fourier transform ion cyclotron resonance mass spectrometry: the teenage years. *Anal Chem* **63**:215A–229A.
16. McIver RT, McIver JR. 2006. *Fourier transform mass spectrometry principles and applications*. Lake Forest, IL: IonSpec.
17. An HJ, Miyamoto S, Lancaster KS, Kirmiz C, Li B, Lam KS, Leiserowitz GS, Lebrilla CB. 2006. Profiling of glycans in serum for the discovery of potential biomarkers for ovarian cancer. *J Proteome Res* **5**:1626–1635.
18. de Leoz ML, An HJ, Kronewitter S, Kim J, Beecroft S, Vinall R, Miyamoto S, de Vere White R, Lam KS, Lebrilla C. 2008. Glycomic approach for potential biomarkers on prostate cancer: profiling of N-linked glycans in human sera and pRNS cell lines. *Dis Markers* **25**:243–258.
19. Kirmiz C, Li B, An HJ, Clowers BH, Chew HK, Lam KS, Ferrige A, Alecio R, Borowsky AD, Sulaimon S, Lebrilla CB, Miyamoto S. 2007. A serum glycomics approach to breast cancer biomarkers. *Mol Cell Proteomics* **6**:43–55.
20. Li B, An HJ, Kirmiz C, Lebrilla CB, Lam KS, Miyamoto S. 2008. Glycoproteomic analyses of ovarian cancer cell lines and sera from ovarian cancer patients show distinct glycosylation changes in individual proteins. *J Proteome Res* **7**:3776–88.
21. Niñonuevo MR, Ward RE, LoCascio RG, German JB, Freeman SL, Barboza M, Mills DA, Lebrilla CB. 2007. Methods for the quantitation of human milk oligosaccharides in bacterial fermentation by mass spectrometry. *Anal Biochem* **361**:15–23.
22. Seipert RR, Barboza M, Niñonuevo MR, LoCascio RG, Mills DA, Freeman SL, German JB, Lebrilla CB. 2008. Analysis and quantitation of fructooligosaccharides using matrix-assisted laser desorption/ionization Fourier transform ion cyclotron resonance mass spectrometry. *Anal Chem* **80**:159–165.
23. Seipert RR, Dodds ED, Clowers BH, Beecroft SM, German JB, Lebrilla CB. 2008. Factors that influence fragmentation behavior of N-linked glycopeptide ions. *Anal Chem* **80**:3684–3692.
24. Seipert RR, Dodds ED, Lebrilla CB. 2009. Exploiting differential dissociation chemistries of O-linked glycopeptide ions for the localization of mucin-type protein glycosylation. *J Proteome Res* **8**:493–501.
25. Clowers BH, Dodds ED, Seipert RR, Lebrilla CB. 2007. Site determination of protein glycosylation based on digestion with immobilized nonspecific proteases and Fourier transform ion cyclotron resonance mass spectrometry. *J Proteome Res* **6**:4032–4040.
26. Dodds ED, Clowers BH, Hagerman PJ, Lebrilla CB. 2008. Systematic characterization of high mass accuracy influence on false discovery and probability scoring in peptide mass fingerprinting. *Anal Biochem* **372**:156–166.
27. Dodds ED, German JB, Lebrilla CB. 2007. Enabling MALDI-FTICR-MS/MS for high-performance proteomics through combination of infrared and collisional activation. *Anal Chem* **79**:9547–9556.
28. Dodds ED, Hagerman PJ, Lebrilla CB. 2006. Fragmentation of singly protonated peptides via a combination of infrared and collisional activation. *Anal Chem* **78**:8506–8511.
29. Marshall AG, Rodgers RP. 2004. Petroleomics: the next grand challenge for chemical analysis. *Acc Chem Res* **37**:53–59.
30. Marshall AG, Rodgers RP. 2008. Petroleomics: chemistry of the underworld. *Proc Natl Acad Sci USA* **105**: 18090–1805.
31. Park Y, Lebrilla CB. 2005. Application of Fourier transform ion cyclotron resonance mass spectrometry to oligosaccharides. *Mass Spectrom Rev* **24**:232–264.
32. Stephens WE. 1946. A pulsed mass spectrometer with time dispersion. *Phys Rev* **69**:691.
33. Mamyrin BA, Karataev VI, Shmikk DV, Zagulin VA. 1973. Mass reflection: a new nonmagnetic time-of-flight high resolution mass-spectrometer. *Soviet Phys JETP* **37**:45–48.
34. Yin H, Killeen K. 2007. The fundamental aspects and applications of Agilent HPLC-Chip. *J Sep Sci* **30**:1427–1434.
35. Yin H, Killeen K, Brennen R, Sobek D, Werlich M, van de Goor T. 2005. Microfluidic chip for peptide analysis with an integrated HPLC column, sample enrichment column, and nanoelectrospray tip. *Anal Chem* **77**:527–533.

36. Mohammed S, Kraiczek K, Pinkse MW, Lemeer S, Benschop JJ, Heck AJ. 2008. Chip-based enrichment and NanoLC-MS/MS analysis of phosphopeptides from whole lysates. *J Proteome Res* **7**:1565–1571.
37. Chu CS, Niñonuevo MR, Clowers BH, Perkins PD, An HJ, Yin H, Killeen K, Miyamoto S, Grimm R, Lebrilla CB. 2009. Profile of native N-linked glycan structures from human serum using high performance liquid chromatography on a microfluidic chip and time-of-flight mass spectrometry. *Proteomics* **9**:1939–1951.
38. Niñonuevo M, An H, Yin H, Killeen K, Grimm R, Ward R, German B, Lebrilla C. 2005. Nanoliquid chromatography-mass spectrometry of oligosaccharides employing graphitized carbon chromatography on microchip with a high-accuracy mass analyzer. *Electrophoresis* **26**:3641–3649.
39. Niñonuevo MR, Perkins PD, Francis J, Lamotte LM, LoCascio RG, Freeman SL, Mills DA, German JB, Grimm R, Lebrilla CB. 2008. Daily variations in oligosaccharides of human milk determined by microfluidic chips and mass spectrometry. *J Agric Food Chem* **56**:618–626.
40. Tao N, DePeters EJ, Freeman S, German JB, Grimm R, Lebrilla CB. 2008. Bovine milk glycome. *J Dairy Sci* **91**:3768–3778.
41. Zhang J, Xie Y, Hedrick JL, Lebrilla CB. 2004. Profiling the morphological distribution of O-linked oligosaccharides. *Anal Biochem* **334**:20–35.
42. Zhang J, Lindsay LL, Hedrick JL, Lebrilla CB. 2004. Strategy for profiling and structure elucidation of mucin-type oligosaccharides by mass spectrometry. *Anal Chem* **76**:5990–6001.
43. Li B, An HJ, Hedrick JL, Lebrilla CB. 2009. Collision-induced dissociation tandem mass spectrometry for structural elucidation of glycans. *Methods Mol Biol* **534**:1–13.
44. Penn SG, Cancilla MT, Lebrilla CB. 1996. Collision-induced dissociation of branched oligosaccharide ions with analysis and calculation of relative dissociation thresholds. *Anal Chem* **68**:2331–2339.
45. Xie Y, Lebrilla CB. 2003. Infrared multiphoton dissociation of alkali metal-coordinated oligosaccharides. *Anal Chem* **75**:1590–1598.
46. Xie Y, Schubothe KM, Lebrilla CB. 2003. Infrared laser isolation of ions in Fourier transform mass spectrometry. *Anal Chem* **75**:160–164.
47. Zhang J, Schubothe K, Li B, Russell S, Lebrilla CB. 2005. Infrared multiphoton dissociation of O-linked mucin-type oligosaccharides. *Anal Chem* **77**:208–214.
48. Lancaster KS, An HJ, Li B, Lebrilla CB. 2006. Interrogation of N-Linked oligosaccharides using infrared multiphoton dissociation in FT-ICR mass spectrometry. *Anal Chem* **78**:4990–4997.
49. Li B, An HJ, Hedrick JL, Lebrilla CB. 2009. Infrared multiphoton dissociation mass spectrometry for structural elucidation of oligosaccharides. *Methods Mol Biol* **534**:1–13.
50. Adamson JT, Hakansson K. 2006. Infrared multiphoton dissociation and electron capture dissociation of high-mannose type glycopeptides. *J Proteome Res* **5**:493–501.
51. Adamson JT, Hakansson K. 2007. Electron capture dissociation of oligosaccharides ionized with alkali, alkaline earth, and transition metals. *Anal Chem* **79**:2901–2910.
52. Hakansson K, Chalmers MJ, Quinn JP, McFarland MA, Hendrickson CL, Marshall AG. 2003. Combined electron capture and infrared multiphoton dissociation for multistage MS/MS in a Fourier transform ion cyclotron resonance mass spectrometer. *Anal Chem* **75**:3256–3262.
53. Hakansson K, Cooper HJ, Emmett MR, Costello CE, Marshall AG, Nilsson CL. 2001. Electron capture dissociation and infrared multiphoton dissociation MS/MS of an N-glycosylated tryptic peptic to yield complementary sequence information. *Anal Chem* **73**:4530–4536.
54. Hakansson K, Hudgins RR, Marshall AG, O'Hair RA. 2003. Electron capture dissociation and infrared multiphoton dissociation of oligodeoxynucleotide dications. *J Am Soc Mass Spectrom* **14**:23–41.
55. Zubarev RA, Kelleher NL, McLafferty FW. 1998. Electron capture dissociation of multiply charged protein cations: a nonergodic process. *J Am Chem Soc* **120**:3265–3266.
56. Cooper HJ, Hakansson K, Marshall AG. 2005. The role of electron capture dissociation in biomolecular analysis. *Mass Spectrom Rev* **24**:201–222.
57. Zubarev RA. 2004. Electron-capture dissociation tandem mass spectrometry. *Curr Opin Biotechnol* **15**:12–16.
58. Zubarev RA, Zubarev AR, Savitski MM. 2008. Electron capture/transfer versus collisionally activated/induced dissociations: solo or duet? *J Am Soc Mass Spectrom* **19**:753–761.
59. Baba T, Hashimoto Y, Hasegawa H, Hirabayashi A, Waki I. 2004. Electron capture dissociation in a radio frequency ion trap. *Anal Chem* **76**:4263–4266.
60. Coon JJ, Syka JE, Schwartz JC, Shabanowitz J, Hunt DF. 2004. Anion dependence in the partitioning between proton and electron transfer in ion/ion reactions. *Int J Mass Spectrom* **236**:33–42.

61. Syka JE, Coon JJ, Schroeder MJ, Shabanowitz J, Hunt DF. 2004. Peptide and protein sequence analysis by electron transfer dissociation mass spectrometry. *Proc Natl Acad Sci USA* **101**:9528–9533.
62. Mikesh LM, Ueberheide B, Chi A, Coon JJ, Syka JE, Shabanowitz J, Hunt DF. 2006. The utility of ETD mass spectrometry in proteomic analysis. *Biochim Biophys Acta* **1764**:1811–1822.
63. Coon JJ, Syka JE, Shabanowitz J, Hunt DF. 2005. Tandem mass spectrometry for peptide and protein sequence analysis. *Biotechniques* **38**:519, 21, 23.
64. Coon JJ, Ueberheide B, Syka JE, Dryhurst DD, Ausio J, Shabanowitz J, Hunt DF. 2005. Protein identification using sequential ion/ion reactions and tandem mass spectrometry. *Proc Natl Acad Sci USA* **102**:9463–9468.
65. Coon JJ. 2009. Collisions or electrons? Protein sequence analysis in the 21st century. *Anal Chem* **81**:3208–3215.

7

Transmission Electron Microscopy and Computer-Aided Image Processing for 3D Structural Analysis of Macromolecules

Dominik J. Green and R. Holland Cheng
Department of Molecular and Cellular Biology,
University of California Davis

7.1. THE TRANSMISSION ELECTRON MICROSCOPE

The Transmission Electron Microscope (TEM) is an electron-based imaging system used to reveal the atomic or molecular details of a specimen. High-energy electrons are used to probe the object in question, resulting in the generation of a two-dimensional (2D) image of the object's three-dimensional (3D) information. The electrons in use, with typical energies of greater than 100 kV, have wavelengths less than a tenth of an Ångström, theoretically allowing for imaging resolution far below the sub-Ångström range. However, due to the presence of imperfect imaging conditions such as lens aberrations and sample irradiation, the information transferred from biological specimens via TEM has yet to reach the sub-Ångström limit.

The electrons used for TEM imaging are generated at the top of the EM column by a variety of methods. Early electron microscopes utilized a tungsten filament as an affordable electron source. By heating the filament to high temperatures (2700 K), electrons of the desired energy can be generated and sent down the column. The development of lanthanum hexaboride (LaB_6) crystals was an improvement over the use of tungsten, as its lower work function of 2.5 eV (vs. 4.5 eV), and its magnitude increase in brightness (current density per unit emission angle) allowed for much better beam generation. Unfortunately, thermionic filament sources such as tungsten and LaB_6 share the common limitation of having high electron energy spreads (1.5–3.0

Address correspondence to R. Holland Cheng, Department of Molecular & Cellular Biology, 007 Briggs, CBS, University of California Davis, Davis, California 95616-8536, USA, 530 752-5659, <rhch@ucdavis.edu>.

T. Jue (ed.), *Biomedical Applications of Biophysics*,
Handbook of Modern Biophysics 3, DOI 10.1007/978-1-60327-233-9_7,
© Springer Science+Business Media, LLC 2010

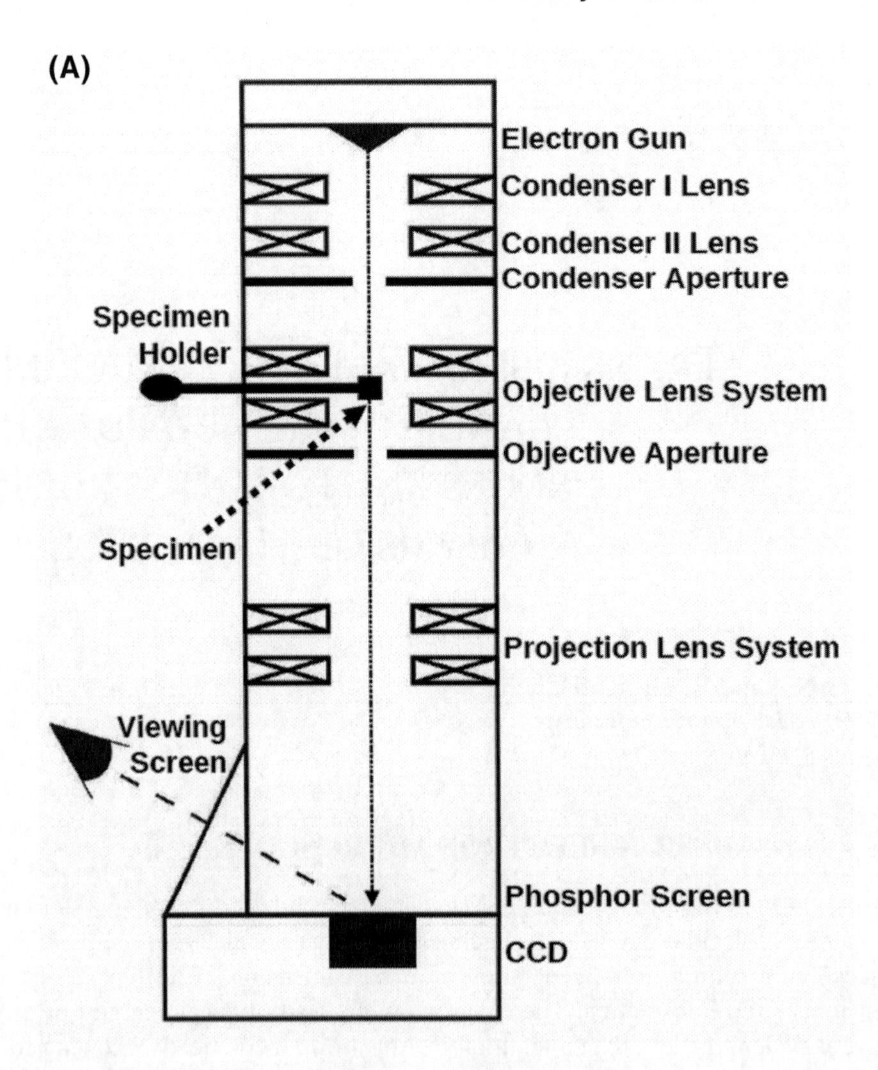

Figure 7.1. **Schematic cross-section diagram and effect of lenses on the electron beam in TEM**. (A) The electron beam generated from the gun passes through a series of lenses and apertures, as well as the specimen, before a magnified image is formed by striking a phosphorescent screen or charged-coupled device (CCD). (B) (facing page) Image formation in a lens can be described as a double-diffraction process. Incoming parallel beams interact with the sample and undergo a corresponding phase shift. These phase-shifted electrons are then focused by the objective lens and screened by the objective aperture. The objective lens generates an optical Fourier transform of the specimen, which can be seen as a diffraction pattern at the back focal plane. Subsequent lenses perform an inverse Fourier transform of the diffraction pattern, producing a magnified image of the specimen. Please visit http://extras.springer.com/ to view a high-resolution full-color version of this illustration.

eV), which will eventually contribute to chromatic aberration with the lens system. The modern TEMs are equipped with a field emission gun (FEG), which generates extremely bright (1000 times greater than tungsten or LaB_6) and low-energy spread (0.3 eV) electrons through the application of an extraction voltage at the cathode tip under ultrahigh vacuum. Usage of FEGs has

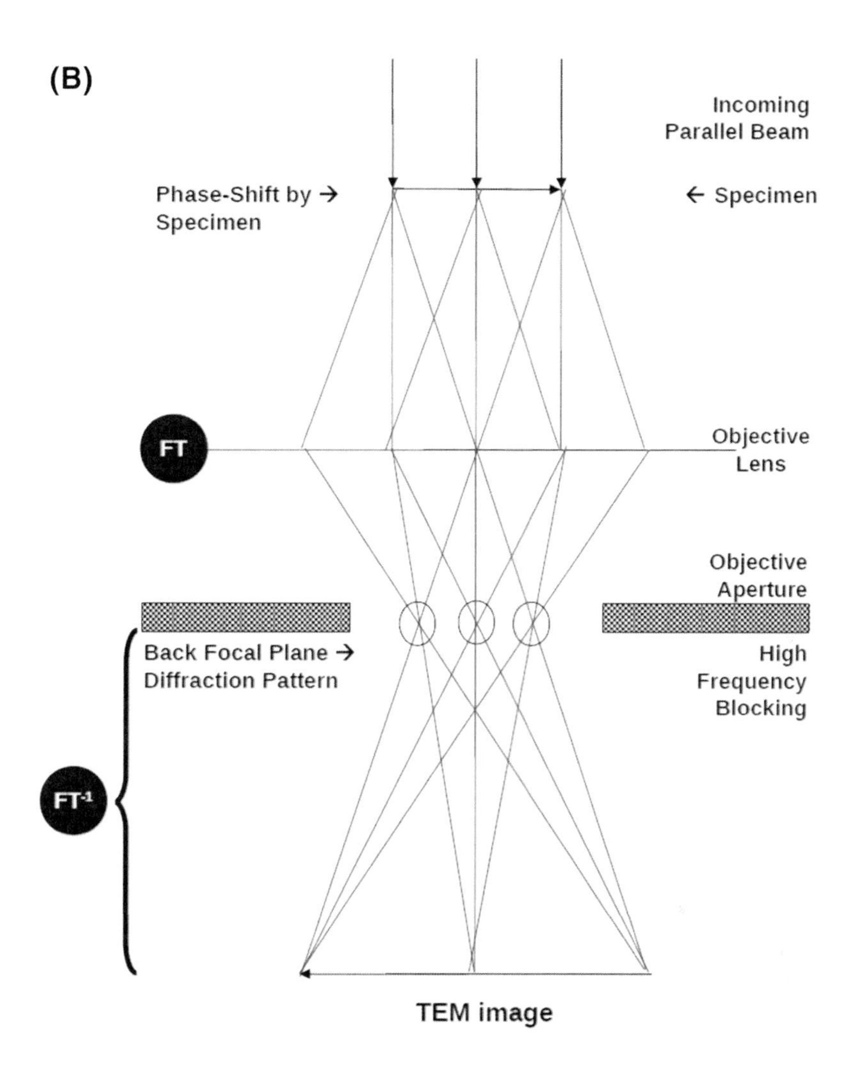

risen considerably since their introduction, as the highly coherent electrons that are produced allow for the highest possible resolution. Its main drawback, however, is a requirement for vacuums of less than 10^{-8} Pa, which can be expensive to generate.

After electrons are generated, they must be sent down the column to impinge upon the sample. Along the length of the column, various magnetic lens systems are used to focus the electrons (Fig. 7.1A). The first of these systems is the condenser lens system, which determines the physical size of the beam as it impinges upon the specimen, as well as the beam's angular convergence, which affects the coherence of the electrons. Typically, the condenser lens system is composed of two lenses: the first lens, Condenser I, demagnifies the image of the source, while the second lens, Condenser II, controls the beam spot size and angular convergence. A condenser aperture may be used to screen out highly divergent source electrons from sample interaction.

The next set of lenses comprises the objective lens system and is positioned around the specimen. The objective lenses are the most important lenses in the EM, as they are responsible for image formation of the specimen. Image formation, according to Abbe's theory, is a two-stage, double-diffraction process (Fig. 7.1B). That is, an image is the diffraction pattern of the diffraction pattern of an object. In the first stage of image formation, a parallel beam of rays incident on the object is scattered and the interference (diffraction) pattern is brought to focus at the back focal plane (BFP) of the lens. This stage is sometimes referred to as the forward Fourier transformation. A lens, essential for image formation, also acts to focus the diffraction pattern at a finite distance from the object at the BFP of the lens. If the lens is removed from behind the object, no image forms, but instead Fresnel diffraction patterns form at finite distances from the object. The second stage of image formation occurs when the scattered radiation passes beyond the BFP of the lens and recombines to form an image. This stage is often referred to as back or inverse Fourier transformation. Image formation is analogous to Fourier analysis in the first stage, and Fourier synthesis in the second stage. To be discussed further below, Fourier image analysis is a powerful method, as it separates the processing of TEM imaging into two stages, where the transform may be manipulated and back-transformed similarly in the second stage to generate a noise-filtered image in the subsequent analysis [1].

Due to the current limitation of magnetic lenses, the objective lenses are responsible for most of the aberrations present in the EM, so its design during engineering and operation during imaging must be carefully optimized to limit these effects. The electrons that pass through the sample can be screened out using an objective aperture located at the BFP. Removal of electrons by the condenser aperture functions as a lowpass filter by blocking out high-angle scattering. The contrast that is generated in an EM image as a result of blocking out these scattered electrons is known as amplitude contrast. Contrast from the EM may also be generated via phase contrast. Such contrast is generated by electrons that are phase shifted due to elastic interactions within the specimen. Phase contrast contributes significantly less in the image than does amplitude contrast, but its effects can be varied in a spatial frequency-dependent manner through the intentional introduction of imaging defocus through the objective lens. Usage of TEM for biological sample analysis primarily operates through phase contrast and will be discussed more below.

The effects of aberrations due to the objective lens can severely limit the attainable resolution of an EM. Due to the differential focusing of electrons along the length of the objective lens, spherical aberrations result, turning the point source of the focused beam into a finite-sized disc (disc of least confusion; Fig. 7.2). This effect is purely a consequence of the imperfect functionality of the objective lens and cannot be eliminated. However, its negative effects may be reduced by the addition of a spherical aberration corrector into the column, which compensates for differential focusing. Additionally, electrons that arrive at the objective lens with different energies will experience different amounts of focus, resulting in chromatic aberration. Thermionic electron sources generate electrons with sufficiently large energy spreads that lead to chromatic aberration. Notably, the effect has been drastically reduced with the introduction of the aforementioned FEG electron sources, which are able to generate electrons with a much narrower energy range. With an FEG electron source, chromatic aberration is no longer a resolution-limiting factor in biological TEM. A final aberration introduced by the objective lens results from non-isotropic focusing effects and is termed astigmatism. The presence of astigmatic focusing results in the unequal emphasis of spatial frequencies in a direction-dependent manner. An astigmatism can be compensated for by introduction of astigmatic correctors positioned after the objective lens, which are able to restore the correct roundness of the electron beam.

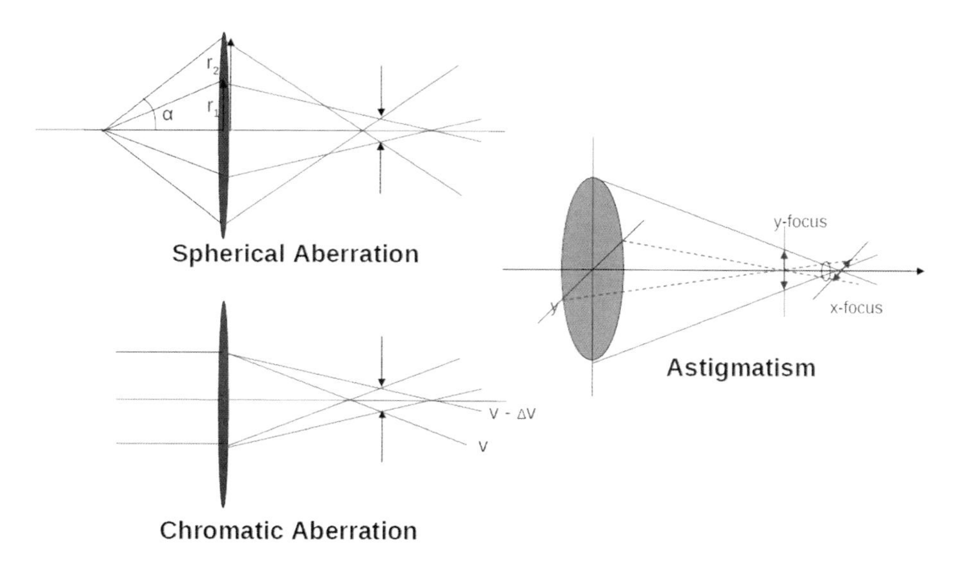

Figure 7.2. Schematic of the aberrations contributed by the objective lens. (A) Differential focusing of the electrons by the lens along its diameter results in spherical aberration. (B) Similarly, differential focusing of electrons of varying energies generates a chromatic aberration. (C) Finally, non-isotropic directional focusing leads to astigmatic effects. Please visit http://extras.springer.com/ to view a high-resolution full-color version of this illustration.

The projector lens system comprises the final lens system of the TEM. The goal of these lenses is to perform any necessary magnifications of the specimen and to project the final image of the specimen onto the imaging screen or collection device. To allow for visualization of the electrons, the imaging screen is coated with a phosphorescent compound that emits faint green photons when struck by the impinging beam. To record an image of the specimen, either photographic film or a charged-coupled device (CCD) can be used. The film or CCD is located beneath the imaging screen, and the amount of beam exposure to these elements can be controlled to allow for optimal exposure times.

The operation of a TEM requires high vacuum conditions. For thermionic electron sources vacuums of 10^{-2}–10^{-4} Pa are required, and as stated previously, 10^{-8} Pa is required for FEG sources in order to keep the FEG cathode free from any molecular contamination. Additionally, any gaseous molecules present within the column are free to interact with the electron beam, so minimizing their presence decreases the likelihood of spurious electron scattering by non-specimen atoms. Significant amounts of time can be lost in order to generate the vacuum pressures necessary for TEM operation, so it is important to work extremely carefully during specimen loading and removal to prevent vacuum failure.

To image the specimen, it must first somehow be positioned within the path of the beam and within the polepiece of the objective lens. Microscopes are available that allow for either top or side entry of the specimen into the EM column. Side-entry columns are most commonly used and allow for tilting of the specimen; however, they also tend to be less mechanically and temperature stable than the top-loading columns. For side-entry columns the specimens are loaded onto metal grids (usually copper or nickel) and are placed into a specimen holder. The holder is then inserted directly into the column perpendicular to the beam path. Top-loading columns place the specimen grid into a mechanical holder that gets lowered down into the EM column along the beam axis. In both scenarios, the height of the specimen relative to the objective lenses must be accurately calibrated for optimal imaging.

Figure 7.3. Schematic of the incoming and outgoing electron wave fronts during imaging. The incoming electrons can be approximated as a planar wave with constant phase. Upon interaction with the sample, the transmitted wave experiences a phase shift relative to its incoming phase, generating phase-based contrast.

7.2. PHASE-CONTRAST IMAGING

The electron beam that impinges upon the sample can be approximated by the following wave equation, with x being the 2D vector corresponding to position in space, k_0 the wave momentum vector, and z the particular plane in the direction of the incident wave (Fig. 7.3):

$$\psi_{\text{incident}}(x) = \exp^{ik_0 \cdot z},$$

As the electrons pass through the sample and interact with its force potentials, they experience a phase shift of $e^{i\eta(x)}$ relative to non-interacting electrons, giving a wave transmission modeled by

$$\psi_{\text{transmitted}}(x) = \exp^{i\eta(x)} \exp^{ik_0 \cdot z}.$$

The phase shift experienced by the passing electrons is influenced by the atomic composition of the specimen. The intensity of the diffracted beam can be approximated as the squared magnitude of the object's structure factor, F_k, given by (with f_i the atomic scattering power or atomic form factor and r_i the position of an atom within the object)

$$F_k = \sum_i f_i \exp^{-2\pi \kappa \cdot r_i}.$$

Atoms with larger nuclei will have larger atomic form factors, which allows them to scatter electrons to a greater degree. Lighter elements present within biological materials will therefore scatter electrons significantly less than the heavier elements found within non-organic materials.

The superposition of the phase-shifted or scattered electrons with the non-phase-shifted or unscattered electrons generates an image with amplitude contrast. Contrary to conventional light microscopy, imaging of biological samples within the TEM is not performed at focus but rather with varying amounts of underfocus or defocus. As will be discussed in the next section, the intentional introduction of defocus can emphasize or deemphasize specific features of the specimen in a frequency-dependent manner. Relating to phase contrast microscopy, biological macromolecules are treated as what is termed "weak phase objects" due to their relatively thin nature (less than tens of nanometers in thickness). Because of their thinness, the electrons that are elastically scattered by the specimen tend to undergo only a *single* scattering event. Increasing the thickness of a sample would increase the probability of the passing electrons interacting multiple times with the sample, causing the weak-phrase approximation to be invalid. These multiple scattering events are undesirable in biological TEM because their interactions with the sample are immensely more difficult to quantify.

Finally, what is recorded by the CCD or photographic film is the intensity of the incoming electron wavefront and is given by

$$f(x) = |\psi(x)|^2 .$$

7.3. THE CONTRAST TRANSFER FUNCTION

Like all optical systems, the TEM is not a perfect imaging system. Due to the perfected formation of glass lenses, modern non-superresolution optical microscopes are now only diffraction limited. However, for TEM, this is not the case as TEM images come nowhere close to the diffraction limit of the electrons that are used. Instead, resolution is limited by the aberrations imparted by the lens system, the mechanical instability of the microscope, and the functionality of image collection devices. The net effect of these imperfections is the contribution of a point spread function (PSF) during imaging, which manifests itself in real space as a blurring of a point source into an Airy disc, or as a frequency-dependent contrast transfer modulation of the image. Similar to optical microscopy, the image that is generated by the electron beams is a convolution of the object with the PSF of the TEM as shown by the Contrast Transfer Function (CTF) (PSF represented by $h(x)$, CTF by $H(s)$, and $\Psi(s)$ as the FT of the object wave equation):

$$\psi_{\text{object}}(x) * h(x) = \left\{ \mathcal{F}^{-1} \Psi_{\text{object}}(s) \cdot H(s) \right\} = \psi_{\text{image}}(x) .$$

Alternatively, the FT of the image wave equation can be generated by multiplication of the FT of the object with the microscope CTF as shown:

$$\mathcal{F}\left\{ \psi_{\text{image}}(x) \right\} = \Psi_{\text{image}}(s) = \Psi_{\text{object}}(s) \cdot H(s) ,$$

$$H(s) = A(s)E(s)\gamma(s) .$$

The term $A(s)$ describes the high-frequency attenuation of the transmitted wave that is removed through the objective aperture, which functions as a binary frequency-dependent mask at the BFP by allowing only the transmission of spatial frequencies within the aperture opening. The sinusoidal-dependent term of the CTF, $\gamma(s)$, is modulated by the amount of defocus (Δz) applied to the specimen, the wavelength of the electrons used in the imaging (λ), and the amount of spherical aberration that is present due to the objective lens (C_s). Since the electron voltage and spherical aberration are constant during operation, the main contribution of CTF modulation is imparted by the operator of the microscope in the form of image defocus:

$$\gamma(s) = -\sin\left[\frac{\pi}{2} C_s \lambda^3 s^4 + \pi \Delta z \lambda s^2 \right] .$$

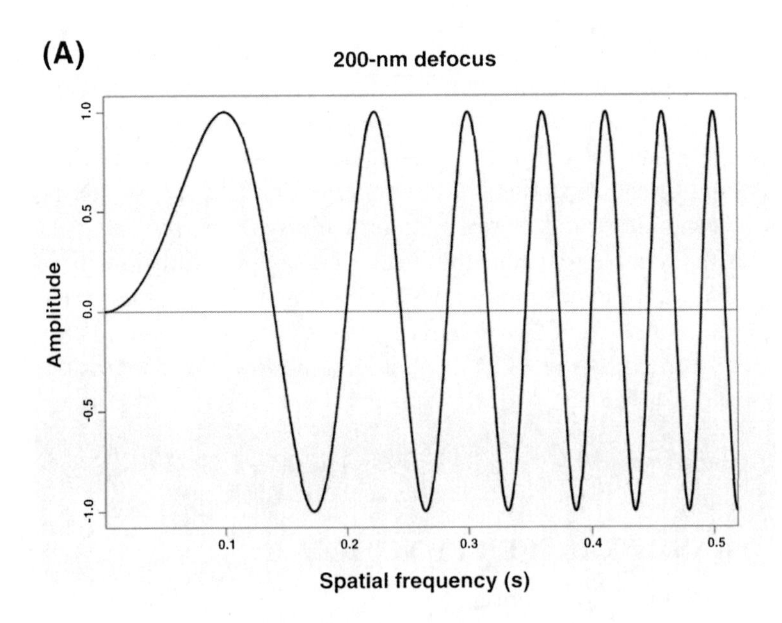

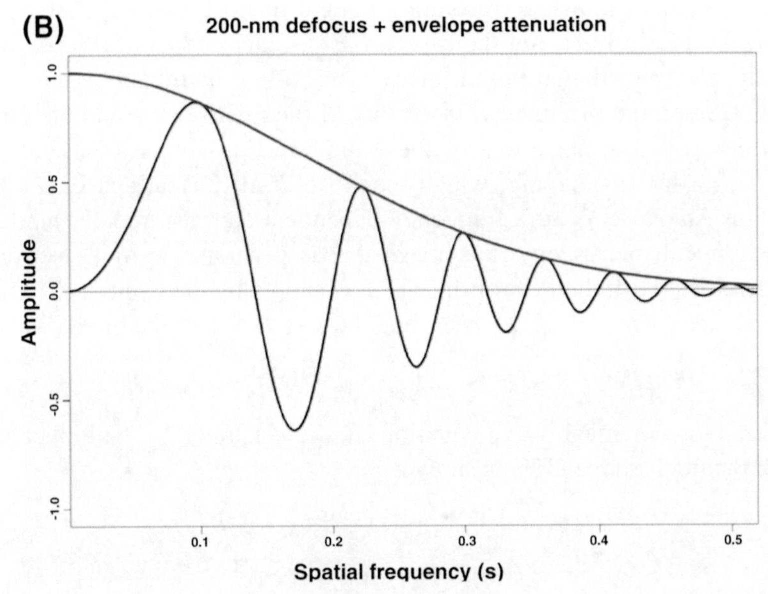

Figure 7.4. Plots of simulated contrast transfer functions (CTFs). (A) CTF at 2000-nm defocus without suppression from an envelope function by having spatial frequency as the variable of the function. (B) Spatial attenuation at high frequencies due to the envelope function. (C) (facing page) CTFs at 1500- (purple) and 2500-nm (green) defocus reveal slower CTF oscillations and a higher point-to-point resolution for the lower defocus and faster CTF oscillations and a lower point-to-point resolution for the higher defocus. Please visit http://extras.springer.com/ to view a high-resolution full-color version of this illustration.

The net effect of $\gamma(s)$ is a sinusoidal contrast inversion of the object's spatial frequencies. The point-to-point resolution of a TEM is determined by the first zero-crossing of the CTF (Fig. 7.4). Higher amounts of defocus result in a lower point-to-point resolution and impart more

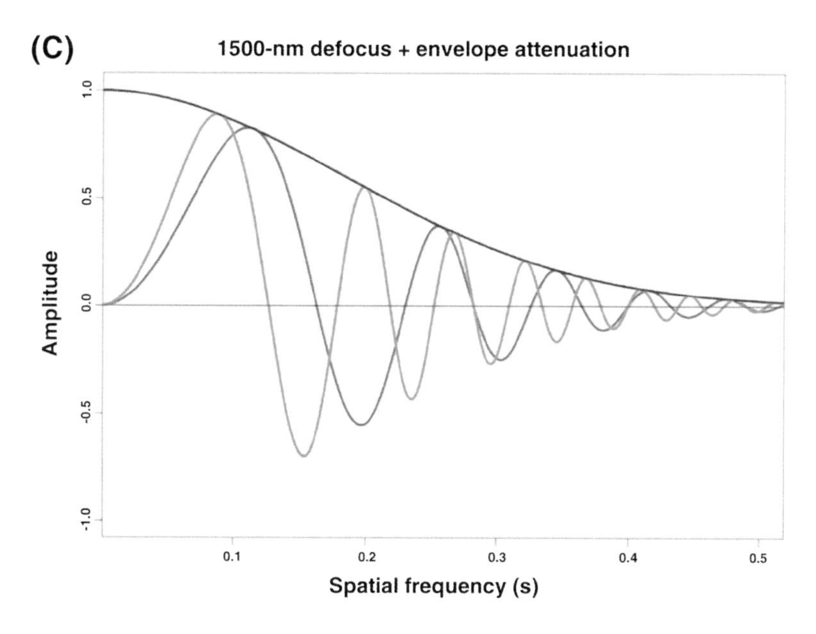

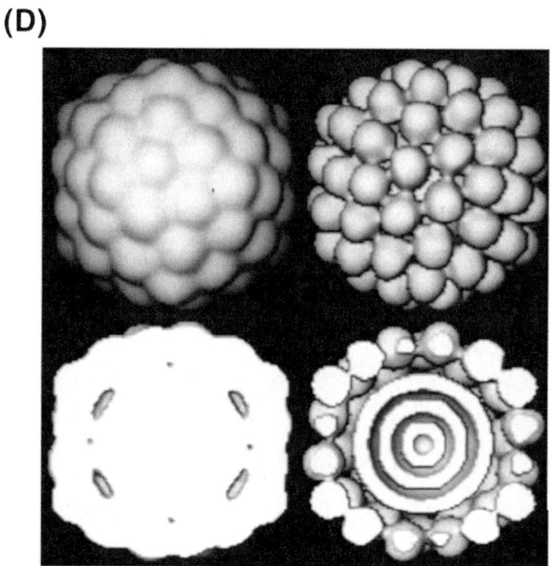

rapid contrast inversions, whereas lower amounts of defocus lead to greater point-to-point resolution and generate less frequent inversions of contrast. In order to obtain a true image of the specimen, the CTF must be corrected for. Additionally, since particular defocus values have unique zero-crossings where no contrast information is transferred, datasets must be collected across a range of defoci in order to compensate for the lack of information at these spatial frequencies [2,3].

The CTF is also modulated by the TEM's own transfer function, or envelope function, $E(s)$, which can be approximated as a Gaussian falloff of information transfer at higher-spatial frequencies. These effects are due to spherical aberration, $E_s(s)$, chromatic aberration, $E_c(s)$ (which is dependent upon the chromatic aberration coefficient, C_c), specimen drift, $E_d(s)$, and vibration, $E_v(s)$, as well as detector contributions, $E_D(s)$, and are the limiting factors in TEM resolution:

$$E(s) = E_S(s)E_C(s)E_d(s)E_v(s)E_D(s),$$

$$E_S(s) = \exp^{-\left(\frac{\pi\theta_0}{\lambda}\right)(C_S\lambda^3 s^3 - \Delta z\lambda s)^2}$$

$$E_C(s) = \exp^{-\frac{\pi^2}{4}\left(C_C\frac{\delta E}{E}\right)^2\lambda^2 s^4}.$$

The effect of the envelope transfer function on biological structural analysis therefore requires a great number of sample images to be generated in order to achieve an appreciable amount of information transfer at higher-spatial frequencies.

Since the PSF modulates the actual features of the sample under study in a frequency-dependent manner, it would be desirable to compensate for its effects to obtain a final image that most closely resembles the specimen. As the CTF is the FT of the PSF, it can be divided out in Fourier space if the correct defocus can be determined as follows:

$$\psi_{\text{object}}(s) = \frac{\Psi_{\text{image}}(s)}{H(s)}.$$

In practice, generation of the power spectrum of recorded image intensity can reveal the frequency-dependent oscillations of the CTF. The oscillations within a micrograph will correspond to a unique defocus that can then be used to correct for the CTF in a spatial-frequency-dependent manner (Fig. 7.4). Once the CTF is accounted for, a more accurate interpretation of the specimen can be determined. If the CTF is not corrected during an analysis, one can expect the effective point-to-point resolution of the data to be reliable only to the first zero-crossing of the CTF.

7.4. THE PROJECTION THEOREM AND SINGLE-PARTICLE RECONSTRUCTION

The goal of single particle reconstruction (SPR) is to determine the 3D structure of a desired biological macromolecule from a set of its 2D projections. To reconstruct a 3D structure from a 2D projection set, the Projection Theorem comes into play (Fig. 7.5). This theorem states that an arrangement of the 2D FTs of various 2D real-space projections of an object can be oriented and combined in 3D Fourier space, wherein a subsequent back Fourier transform (FT^{-1}) of the 3D Fourier information can generate a 3D representation of the original object in real space. Usage of the TEM automatically generates the 2D projections required by the projection theorem. The determination of the orientation of each projection and their summed conversion into a 3D model are performed computationally through a variety of related algorithms. The orientation of a particular projection or class average is determined by five parameters: two Cartesian centers, x and y, and three Euler rotations, out-of-plane θ and ϕ, and in-plane ψ. Depending on the reconstruction algorithm that is used, the order of parameter determination can vary and even may involve determination of all five parameters simultaneously [4–6].

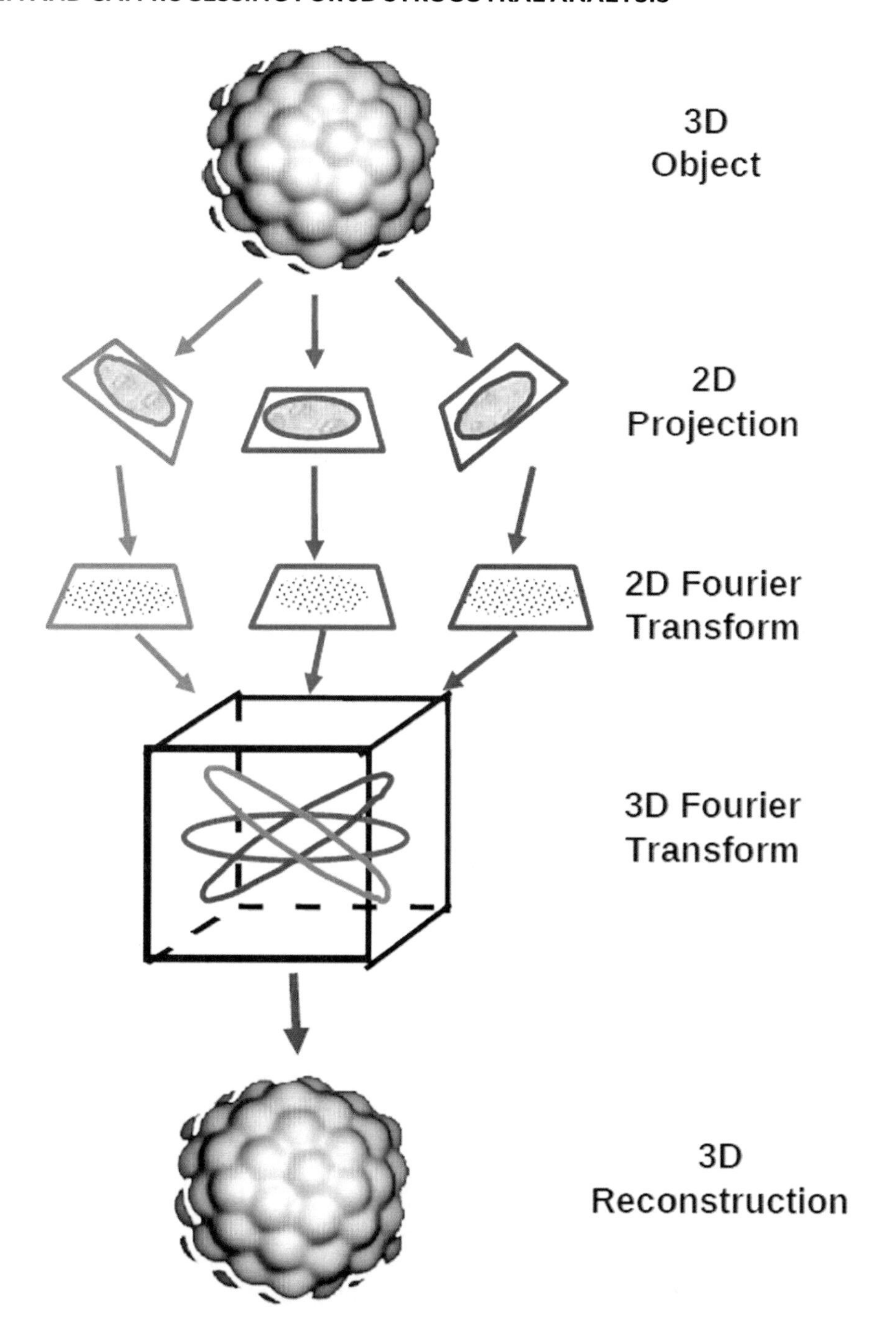

Figure 7.5. **Projection theorem at a glance**. A 3D object is imaged via TEM in varying orientations, generating 2D projections of itself. FTs of the 2D projections are oriented relative to one another in 3D Fourier space, followed by a Fourier back transform that regenerates a representation of the starting object. Please visit http://extras.springer.com/ to view a high-resolution full-color version of this illustration.

Due to the low signal-to-noise ratios (SNRs) present in TEM and the fact that 3D information is superimposed into a 2D projection, the usage of a single projection image is insufficient to obtain an accurate 3D representation of a macromolecule. To overcome these problems, multiple projection images with known orientations must be combined to generate a more accurate 3D representation. If enough particles are used, the SNR of the higher-spatial frequency terms is increased and higher-resolution structures can be attained. Additionally, by ensuring that the projection images are randomly oriented, all regions of Fourier space can be adequately filled for accurate reconstruction back into a real-space volume. If projections of particular views are not included in a reconstruction process, then their corresponding spatial frequencies in Fourier space will be missing and the real space reconstruction will suffer as a result. Such an occurrence is common in tomography, where tilt angle limitations generate a missing wedge or cone in Fourier space but can be minimized in SPR by ensuring that all possible projected orientations are present in the reconstruction process. As just stated, single-particle reconstruction does not rely on a *single, lone* molecule to determine a structure, but rather a collection of *isolated, identical* molecules.

7.5. ADVANTAGES OF SPR OVER X-RAY CRYSTALLOGRAPHY

SPR was developed as a means to study proteins and protein complexes that were not amenable to study via X-ray diffraction (XRD) techniques. The first requirement for XRD is a suitable 3D crystal of a purified protein. Generation of these crystals is often the limiting step in structure determination, and for many proteins, crystals may be unable to form, too small for analysis, or unable to diffract to an appreciable resolution. Due to the ability of the EM to visually identify individual protein complexes, 3D crystals do not need to be grown. By taking an individual biomacromolecule and averaging various projections together in an orientation-dependent manner, SPR is able to sidestep the requirement for an ordered array of protein. Another significant advantage related to this is the fact that SPR is able to analyze heterogeneous metastable complexes. Whereas a crystal used for XRD usually contains a highly stabilized macromolecule in a fragile packing arrangement, SPR allows proteins to easily exist in a variety of conformations. It is therefore easy to image samples using EM with a wide range of substrate and buffer conditions that are not amenable to particular crystal analyses. A common problem frequently encountered by crystallography is the growth of crystals of large macromolecular complexes. Such complexes have immense amounts of surface area that often are not capable of forming favorable crystal contacts. Additionally, these complexes tend to be somewhat fragile and can easily dissociate into subcomplexes during the crystallization process. SPR is able to handle such large complexes, again by treating them as individual components in the data-collection and structural-refinement process. Another potential drawback of crystallography is the influence of crystal packing upon protein structure. Not only are the crystallized proteins present in only a partially hydrated environment, they also exhibit crystal contacts that may not be indicative of their structure while free in solution. These hydration and crystal packing effects are not present in SPR because the protein complexes can be imaged in a completely hydrated environment and outside of the crystal context. A final strength of SPR over crystallography is the amount of pure protein that is required for a structural analysis. Usage of SPR can determine the structure of macromolecules with as little as 0.5 mg/mL of sample, whereas some crystal growth conditions can require as much as 10–30 mg/mL of purified sample. Additionally, the protein preps for SPR can contain modest levels of background impurities as their low molecular weights will only contribute to background signal, whereas similar background contamination in crystallization attempts can prevent successful crystal growth [2,7].

However, SPR does have drawbacks when compared to XRD. First and foremost is the resolution limit on SPR due to SNR attenuation at higher-spatial frequencies. For high-resolution structures to be generated, thousands to tens of thousands of images need to be averaged together, which requires significant computational effort. XRD does not have this problem and is able to achieve high resolution because of the SNR-amplifying effect of an ordered crystal. Additionally, limitations on the attainable resolution of imaging data from biological samples has yet to pass the 3.5-Å barrier, keeping it significantly behind the sub-Ångström resolution capability of XRD. Finally, there is a size limitation on SPR usage. Currently, the smallest structures solved to date (and only to a very modest resolution) have been on the order of 150–250 kDa. Structures with molecular weights smaller than this do not have enough mass to generate significant electron scattering events, making them nearly invisible above background image levels.

7.6. SAMPLE PREPARATION

To generate data suitable for SPR, the macromolecule under study must be prepared in a suitable fashion. Any purified protein, whether produced through recombinant means or purified directly, can be suitable for SPR if it is relatively homogeneous in composition. Significant amounts of contaminating background proteins with sizes similar to the sample under study can interfere with the reconstruction process if they are not removed prior to imaging. It is therefore advisable to run protein preparations through multiple purification columns to ensure the removal of contaminants similar in shape or size. Additionally, high levels of salt, glycerol, or other cryoprotectants will result in the degradation of image quality due to their interaction with the electrons, so dialysis may be required to reduce their concentrations when these compounds are present [1,8].

Several preparation techniques are used to image macromolecules within the EM. The most common practice is negative staining and involves the use of heavy metal salts such as uranyl acetate, lead citrate, or ammonium molybdate to generate contrast. Briefly, the sample is applied to a carbon-coated EM grid and blotted dry. The negative stain is then applied to the sample-containing grid and is blotted dry as well. The metal salts orient themselves around the periphery of the sample, providing a high-contrast surface detail of the specimen. Electrons more prevalently interact with the stain's heavy metal atoms due to their larger scattering cross-section than with the lower atomic mass atoms within the biological specimen, causing a greater degree of contrast enhancement compared to a biological sample alone. Although use of negative stain can provide high-contrast images, surface tension effects upon blotting/drying can lead to distortions of the sample that alter its true structure. Also, since the negative stain particles are of finite size, they do not always penetrate into the cavities or crevices of biological macromolecules, leaving these structural features irresolvable in the final image. Recent techniques have involved embedding the biological sample within an intact layer of heavy metal salts. Such stain-embedding techniques have the advantage of high image contrast but are able to circumvent most of the effects of particle distortion due to drying. The most significant development in sample preparation, however, has involved the embedding of sample within a thin-layer amorphous ice instead of metal salts. Such attempts are able to successfully preserve the native water-like environment of the biological sample and do not suffer from the distortion artifacts of negative staining. By placing the sample onto a grid coated with holey carbon, the specimen is able to migrate into a meniscus-like layer of water within the carbon holes. Rapid plunging of the sample into a cryogen with a high specific heat capacity such as liquid ethane causes the water to freeze rapidly enough to prevent an ordered water phase from forming, which would distort the sample's liquid environment and potentially alter its native structure. The sample can then be kept at extremely low "cryo" conditions during data collection with specialized liquid nitrogen–cooled cryoholders. The process of freezing and imaging a frozen biological sample is collectively known as cryo-electron microscopy (cryoEM) and is currently the gold-standard in SPR for native, high-resolution structure determination.

7.7. IMAGING CONDITIONS

The energy of the electrons imparted to a biological sample are immense and are enough to quickly destroy the specimen before adequate data can be collected. To help reduce the effects of electron beam–induced damage, the EM must be operated in a manner that reduces the total dose of electrons that impinge upon the sample. Common techniques involve reduction of either beam brightness or beam spot sizes, which limit the number of electrons that reach the sample. Exposures of 15–25 $e^-/Å^2$ are enough to alter the physical and chemical structure of a biological specimen, so reducing the amount of electron exposure to 5–10 $e^-/Å^2$ is desirable. However, such reduced beam exposure drastically affects the contrast in the resulting image, which has a pronounced effect on the ability to detect a particle above background due to the extremely low image SNR across all spatial frequencies. Development of cryoEM techniques revealed the preserving nature of imaging at liquid nitrogen temperatures, allowing the frozen specimen to tolerate a modest amount of sample radiation before becoming irreversibly damaged. As previously stated, the use of cryoEM requires specialized specimen holders that can accommodate a cryogen such as liquid nitrogen, which serves to keep the specimen frozen throughout image collection. Additionally, anti-contaminator blades may be placed within the EM column surrounding the specimen in order to adsorb stray gaseous water molecules present in the column. The readsorption and subsequent freezing of water molecules onto an extremely cold grid can generate cubic or hexagonal ice crystals that can hinder optimal data collection. Recent advances in cryoEM have involved the use of liquid helium as a cryogen, with purported increases in radiation damage over that of liquid nitrogen.

7.8. DATA VALIDATION

Once micrographs (film or CCD datasets) have been collected for use in SPR, they must be checked for quality. By computing the power spectrum of a micrograph either with an optical diffractometer or via a computerized fast Fourier transform, one is able to determine if any specimen drift has occurred, and if astigmatism or charging effects are present. Specimen drift is the result of mechanical instability of the specimen holder during image recording. It can be easily characterized by the directional absences of CTF oscillations (also called Thon rings) in the direction of drift. Astigmatism can be identified by elliptical Thon rings present in the power spectrum, indicating a direction-dependent difference in focus in the micrograph. Charging effects occur when localized charges are present on the specimen or grid. These charges are able to interact with the electron beam, generating direction-independent imaging artifacts. Micrographs that exhibit any of these three phenomena should not be used in subsequent structural analysis. Additionally, by looking at the extent of the Thon rings present within a power spectrum, one is able to determine a lower-level estimate of the resolution limit of a particular micrograph. As imaging conditions are perfected, a skilled electron microscopist will be able to maximize the number of Thon rings present within a power spectrum. The more Thon rings that are present, the further out in reciprocal space useful information has been transferred on that micrograph, indicating higher-resolution content [9,10].

7.9. DATA SELECTION & PREPARATION

SPR requires the extraction of a single-protein complex for structural elucidation. To perform this, the user must either interactively box out desired macromolecule image projections from a micrograph or automatically extract the images computationally. Interactive boxing is an extremely time-consuming process, but it may be necessary for highly heterogeneous preparations or if automated boxing has proven unsuccessful. A benefit of manual particle selection is that it allows the user to get familiar with orientations or heterogeneity present in the dataset that would not happen with automatic particle selection. Many techniques have been developed to perform automated particle selection, ranging from correlation-based boxing from a reference dataset to neural network–based boxing refinement. Depending on the size of the molecule present and the quality of the data, automatic boxing may or may not be a viable option for particle selection [11].

Once the particles required for analysis have been selected, they are usually masked along their periphery to remove any unwanted background signal. It is also desirable to apply a real space- or Fourier space-based filter to the selected images. A particularly effective frequency-based filter used in many SPR experiments is the bandpass filter. The bandpass filter applies both a highpass filter to remove any low-resolution background ramping that is commonly found in images and a lowpass filter to remove high-resolution noise components that may negatively bias the reconstruction process. Systematic extension of the lowpass filter to subsequently higher and higher-spatial frequencies can be performed to slowly provide the refinement process with increasing amounts of higher-resolution features for later stages of image alignment (Fig. 7.6). Other filters used in SPR include median, wavelet, and bilateral filters, the use of which depend on the desired filtering characteristics.

7.10. MULTIVARIATE STATISTICAL ANALYSIS

Once the particle images have been preprocessed, it is beneficial to perform a preliminary analysis to familiarize oneself with the data. Individual particle images contain extremely low SNRs, so the amount of information one can glean from them is very limited. However, by

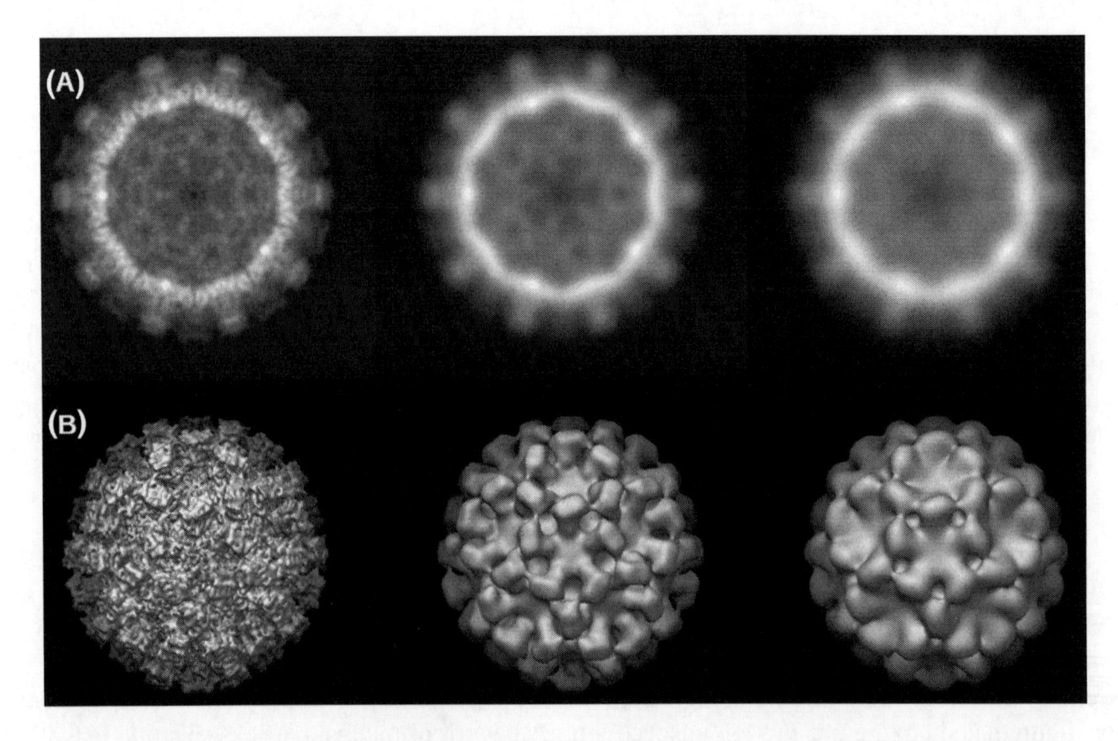

Figure 7.6. The effect of lowpass filtration. (A) 2D reprojections down the 5-fold axis of Norwalk virus (1IHM). (B) 3D surface renderings of Norwalk virus. Lowpass filters of spatial frequency 1/(10 Å) (Column #1), 1/(30 Å) (Column #2), and 1/(50Å) (Column #3). Structural detail is gradually lost as the lowpass filter is extended to lower spatial frequencies.

grouping together similar particles (based on projection orientation or conformational orientation), the SNR of those particular views can be increased, allowing more meaningful structural details to become apparent (Fig. 7.7). Statistical methods such as multivariate statistical analysis (MSA) have been devised in order to analyze the variation present within an image dataset. Variation can relate to a number of factors, including the variation present due to differences in the projection orientation dataset and conformational variation due to component variability or structural flexibility.

The overall goal of MSA is dimension reduction of the dataset and is accomplished by determining the fewest number of components required to describe the dataset as a whole. This operation is performed by determining the eigenvectors of the covariance matrix from vectorial representations of all projection images. By analyzing the eigenvectors that result, unique contributors of variance within the dataset can be identified and can be subsequently used for classification. Linear combination of the various eigenvectors with one another can be performed, resulting in an accurate approximation of the original dataset. Through MSA application, instead of representing the full dataset as an array of hundreds to thousands of images each with tens of thousands of pixel components, it can now be interpreted as a linear combination of a very small subset of eigenvectors and a datafile that describes the relative contribution of each eigenvector to the original images.

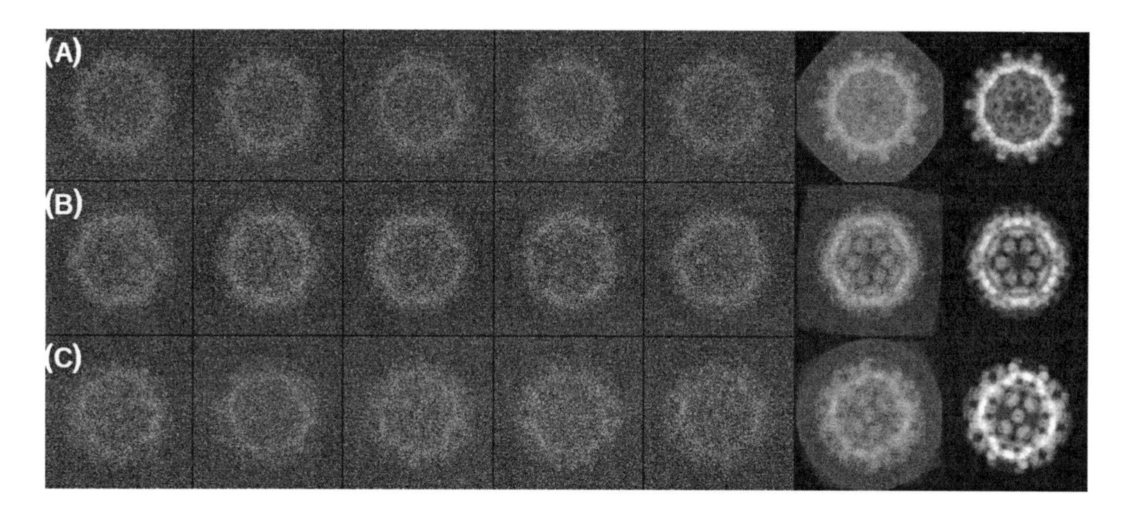

Figure 7.7. Classification of orientation subsets of Norwalk virus reprojections. Orientation of the reprojections corresponding to (A) 5-fold, (B) 3-fold, and (C) 2-fold axes. Columns 1–5 are noisy representations with random in-plane rotations of their particular class. Column 6 corresponds to the aligned average of the particles in columns 1–5. Column 7 is a reprojection of a Norwalk virus model in a corresponding orientation to column 6. The class averages in column 6 reveal much more structural information than the individual images in columns 1–5 and bear a strong resemblance to their parent reprojection in column 7.

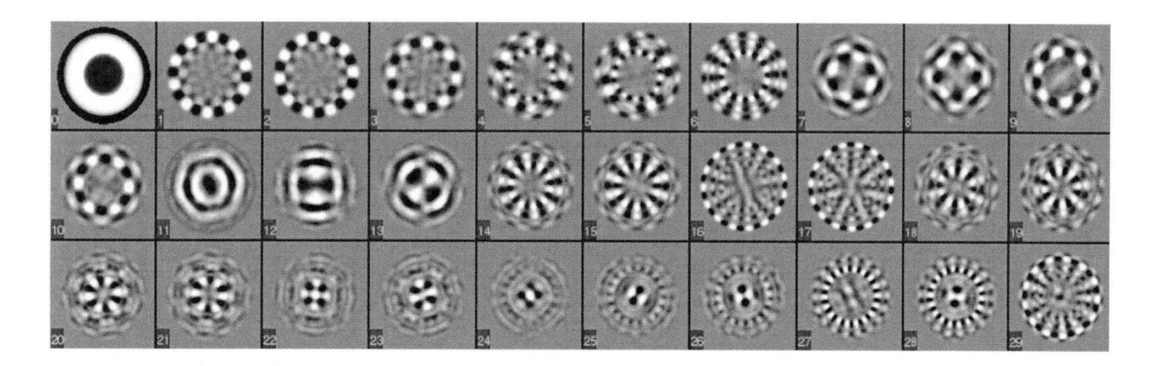

Figure 7.8. First thirty eigenimages generated from a mock dataset of reprojected lumazine synthase capsids lowpass filtered to 5 Å. The first eigenimage corresponds to the center of mass of the full datacloud and represents the average of the dataset. The early eigenimages reveal lower-order symmetry modulations, while subsequent ones reveal higher-order symmetry modulations that characterize finer capsid details. Please visit http://extras.springer.com/ to view a high-resolution full-color version of this illustration.

The eigenvectors that are determined by MSA can be directly interpreted as eigenimages (images that are pictorial representations of the eigenvectors; Fig. 7.8). By analyzing the variations present within the dataset in a visual manner, one may be able to better understand inherent differences within the dataset as a whole. As the eigenvectors are determined, the most predominant variations within the dataset are found to exhibit the highest eigenvalues. The higher an eigenvector's eigenvalue, the more variation that particular eigenvector describes within the whole of the dataset. The most significant eigenvectors within SPR data tend to characterize low-frequency structural modulations. These include the orientation of the macromolecule, its overall shape and size, symmetry arrangements, and large-scale structural variations such as changes in subunit binding stoichiometry or large protein domain movements. Less significant eigenvectors tend to have smaller eigenvalues but are useful in specifying higher-frequency modulations present in the data such as small-scale structural rearrangements and fine changes in adjacent orientation parameters. The least significant eigenvectors contain the lowest eigenvalues and tend to describe the noise present within the data. These eigenvectors are not useful in describing actual variations present within the images and can be disregarded during eigenanalysis [12].

7.11. CLASSIFICATION

As mentioned above, classification of individual image projections based on similarity is a powerful technique used to overcome the inherently low SNR present in TEM datasets. Classification can be performed in two different manners: unsupervised or supervised. Unsupervised or reference-free classification classifies images within a dataset without the use of outside references, while supervised or non-reference-free classification classifies the images with respect to outside references [13,14].

Unsupervised classification is a computationally intensive technique, as it seeks to find relations between images with zero outside guidance. Classical techniques for performing unsupervised classification involve *K*-means clustering and Hierarchical Ascendant Classification (HAC). Both of these techniques require previous application of MSA and eigenvector generation before being performed. By analyzing the distribution of images in multidimensional eigenspace performed by MSA, *K*-means and HAC are able to classify similar images with one another in an eigenvector-dependent fashion. *K*-means clustering is accomplished by placing a set number of seed class markers into the eigenspace datacloud (Fig. 7.9). These seeds correspond to gravity centers of individual image classes. With each iteration, new classes are generated by migrating the gravity centers closer to the centers of clustered images within the datacloud. *K*-means clustering is limited due to the fact that the location of starting class seeds can greatly influence the final classification results. HAC is performed by successively grouping like images together until a certain classification threshold is reached. To perform the grouping, HAC through Ward's criterion seeks to maximize the interclass variance (variance between different classes) while minimizing the intraclass variance (variance within a single class). Doing so helps to encourage the formation of classes that are quite dissimilar from one another but are made up of individual images that are very alike. Non-reference-based classes can also be generated through the use of Self-Organizing Maps (SOMs). An SOM implements dual-layer neural networks to iteratively classify molecules according to similarity, in addition to providing a 2D layout that indicates which classes are most similar to one another. Classes adjacent to one another on an SOM are most similar to one another, while classes found at the corners of the layout tend to represent extrema of the classified variability.

With supervised classification, images within a dataset are not compared to one another but instead are compared to a reference dataset. Performance of reference-based classification is

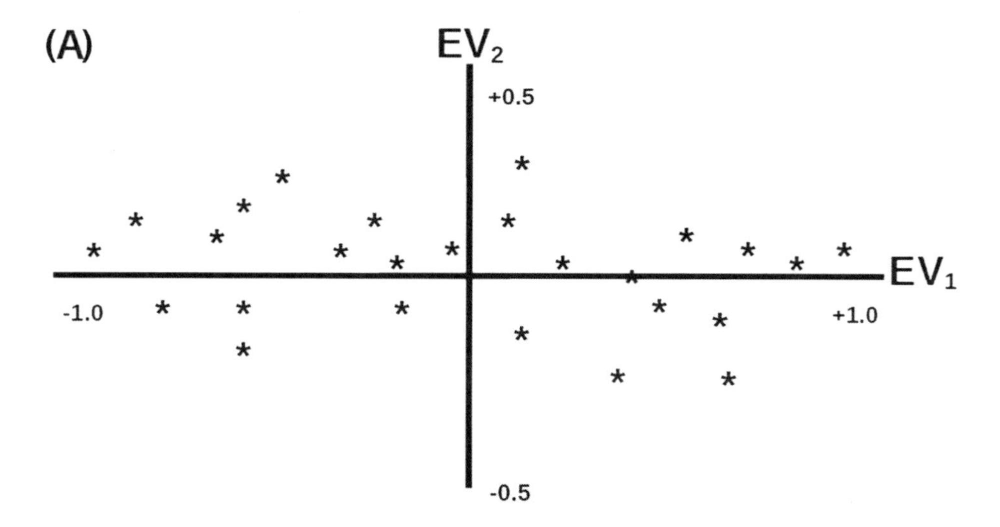

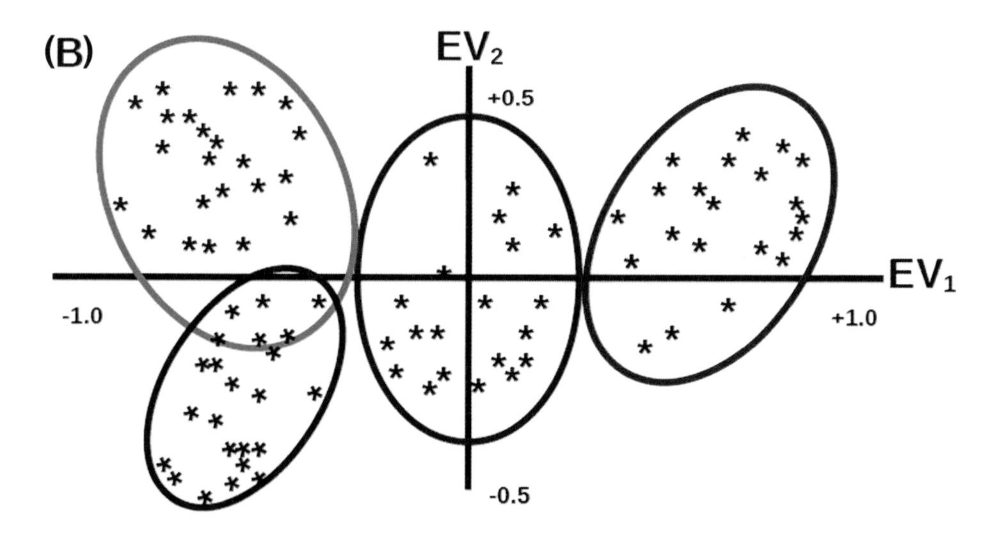

Figure 7.9. **Images distribution of in multidimensional eigenspace**. (A) Example of a 2D projection of image vectors in eigenspace onto the first and second eigenvectors. Principal variation occurs across eigenvector 1, with eigenvalues spanning from –1 to +1. Less variation is seen along the direction of eigenvector 2, with eigenvalues only ranging from –0.5 to +0.5. (B) Classification of a dataset projected onto the first and second eigenvectors, as determined by *K*-means clustering. Images within a class share similar eigenvalues with one another across both eigenvectors and can be grouped into nearly homogeneous classes. The classes in green and purple show some overlap, as there is not a complete distinction between them based solely on two eigenvectors. Please visit http://extras.springer.com/ to view a high-resolution full-color version of this illustration.

less computationally demanding than non-reference-based classification as class seeds are already generated and do not need to be artificially derived through iterative procedures. The main drawback of supervised classification, however, is the bias that is introduced by the reference projections. The classification of an input dataset will be guided by the references, and if there is a significant difference between the data and reference datasets, inaccurate or inconsistent classification may result. Despite this limitation, reference-based classification is a powerful technique for use in refinement cycles as proper reference projections can effectively seed the classification of projection images that are limited in popular number and would not likely arise in unsupervised techniques.

7.12. ANGULAR RECONSTITUTION

In the method of angular reconstitution, particles are grouped into related class averages via MSA and subsequent classification. To determine the orientation of the class averages relative to one another, angular reconstitution is performed. This method relies on the determination of common projection lines that are found from 2D Radon transforms by comparing all generated class averages against one another (Fig. 7.10). By iteratively refining the orientations determined for each class average with the goal of minimizing the orientation error across all classes, a reliable 3D model can be generated [15].

7.13. REFERENCE-BASED REFINEMENT

Reference-based refinement is a powerful technique that can be used to determine the structure of a macromolecule. To start, a set of reference projections is generated from an initial starting model that is projected across an asymmetric unit (Fig. 7.11). The images within the dataset are then compared to the set of reference projections in order to determine their relative orientations. Once the orientation of each image has been determined, a 3D model can be generated. This procedure is iterated until no further changes occur within the outputted volumes between iterations.

The most common flavor of reference-based refinement used in SPR involves the classification of individual projection images to a particular reference projection. All particles assigned to a specific reference orientation are aligned to one another and then averaged, generating a class average for that particular orientation. Noise-reduced class averages for each reference projection result are subsequently used to generate a 3D model. A variation on this theme exists that involves the generation of class averages *before* comparison to reference projections. Here, signal-strengthened class averages can be very accurately correlated to a reference, as opposed to correlation with a noise-weakened individual projection. A final variation avoids the usage of class averages altogether. In this case, individual images are assigned an orientation through reference matching, and are then directly used to generate a volume.

The correlations used to determine the level of matching between the image dataset and reference projections are extremely varied and depend upon the single-particle reconstruction package in use. Some packages use direct real-space correlations between the images, while others rely on Fourier space–based metrics. The Polar Fourier Transform (PFT) is noteworthy for its use in spherically symmetric objects such as viral capsids. The image and reference projections are converted into polar coordinates followed by FT of the polar transforms generating a PFT. The correlation of the PFTs allows for very accurate determination of a capsid's orientation because it maximizes the usage of symmetry-containing information. Recently, advanced

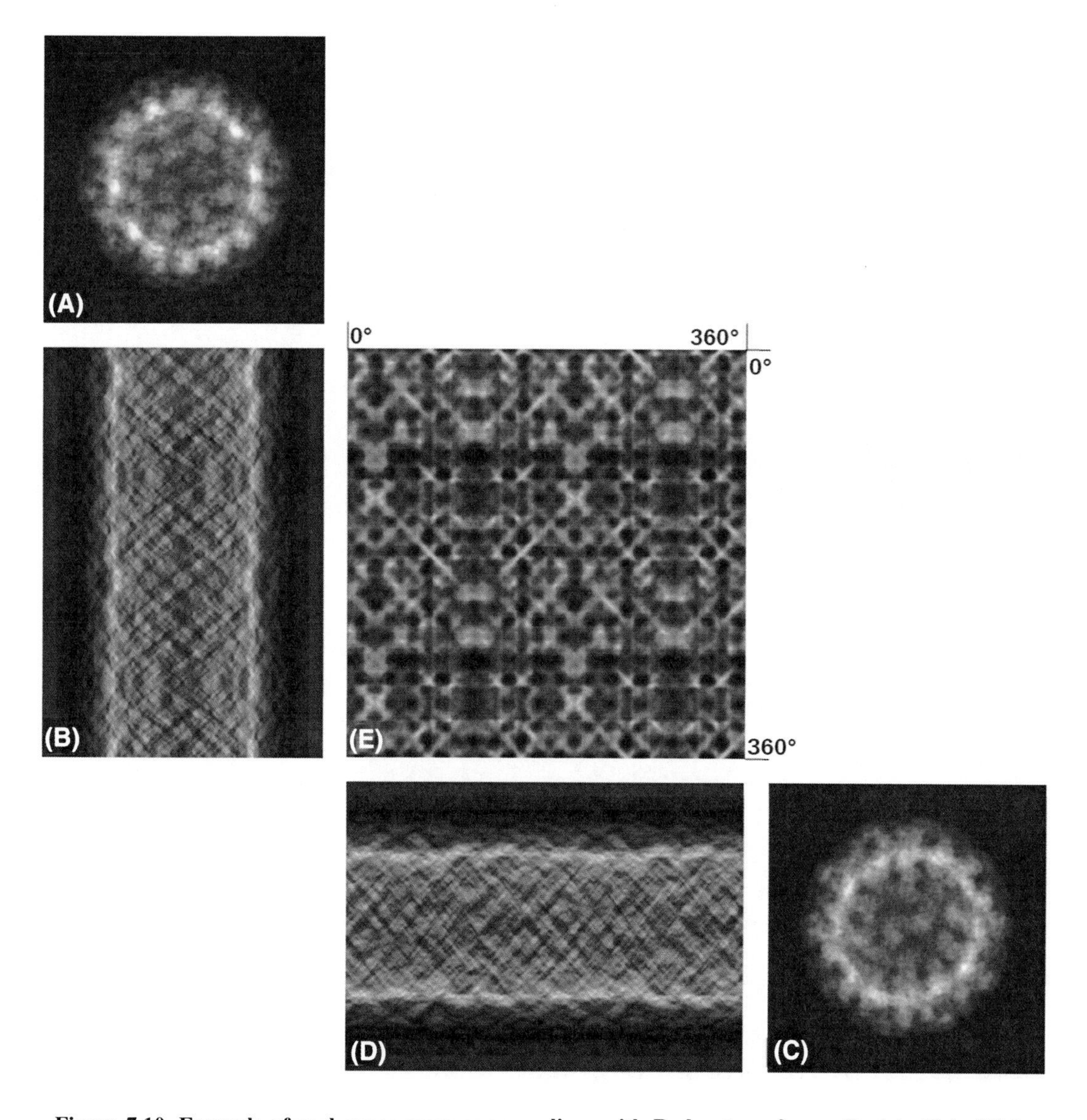

Figure 7.10. Example of real-space cross-common lines with Radon transforms. Particle #1 in (A) is compared to particle #2 in (C) by correlating their respective Radon transforms in (B) and (D), generating a correlation map (E). The correlation map reveals the orientation between the two projections through the location of its correlation peaks.

orientation determination techniques involving neural networks or Monte Carlo methods with simulated annealing have been applied. By turning random initial guesses into iterative guided searches, highly accurate orientations can be derived from random starting inputs [3,6,16].

A common technique used in orientation determination involves the usage of focal pairs. During data collection, two images of a specimen field are taken: a first image at low defocus followed by a second at high defocus. The high-defocus image contains higher contrast and better-resolved lower-spatial frequency information, which is beneficial for accurate determination

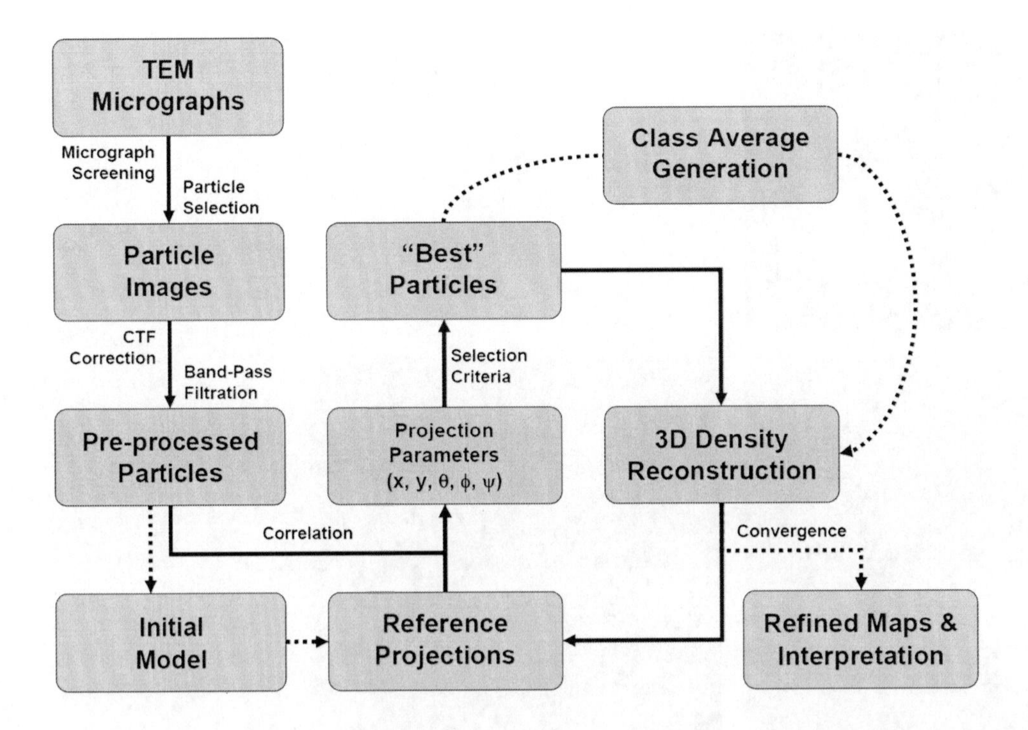

Figure 7.11. Workflow of reference-based refinement. Please visit http://extras.springer.com/ to view a high-resolution full-color version of this illustration.

of orientation (Fig. 7.12). The orientation determined from the far-defocus image is then applied to the projection image from the close-to-focus image. By utilizing focal pairs in the refinement process, orientations determined from close-to-focus images whose low-spatial frequency information is limited are circumvented [2].

7.14. INITIAL MODEL GENERATION

Initial models for reference-based refinement can be generated in numerous ways. The easiest method, which mimics the molecular replacement technique in X-ray crystallography, utilizes a lowpass-filtered density map from a previously determined macromolecule. If the model chosen is close in overall shape, size, and subunit composition, it can usually serve as a successful initial model for orientation refinement. Being that the dataset should drive the overall refinement procedure, the structural details present within the dataset are expected to reveal themselves during refinement, better approximating the actual structure after each successive iteration. However, some concern for initial model bias exists. By using an inappropriate starting model, a refinement may fail to converge to its true structure as it is too strongly influenced by the model at hand and correct orientation determinations cannot be achieved. To prevent such a result, initial model bias can be circumvented by employing additional initial model generation techniques [17].

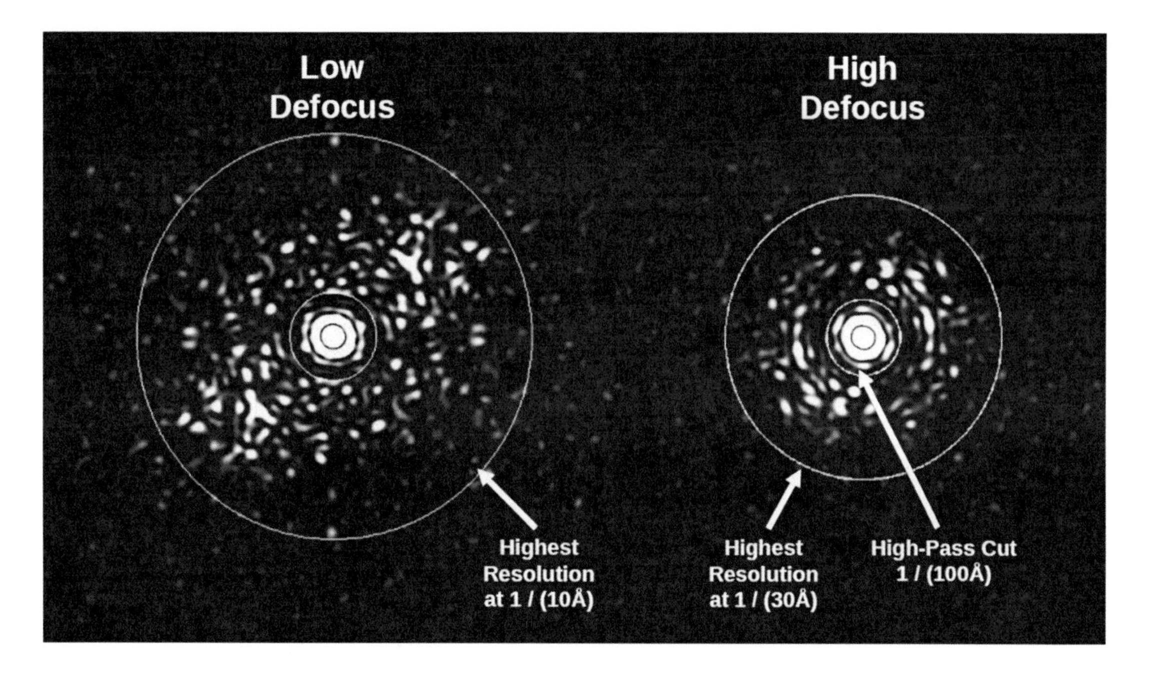

Figure 7.12. Fourier transforms derived from a focal pair of micrographs. The FT on the left was recorded at low defocus and contains high-resolution information out to 1/(10 Å), while the FT on the right was recorded at high defocus and only contains high-resolution information out to 1/(30 Å). The high defocus image will not be influenced by high-resolution noise during orientation refinement and should result in a more accurate orientation determination than use of the close-to-focus image.

The use of self-common lines and cross-common lines makes it possible to generate an initial model that is free of bias but contains relevant information about the structure in hand. The orientation of highly symmetric objects, such as 60-fold symmetric viral capsids, can be determined using self-common lines. Self-common lines are present within the Fourier transform of a projection image and are the result of the multiple symmetry elements present within a symmetric macromolecule (Fig. 7.13). The Fourier components along these lines will be identical to one another. An icosahedron will have 37 such pairs of self-common lines, and by applying all possible orientations to a particular image and determining the self-common lines for each, an optimal orientation can be determined for that particle. This solution can then be used to perform a single-image reconstruction to generate a starting model, or can be combined with other images of known orientation to form a noise-suppressed initial model. Unfortunately, if the particle in question does not possess a high degree of symmetry, self-common lines cannot be effectively used to determine an orientation de novo and cross-common lines must be used instead by determining orientations between particle images. Similar to self-common lines, the 2D FT of a projection image will share exactly one line in common with the 2D FT of another projection image. Along this line, all values of amplitude and phase will be similar between the two projections, so by searching orientation space for such a line, one can determine the relative orientation between two projection views. If multiple views are input, the overall error among all orientations is minimized in an iterative fashion. Once the orientations of these images are

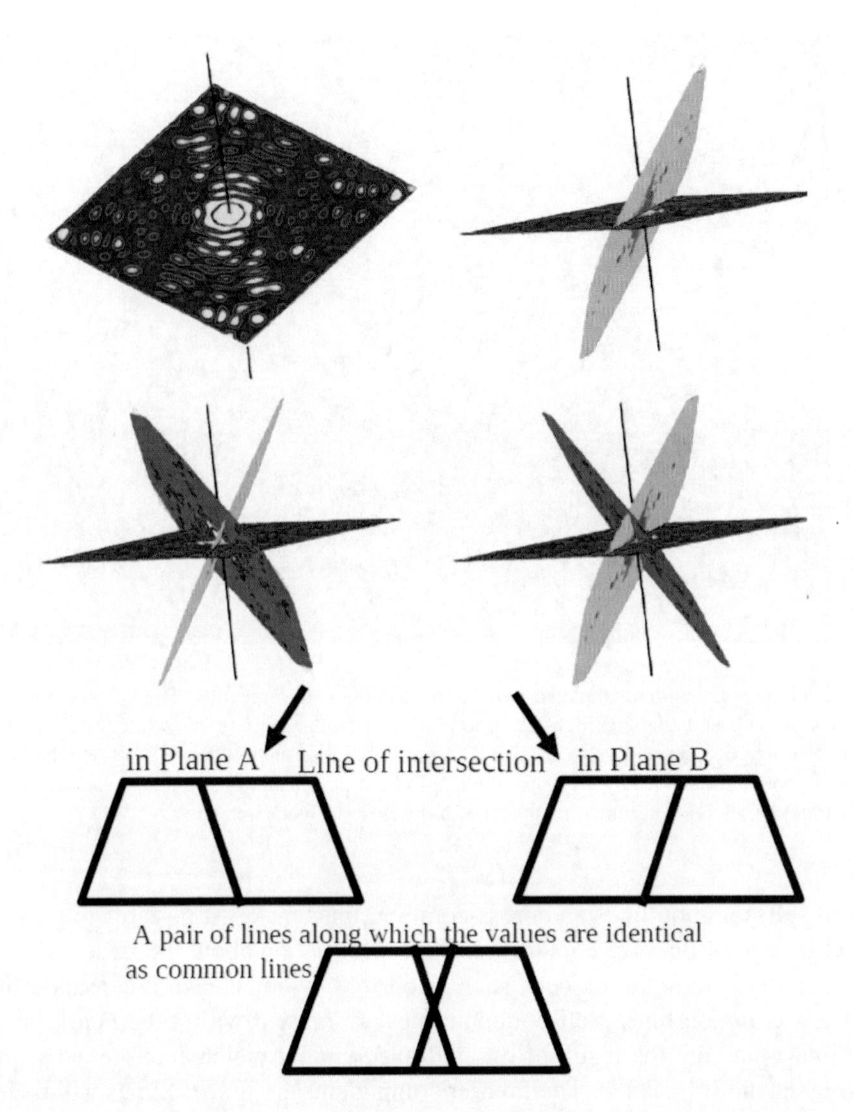

Figure 7.13. Self-common lines of a viral capsid. (A) FT of a projection oriented near the 3-fold axis. (B) Introduction of a symmetry-related projection generates a common line between the two projections. (C) Introduction of a third symmetry-related projection generates a set of three common lines, with each projection containing only two of the three possible common lines. The Fourier amplitude and phase values along these pairs of lines are identical to one another. Please visit http://extras.springer.com/ to view a high-resolution full-color version of this illustration.

known, they can then be used to generate an initial model for refinement. To improve the SNR of the 2D FTs, instead of using the FT of individual images, class averages can be used. By combining the signal of numerous projection images, a more accurate cross-common line orientation can be determined [5].

A third way to generate an initial model involves the generation of a volume with a randomized Gaussian density distribution. By approximating the overall shape and size of the macromolecular complex with a set of spheres or ellipsoids, a rough starting model can be generated to seed a refinement attempt. Again, the images within the dataset should drive the refinement cycle toward the correct structure despite the fact that a very vague starting volume is used. By doing so, model bias is nearly completely eliminated, and one can be confident in their structure if consistent results are obtained from numerous random Gaussian models. This technique may fail to work if the macromolecule in question does not possess unique or significant surface features for the refinement cycle to lock on to.

Another very successful method for initial model generation is the random conical tilt. An image dataset is recorded at zero tilt within the microscope. The sample grid is then tilted and another image is recorded, albeit with the particles in different projection orientations due to the tilting. Since the angle of tilt between the two image recordings is known, the exact relationship between them can be determined, thereby defining the orientation of the images. By doing so, a relatively trustworthy and data-driven initial model is generated. A bonus with the random conical tilt is that it is able to correctly determine the handedness of the complex being studied.

7.15. RECONSTRUCTION TECHNIQUES

Once the orientations of the image projections are known, they must be combined to generate a 3D reconstruction of the object. Many techniques have been devised to accomplish this task and fall into two main categories of either transform-based or series expansion–based methods. Transform methods include weighted back projection, Fourier-Bessel inversion (common in reconstruction of icosahedral objects), and direct Fourier inversion. Transform algorithms are relatively quick to execute and are well suited to parallel processing but may suffer from Fourier-based artifacts when a solution is undersampled at a particular spatial frequency. Examples of series expansion include algebraic reconstruction techniques (ARTs) and simultaneous interactive reconstruction techniques (SIRTs). The series expansion algorithms are extremely computationally intensive, as they require the iterative solving of a large system of equations; however, they also produce the most reliable results by generating reconstruction volumes that are most consistent with the data that are input. Significant improvements have been made in the parallelization of series expansion methods, and it is likely that these techniques will become the predominant choice for volume reconstruction in SPR [18,19].

7.16. RELIABILITY ASSESSMENT

The reliability of a 3D model generated through SPR must be determined before it can be analyzed in detail. Simple checks can be performed to ensure that the reconstruction process was self-consistent. For classification-based methods, the particles within a class average should resemble the class average and their class partners. Additionally, the class average that a reference projection is matched to should reveal high similarity. For reference-based refinements, a useful trick is to perform non-reference-based classification. If non-reference-based classes match reasonably well to the classes determined through reference-based refinement, one can be confident of the consistency of the reconstructions. To eliminate the worry about initial model bias, separate reconstructions using various starting models can be performed. If a consistent solution is reached among the various starting models, it can be assured that model bias was not a factor in the reconstruction process. Also, one may track the orientation determined for a particle or set of particles throughout the refinement procedure. If a particle jumps among many orientations and never stabilizes, it is a candidate for removal from the dataset due to its unreliable behavior. Finally, it is important to determine the resolution of the final output volume. If a reconstruction is not filtered to the proper resolution, spurious details can interfere with the structural analysis and may even lead to the identification of noise-based features that are not characteristic of the actual sample. To perform a resolution check, SPR typically employs the Fourier shell correlation (FSC):

$$\mathrm{FSC}(r) = \frac{\Sigma_{r_i \in r} F_1(r) \cdot F_2(r)^*}{\left\{ \Sigma_{r_i \in r} \left| F_1(r) \right|^2 \cdot \Sigma_{r_i \in r} \left| F_2(r) \right|^2 \right\}^{1/2}} .$$

This metric determines the summed correlation between successive shells of radial spatial frequency, r, in Fourier space of half-datasets compared against one another. The extent of their similarity in Fourier space to a correlation cutoff value of 0.5 gives a rough estimate of the resolution achieved in the reconstruction. Unfortunately, computation of the FSC from half-datasets will lead to an underestimation in achieved resolution due to halving of the dataset. Additionally, the FSC can be quite sensitive to aligned noise components or filtering artifacts, so care must be taken when assigning a reconstruction's final resolution cutoff. Debates in the field have existed over the actual numerical cutoff to use for resolution determination. Instead of the 0.5 cutoff criterion, some have suggested the use of a 3σ cutoff, where signal consistently emerges higher than random noise fluctuations, or the half-bit criterion, which is the resolution where enough information has been collected to interpret a volume. Other metrics have been developed, including Differential Phase Residual (DPR) and Spectral Signal-to-Noise Ratio (SSNR), and may be seen in publications in conjunction to an FSC curve [3,20].

7.17. ANALYSIS OF HETEROGENEITY

Several computational techniques have been developed to determine the structure of sample preparations or macromolecular complexes that are heterogeneous in nature. The first of these is termed focused classification. An initial reconstruction is performed with all of the data, followed by a 3D variance calculation. This step is able to determine which regions of the reconstructed volume exhibit the most density variance and, likely, pinpoints a region of structural flexibility within the macromolecule. By focusing in on these regions with a 3D mask and performing subsequent 2D MSA, variability in this focused region can be sorted via classification into more homogeneous datasets. These datasets are then used to generate parallel reconstructions that should characterize the structural heterogeneity present in homogeneous volumes. An offshoot of this approach is the use of 3D MSA. By generating a starting anchor set of numerous 3D reconstructions followed by 3D MSA, the structural variations present can be classified in a fashion similar to 2D MSA, generating more homogeneous subpopulations in the process. These subpopulations are then refined independently to produce a series of homogeneous reconstructions displaying the unique structural variations. A third, and perhaps the most complicated and computationally intensive approach, involves the use of Maximum Likelihood in the refinement process. A starting anchor set of numerous random models are generated, followed by independent maximum-likelihood refinement of the random models. Through its iterative procedure, the maximum-likelihood algorithm will progressively choose particles that most accurately match one of the 3D references they are being compared to. If structural variations are indeed present, homogeneous datasets will "self-crystallize" themselves computationally from the starting dataset by grouping with one of the various models throughout consecutive iterations [21–23].

7.18. SUMMARY

Transmission electron microscopy has seen a recent boom in its usage for macromolecular structural determination. Advances in electron microscope design, sample preparation techniques, and in refinement and reconstruction algorithms have pushed the resolution capabilities of SPR to nearly the 3-Å range. At this resolution, alpha-helices and beta-strands can be easily resolved. Additionally, de novo peptide backbone traces can be performed with high accuracy, and even bulky aromatic sidechains can be identified. Further efforts in SPR will concentrate on massively parallel refinements involving the usage of tens of thousands to hundreds of thousands of particles to push the resolution limit even further and to truly challenge the resolution achieved by XRD or NMR.

ACKNOWLEDGMENTS

Thanks to Dr. M. Kawano for help with the figure sets.

PROBLEMS

7.1. What fraction of the speed of light is the velocity of an electron experiencing an accelerating voltage of 200 kV? What is the relativistic mass of an electron accelerated to 1000 kV? Why are these values significant? [Adapted from Williams D, Carter C. 1996. *Transmission electron microscopy:bBasics I*. New York: Springer.]

7.2. Given that $\gamma(s) = -\sin\left[\frac{\pi}{2}C_s\lambda^3 s^4 + \pi\Delta z\lambda s^2\right]$, how would the introduction of a Zernike-like phase plate at the back focal plane affect the CTF if the phase plate imparts a $-\pi/2$ phase shift? Why would this be beneficial? [Adapted from Danev R, Nagayama K. 2001. Transmission electron microscopy with Zernike phase plate. *Ultramicroscopy* **88**:243–252.]

7.3. In cylindrical coordinates, a particle's density and its transform are related by $p(r,\phi,z) = \Sigma_n \int g_n \exp^{in\phi} \exp^{2\pi i\zeta Z} dZ$. The Fourier-Bessel transform of g_n is given by $G_n(R,Z) = \int_0^\infty g_n(r,Z) J_n(2\pi Rr) 2\pi R dr$. The representation for transform data with discrete sampling is $F_j = \Sigma_{n=-N}^{N} B_{jn} G_n \quad (n = -N \ldots N)$, with F_j being the available values of the transform independent of the particle density, B_{jn} the sampling matrix of the transform, and G_n the

Fourier transform of the particle density, and the solution being $G_n = \frac{1}{2m}\Sigma_{j=1}^{2m} B_{km}^{*} F_j$ $(n = -N \ldots N)$. Assuming that (1) the data provided are evenly spaced and (2) there are as many observations F_j as unknowns G_n $(2m \geq 2N + 1)$, what is the minimum number of views, *m*, required to reconstruct a particle of diameter *D* to a given resolution *d* (= $1/R_{max}$)? [Adapted from Crowther R, DeRosier D, Klug A. 1970. The reconstruction of a three-dimensional structure from projections and its application to electron microscopy. *Proc R Soc A* **317**(1530):319–240.]

FURTHER READING

Baker T, Olson N, Fuller S. 1999. Adding the third dimension to virus life cycles: three-dimensional reconstruction of icosahedral viruses from cryo-electron micrographs. *Microbiol Mol Biol Rev* **63**(4):862–922.

Cheng RH, Hammar L. 2004. *Conformational proteomics of macromolecular architectures*, Singapore: World Scientific. ISBN 981-238-614-9

Frank J. 2006. *Three-dimensional electron microscopy of macromolecules*. New York: Oxford UP.

Orlov I, Morgan DG, Cheng RH 2006. Efficient implementation of a filtered back-projection algorithm using a voxel-by-voxel approach. *J Struc Biol* **154**:287–296

Scherzer O. 1949. The theoretical resolution limit of the electron microscopy. *J Appl Phys* **20**(1):20–29.

Williams D, Carter C. 1996. *Transmission electron microscopy: basics I*. New York: Springer.

REFERENCES

1. Baker TS, Olson NH, Fuller SD. 1999. Adding the third dimension to virus life cycles: three-dimensional reconstruction of icosahedral viruses from cryo-electron micrographs. *Microbiol Mol Biol Rev* **63**(4):862–922.
2. Cheng RH, Olson NH, Baker TS. 1992. Cauliflower mosaic virus: a 420 subunit (T = 7), multilayer structure. *Virology* **186**(2):655–668.
3. Cheng RH, Reddy VS, Olson NH, Fisher AJ, Baker TS, Johnson JE. 1994. Functional implications of quasi-equivalence in a T = 3 icosahedral animal virus established by cryo-electron microscopy and X-ray crystallography. *Structure* **2**(4):271–282.
4. Crowther RA, Amos LA, Finch JT, De Rosier DJ, Klug A. 1970. Three-dimensional reconstructions of spherical viruses by Fourier synthesis from electron micrographs. *Nature* **226**(5244):421–425.
5. Fuller SD, Butcher SJ, Cheng RH, Baker TS. 1996. Three-dimensional reconstruction of icosahedral particles—the uncommon line. *J Struct Biol* **116**(1):48–55.
6. Baker TS, Cheng RH. 1996. A model-based approach for determining orientations of biological macromolecules imaged by cryoelectron microscopy. *J Struct Biol* **116**(1):120–130.
7. Gong ZX, Wu H, Cheng RH, Hull R, Rossmann MG. 1990. Crystallization of cauliflower mosaic virus. *Virology* **179**(2):941–945.
8. Fujiyoshi Y. 1989. High resolution cryo-electron microscopy for biological macromolecules. *J Electron Microsc (Tokyo)* **38**(Suppl):S97–S101.
9. Booth CR, Jakana J, Chiu W. 2006. Assessing the capabilities of a 4k x 4k CCD camera for electron cryo-microscopy at 300 kV. *J Struct Biol* **156**(3):556–563.
10. Chen DH, Jakana J, Liu X, Schmid MF, Chiu W. 2008. Achievable resolution from images of biological specimens acquired from a 4k x 4k CCD camera in a 300-kV electron cryomicroscope. *J Struct Biol* **163**(1):45–52.
11. Ludtke SJ, Baldwin PR, Chiu W. 1999. EMAN: semiautomated software for high-resolution single-particle reconstructions. *J Struct Biol* **128**(1):82–97.
12. Sines J, Rothnagel R, van Heel M, Gaubatz JW, Morrisett JD, Chiu W. 1994. Electron cryomicroscopy and digital image processing of lipoprotein(a). *Chem Phys Lipids* **67–68**:81–89.
13. Schatz M, van Heel M. 1990. Invariant classification of molecular views in electron micrographs. *Ultramicroscopy* **32**(3):255–264.

14. van Heel M, Harauz G, Orlova EV, Schmidt R, Schatz M. 1996. A new generation of the IMAGIC image processing system. *J Struct Biol* **116**(1):17–24.
15. Serysheva, II, Orlova EV, Chiu W, Sherman MB, Hamilton SL, van Heel M. 1995. Electron cryomicroscopy and angular reconstitution used to visualize the skeletal muscle calcium release channel. *Nat Struct Biol* **2**(1): 18–24.
16. Cheng RH, Kuhn RJ, Olson NH, Rossmann MG, Choi HK, Smith TJ, Baker TS. 1995. Nucleocapsid and glycoprotein organization in an enveloped virus. *Cell* **80**(4):621–630.
17. Wikoff WR, Wang G, Parrish CR, Cheng RH, Strassheim ML, Baker TS, Rossmann NG. 1994. The structure of a neutralized virus: canine parvovirus complexed with neutralizing antibody fragment. *Structure* **2**(7):595–607.
18. Crowther RA, Klug A. 1971. ART and science or conditions for three-dimensional reconstruction from electron microscope images. *J Theor Biol* **32**(1):199–203.
19. Orlov IM, Morgan DG, Cheng RH. 2006. Efficient implementation of a filtered back-projection algorithm using a voxel-by-voxel approach. *J Struct Biol* **154**(3):287–296.
20. van Heel M, Schatz M. 2005. Fourier shell correlation threshold criteria. *J Struct Biol* **151**(3):250–262.
21. Zhou ZH, Liao W, Cheng RH, Lawson JE, McCarthy DB, Reed LJ, Stoops JK. 2001. Direct evidence for the size and conformational variability of the pyruvate dehydrogenase complex revealed by three-dimensional electron microscopy: the "breathing" core and its functional relationship to protein dynamics. *J Biol Chem* **276**(24):21704–21713.
22. Haag L, Garoff H, Xing L, Hammar L, Kan ST, Cheng RH. 2002. Acid-induced movements in the glycoprotein shell of an alphavirus turn the spikes into membrane fusion mode. *EMBO J* **21**(17):4402–4410.
23. Xing L, Casasnovas JM, Cheng RH. 2003. Structural analysis of human rhinovirus complexed with ICAM-1 reveals the dynamics of receptor-mediated virus uncoating. *J Virol* **77**(11):6101–6107.

8

RAMAN SPECTROSCOPY OF LIVING CELLS

Tyler Weeks and Thomas Huser
Department of Internal Medicine, and NSF Center for Biophotonics, University of California, Davis

8.1. INTRODUCTION

Raman spectroscopy is an optical technique for the vibrational spectroscopy of molecules. It is a rather unique and powerful technique, because, similar to magnetic resonance imaging (MRI) or X-ray imaging, it is capable of live cell chemical analysis and imaging without destroying or altering the biological materials under investigation. In contrast to these techniques, however, Raman spectroscopy works really well at the cellular and subcellular levels, because it can easily be integrated with optical microscopy. Also, it does not require exogenous probes, such as in fluorescence microscopy, but instead relies fully on an intrinsically generated signal. Raman spectroscopy is based on the fact that a very small number of photons from an intense light source (typically a laser) can inelastically scatter off molecular bonds in a sample, leading to discrete shifts in the wavelength (or color) of the scattered photons. This chapter will provide a brief introduction to Raman spectroscopy, explain the principles behind it, provide a brief "how-to" guide for interested experimentalists, and discuss applications and recent advancements in the chemical analysis of living cells.

The Raman effect was first experimentally discovered in 1928 by C. V. Raman and K. S. Krishnan [1] by using sunlight as the light source, narrow spectral filters that Raman and Krishnan had developed, and their naked eyes as detectors. Raman was subsequently awarded the Nobel Price in Physics in 1930 "for his work on the scattering of light and for the discovery of the effect named after him." In the following years, mercury arc lamps served as the main sources for Raman spectroscopy, but Raman spectroscopy as a technique underwent a major

Address correspondence to Thomas Huser, NSF Center for Biophotonics, University of California, Davis, 2700 Stockton Blvd., Suite 1400, Sacramento, CA 95817, USA, 916 734-1772, 916 703-5012 (fax), <trhuser@ucdavis.edu>

T. Jue (ed.), *Biomedical Applications of Biophysics*,
Handbook of Modern Biophysics 3, DOI 10.1007/978-1-60327-233-9_8,
© Springer Science+Business Media, LLC 2010

renaissance in the mid to late 1960s, after the laser was invented, which served as the bright monochromatic light source that truly enabled highly sensitive optical spectroscopy. More recently, Raman spectroscopy has benefited greatly from the development of new, inexpensive steep-edge filters and more sensitive detectors. These advancements have enabled the development of compact commercial devices, which are used widely for chemical process analysis by the pharmaceutical industry, or micro-Raman spectrometers, which were initially used by Materials Scientists, but are now gaining significant popularity in the Life Sciences.

The main interest in Raman spectroscopy by Life Scientists is based on the fact that Raman spectroscopy enables the dynamic chemical analysis of living cells with subcellular spatial resolution. In essence, no other technique can claim a similar combination of high sensitivity, non-destructive chemical analysis of molecular constituents, and high spatial resolution. Standard techniques for cellular imaging and analysis, such as biological immunoassays and optical and fluorescence microscopy, have high sensitivity but suffer from a lack of molecular specificity, they perturb the sample by exogenous introduction of fluorescent probes and antibodies, or they require destructive processes such as cell fixation or lysis. Mass spectrometry (MS), in particular matrix-assisted laser desorption/ionization time-of-flight mass spectrometry (MALDI-TOF), can achieve similar spatial resolution, but requires the use of special substrates and destroys cells in the process of the analysis. MS is nonetheless an interesting technique that can be used in combination with Raman spectroscopy to obtain elemental maps of a sample, since Raman spectroscopy provides complementary information about molecular bonds. Secondary Ion Mass Spectrometry, especially when used with tightly focused gallium ion beams to achieve a resolution on the nanometer scale (termed "NanoSIMS"), can obtain even higher spatial resolution and sensitivity, but it also is not compatible with living cells and destroys cells in the process. Nuclear magnetic resonance (NMR) and its imaging analog MRI cannot compete with the superior spatial resolution and sensitivity provided by Raman spectroscopy. Only infrared (IR) absorption spectroscopy can claim to provide similar information, but it does not achieve subcellular spatial resolution, and is generally difficult to perform on living cells because IR wavelengths are highly absorbed by water. The only problem with Raman spectroscopy is that it is a process with fairly low efficiency, requiring high sample concentrations (albeit minute sample volumes), but this issue is currently being addressed by a number of research groups worldwide. Raman spectroscopy is destined to continue its rapid growth and become a major new driver in Life Sciences research for the foreseeable future. Readers interested in a more detailed discussion of biological Raman spectroscopy will find an excellent resource in recent inexpensive textbooks (such as [2]) or in-depth discussions of many modern microscopic imaging techniques in edited books (e.g., [3]).

8.2. RAMAN SCATTERING: INELASTIC LIGHT SCATTERING BY MOLECULAR BONDS

Raman spectroscopy is the detection and dispersion of photons that are inelastically scattered by molecular vibrations. When a monochromatic laser source with frequency ω_{i} (photon energy E_{i}) is used to probe a sample, most of the interacting photons encounter elastic scattering where the frequency of the photons remains unchanged after the scattering event. This process is called

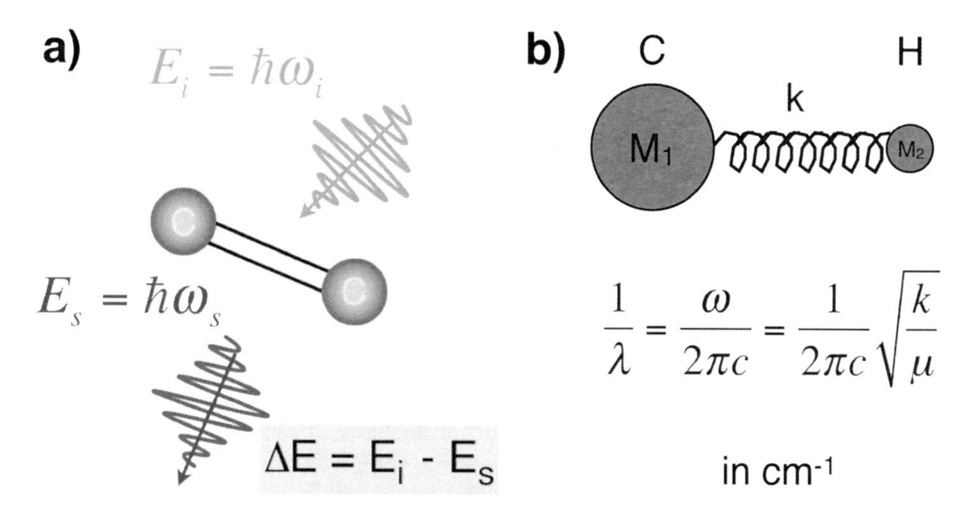

Figure 8.1. (a) Schematics of an inelastic photon scattering event: incoming photons with energy E_i scatter off a molecular bond, thereby depositing energy into the bond vibration. The outgoing, inelastically scattered photons have a slightly lower energy E_s. (b) Model of a molecular bond: two atoms with masses M_1 and M_2 are connected by a spring with spring constant k. The energy of the resulting vibration written in spectroscopic terms is shown below, where c is the speed of light and $\mu = M_1M_2/(M_1 + M_2)$ is the reduced mass of both atoms. Please visit http://extras.springer.com/ to view a high-resolution full-color version of this illustration.

Rayleigh scattering. The small fraction of photons, however, that scatter inelastically off molecular bonds exchange energy with the vibrational energy levels of the molecules in the sample and are *Raman* scattered (see the schematic depiction in Fig. 8.1). These scattered photons, with a new frequency ω_s, will show up as discrete peaks in a Raman spectrum, where each peak corresponds to a distinct type of molecular vibration. Typically, these peaks are represented in terms of a wavelength-independent relative energy unit called "wavenumbers (cm^{-1})," which reflects the amount of energy a photon has exchanged with the molecular vibration during the scattering event. Since molecular vibrations are strongly dependent on the molecular conformation and environment, spectral analysis of the scattered photons can be used not only to identify a molecular bond but also to assess the chemical microenvironment in which it is found. For example, the schematics in Figure 8.1a show a C=C double bond that only allows for a stretching vibration along the axis connecting both carbon atoms. If this vibration is momentarily in the ground state, the incident photon (shown in green) can excite it to the first vibrational state, i.e., termed a stretching vibration with a fixed frequency. The scattered photon will have transferred parts of its energy to this vibration and has a lower energy than the incident photon, which is represented by its red color. Molecular bond vibrations are quantized, so only certain well-defined amounts of energy can be exchanged between photons and molecular bonds, leading to a Raman spectrum with discrete peaks, each of which can be assigned to a specific vibrational mode.

At this point it is useful to remind ourselves that the energy of a photon depends on the frequency (wavelength) of the electromagnetic wave packet that it represents and that it can be written as $E = h\nu = \hbar\omega$. For consistency, we will use either the energy E or frequency ω to describe photons for the remainder of this text. The symbol ν will be reserved for the representation of different vibrational energy levels of the molecules (starting from Fig. 8.2).

Let us briefly consider another analogy. In very simplistic terms, a specific molecular bond can be represented by a harmonic oscillator (see Fig. 8.1b). Here, the two atoms that form the bond have specific masses M_1 (e.g., a carbon atom) and M_2 (e.g., a hydrogen atom), and the bond itself is represented by a spring with spring constant k. Since the energy of a stretching vibration along the axis of the spring is proportional to its frequency (i.e., the inverse of the wavelength), optical spectroscopists have agreed to use a simplified notation and describe molecular vibrations in terms of an inverse wavelength with inverse centimeters (cm^{-1}) as unit. In this case, the energy of the stretching vibration can be calculated quite easily by the reduced mass μ of the two atoms forming the bond, and the spring constant. This simple mechanical model is sufficient for the description of diatomic molecules; however, it is no longer valid for more complex molecules and we have to resort to a somewhat more complex picture, which ultimately for large molecules can only be solved numerically. A number of excellent software packages exist that can calculate the normal mode vibrations of complex molecules within reasonable limits (typically up to the size of amino acids).

In inelastic light scattering there are generally speaking two possible interactions between photons and molecular vibration. A molecule can be momentarily in a vibrational ground state ν_0 from where it can pick up energy from the photon and remain in an exited vibrational state ν_1 after the scattering event. In this case, the photon has transferred part of its energy to the molecular vibration, leading to a shift to longer wavelengths. This event is described as Stokes-shifted Raman scattering (see Fig. 8.2a). Alternatively, the molecule could already be in the exited vibrational state ν_1. In this case, the incident photon can pick up energy from the vibration, returning the molecule to the vibrational ground state, and the energy of the scattered photon is now shifted to shorter wavelengths (anti-Stokes Raman scattering). This is depicted in Figure 8.2b. During the scattering event the molecular vibration and photon are momentarily indistinguishable due to Heisenberg's uncertainty principle. This short-lived state is what we call a "virtual state," and it is depicted by the combined energy level of the photon and the molecule in Figure 8.2.

8.2.1. The Induced Dipole Moment

In order to better understand the scattering process it is important to appreciate how light interacts with molecules. Since simple molecules are much smaller than optical wavelengths of light, a photon can be approximated as a uniform oscillating electric field. Much like the effect of the moon's gravity on oceanic tides, this oscillating electric field can affect the size and shape of a molecule. The result of this interaction is often described on two different scales: the microscopic scale, which describes the response of a single molecule, and the macroscopic scale, which describes the average response of bulk material. On the microscopic scale the induced electric dipole moment vector, **p**, is a measure of how much the electric field "tweaks" a molecule. At low field strengths, **p** is simply proportional to the electric field, **E**:

$$\mathbf{p} = \alpha \mathbf{E}. \tag{8.1}$$

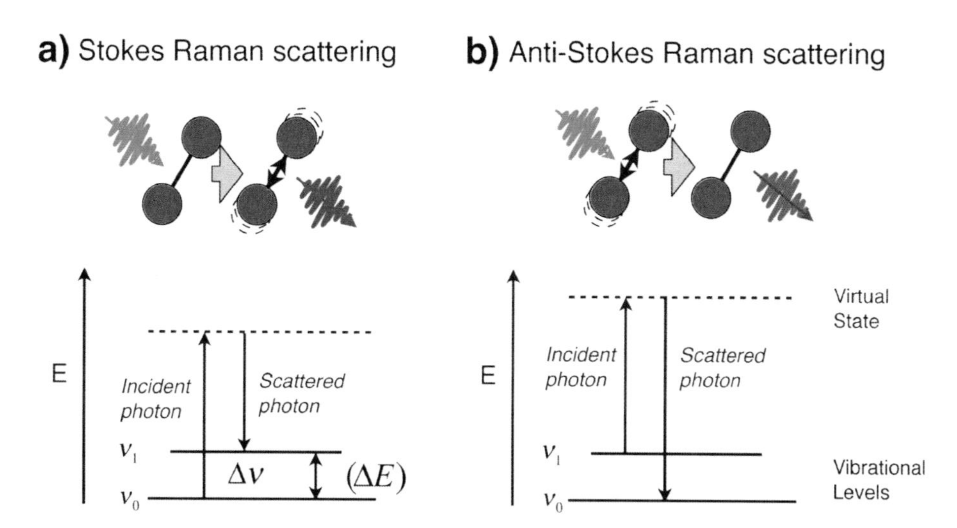

Figure 8.2. (a) Schematics of an inelastic photon scattering event where the photon deposits energy into the bond vibration, and its associated energy diagram. The scattered photon has lost energy relative to the incident photon. This process is called Stokes-shifted Raman scattering. (b) Schematics of an inelastic photon scattering event where the photon scatters off a vibrationally excited bond, and its associated energy diagram. The scattered photon has gained energy relative to the incident photon. This process is called anti-Stokes-shifted Raman scattering. Please visit http://extras.springer.com/ to view a high-resolution full-color version of this illustration.

Here, **p** is linearly dependent on **E** and is related to **E** by a second-rank tensor α, called the polarizability. If you are not familiar with tensors you can think of them as matrices with as many elements as there are possible interactions within a system. In the case of the polarizability tensor the elements correspond to the amount a molecule is deformed in the x, y, and z directions by light that is polarized in some arbitrary direction.

8.2.2. Polarizability and Raman Scattering

The polarizability, α, as defined above (a proportionality factor connecting the induced dipole moment, **p**, of a molecule and an oscillating electric field, **E**) can be thought of as a measure of how difficult it is for an electric field to stretch a particular chemical bond. Just like a rubber band, it is reasonable to imagine that the further the bond is stretched the more difficult it becomes to stretch it, i.e., α may change as a function of nuclear displacement. Nuclear displacement here means the amount by which the nuclei of the atoms that are partners in a chemical bond are further displaced from their normal position due to an incident electric field. In order to more clearly describe the relationship between α and the nuclear displacement we can expand the polarizability tensor α in a Taylor series with respect to the amount the bond has stretched. For the simplest case in which we consider only a single (e.g., stretching) vibration, this expansion can be written as

$$\alpha = (\alpha)_0 + \left(\frac{\partial \alpha}{\partial r}\right)_0 r + \frac{1}{2}\left(\frac{\partial^2 \alpha}{\partial r^2}\right)_0 r^2 + \dots, \qquad (8.2)$$

where $(\,)_0$ indicates the value at equilibrium and r is the coordinate along which the molecule is being stretched. We can further make another simplification by using the mechanical harmonic oscillator approximation, which only retains terms that are linearly dependent on the nuclear displacement, in this case just the first two terms. Dealing with the anharmonic, or higher-order terms, is outside the scope of this treatment. This approximation ensures that we stay within the limits of the simple mechanical model that we introduced in the previous section and the molecule acts like a system of balls and springs (see Figs. 8.1 and 8.2). Since we are assuming that this vibrating molecule will act just like a harmonic oscillator, we can also assume that the displacement along r will vary sinusoidally. Thus we can define an expression for the polarizability as the molecule vibrates:

$$\alpha = (\alpha)_0 + \frac{1}{2}\left(\frac{\partial \alpha}{\partial r}\right)_0 r_0 \sin(\omega_k t + \delta). \tag{8.3}$$

Here, r_0 is the equilibrium displacement, ω_k is the frequency of a normal mode of vibration of the molecule, and δ is a phase shift that accounts for the fact that in a large sample each molecule can vibrate with its own, discrete, but random phase. Assuming that the **E**-field also oscillates sinusoidally, $\mathbf{E} = E_0 \sin(\omega_l t)$, we can use Eqs. (8.1) and (8.3) to obtain an expression for the induced dipole moment:

$$\mathbf{p} = (\alpha)_0 E_0 \sin(\omega_l t) + \frac{1}{2}\left(\frac{\partial \alpha}{\partial r}\right)_0 r_0 E_0 \sin(\omega_k t + \delta)\sin(\omega_l t). \tag{8.4}$$

The first term in Eq. (8.4) describes elastic (Rayleigh) scattering since the frequency of the induced dipole moment will oscillate with frequency ω_l, which is the same as that of the incident **E**-field. The only selection rule for Rayleigh scattering is that the polarizability, $(\alpha)_0$, be nonzero at equilibrium, but this is virtually always the case, so this effect is typically observed. By using a simple trigonometric identity we can see that the second term corresponds to inelastic, or Raman, scattering. Given the identity

$$\sin(A)\sin(B) = \frac{1}{2}\left[\cos(A-B) - \cos(A+B)\right], \tag{8.5}$$

the second term in Eq. (8.4) then becomes

$$p_{\text{Raman}} = \frac{1}{4}\left(\frac{\partial \alpha}{\partial r}\right)_0 r_0 E_0 \left[\cos((\omega_l - \omega_k)t + \delta) - \cos(\omega_l + \omega_k)t + \delta\right]. \tag{8.6}$$

Equation (8.6) clearly shows that the interaction of the sinusoidally oscillating molecule with the sinusoidally varying **E**-field results in two inelastically scattering contributions: $(\omega_l - \omega_k)$, the Stokes shift, in which the incident laser photons are red-shifted by ω_k, and $(\omega_l + \omega_k)$, the anti-Stokes shift, in which the incident photons are blue-shifted by ω_k. Unique molecular modes of vibration will appear as distinct lines in a Raman spectrum that appear at $\omega_l \pm \omega_k$, which can then be used to infer structural and environmental information. The connection between vibrational energy levels and the Raman spectrum is shown schematically in Figure 8.3. Traditionally, however, a Raman spectrum is shown as the intensity of the scattered photons plotted

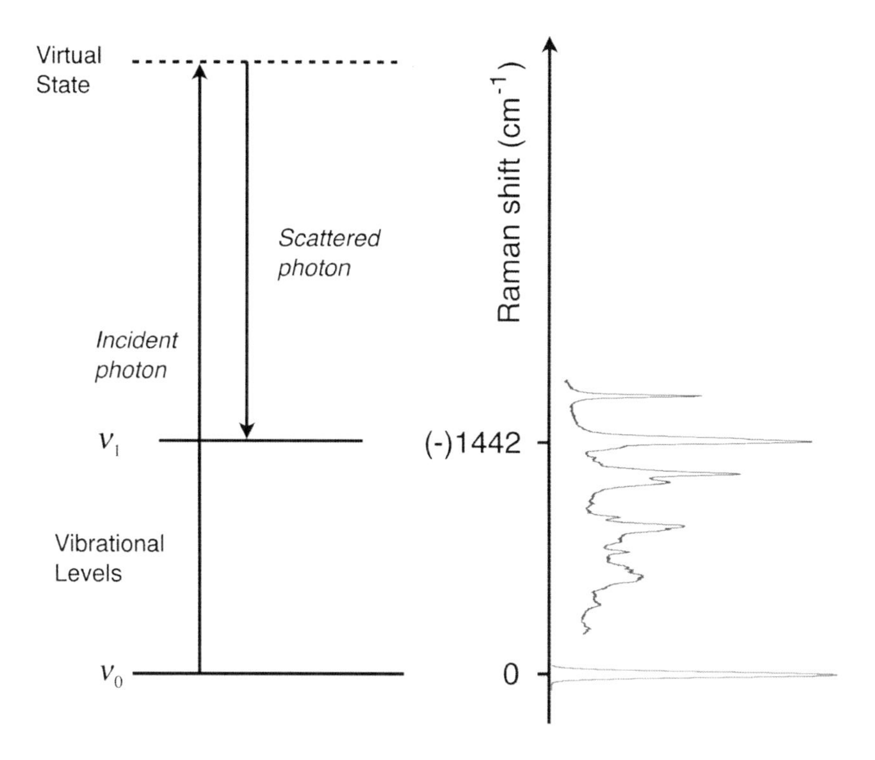

Figure 8.3. The energy diagram describing Raman scattering is directly related to the Raman spectrum of a molecule. Here, the energy of the incoming photons (e.g., from a laser) defines the vibrational ground state, while the scattered photons are shifted to the red (blue) depending on the bond vibration and the scattering process. This is shown schematically for the spectrum (red) obtained from an unsaturated fatty acid (oleic acid), where the lipid CH deformation mode results in a peak at 1442 cm^{-1}. Other peaks are not assigned to keep the energy diagram simple. Please visit http://extras.springer.com/ to view a high-resolution full-color version of this illustration.

against their Raman shift (here, ω_k). This is shown in Figure 8.4 on the example of the fingerprint spectrum obtained from a fatty acid, oleic acid. Here, both the anti-Stokes scattered spectrum as well as the Stokes-shifted spectrum are shown. While these spectra show a wealth of peaks that can all be attributed to specific vibrational modes within the molecule, it has to be noted that in most cases one would obtain even more peaks if all vibrational modes were Raman active. Raman activity, however, as briefly mentioned before, is defined by certain selection rules. Unlike infrared (IR) spectroscopy, which requires a changing dipole moment, the selection rule for Raman scattering requires only that at least one of the tensor terms, $(\partial\alpha/\partial r)_0$, be nonzero.

Although Raman spectroscopy can provide substantial chemical information, the signals are very weak with a differential scattering cross-section of only ~10^{-30} cm^2. By comparison, fluorescence typically has an absorption/excitation cross-section of ~10^{-16} cm^2. Several techniques have been developed to "boost" the signal, including resonance Raman scattering (RRS), surface-enhanced Raman scattering (SERS), and coherent Raman scattering (CRS). We will briefly discuss some of these in more detail later.

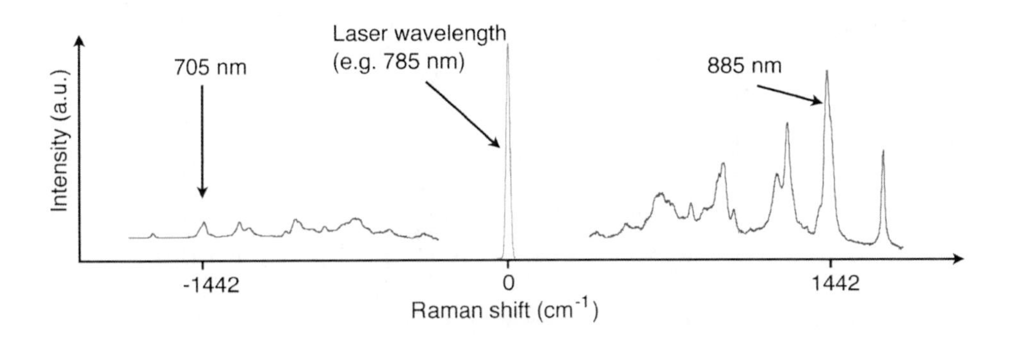

Figure 8.4. The fingerprint Raman spectrum of oleic acid results in both Stokes and anti-Stokes scattered photons. The distribution with which each fraction of photons is scattered to the left or right of the laser line (blue) depends on the temperature. Raman spectra are labeled in relative wavenumbers (cm^{-1}) to describe the spectrum in terms independent of the source wavelength. Please visit http://extras.springer.com/ to view a high-resolution full-color version of this illustration.

8.2.3. Spectroscopy

As mentioned, Raman spectroscopy provides rich chemical information. Most of this information is extracted based on peak assignments and relative intensities but can also involve peak shapes and dependence on the polarization of the laser light. Table 8.1 summarizes some of the main vibrational bands of interest for biomolecules. The strength of the Raman signal is related to the polarizability by the following expresssion:

$$I_{\text{Raman}} = NL\Omega\frac{\hbar}{2m\omega_k}\left(\frac{\partial\alpha}{\partial r}\right)_0^2\frac{\omega_l^4}{c^4}I_l. \tag{8.7}$$

Here, I_{Raman} and I_l are the intensities of the Raman line and the excitation laser, respectively; N is the number of bonds; L is the length of the focal volume; Ω is the solid angle over which the signal is collected; and m is the reduced mass of the vibrating molecule. An important technical note that is highlighted by Eq. (8.7) is the ω^4 dependence of the scattering signal. The result of this dependence is that Raman spectra generated with green (532 nm) or blue (488 nm) laser light are significantly stronger than the ones generated in the red part of the optical spectrum. This is schematically depicted in Figure 8.5, where the Raman spectrum of oleic acid is shown as it would appear when obtained with a green laser source or a near-infrared laser source. This potential advantage can, however, be offset by the excitation of autofluorescence, which can easily overwhelm the Raman scattering signal. This is particularly problematic for biological samples, where some amino acids and certain proteins exhibit particularly strong autofluorescence. Therefore, although shorter excitation wavelengths provide a stronger Raman signal, when analyzing biological samples, red (632.8, 647 nm) or near-infrared (785, 830 nm) laser light is often used to minimize contributions from autofluorescence and also maximize tissue penetration. It should also be noted that the Raman signal intensity is inversely proportional to

Table 8.1

Bond	Energy (cm^{-1})
C-H	2850-2960
=C-H	3020-3100
C=C	1650-1670
≡C-H	3300
C≡C	2100-2260
C-Cl	600-800
C-Br	500-600
C-I	500
O-H	3400-3640
C-OH	1050-1150
⋛C-H	3030
(benzene ring)	1600,1500
N-H	3310-3500
C-N	1030,1230
C=O	1670-1780
COOH	2500-3100
C≡N	2210-2260
NO_2	1540

the molecular vibrational frequency, ω_k, so that the greater the Raman shift the less efficient the scattering will be. This dependence is typically not limiting since most biologically relevant Raman lines are found at wavenumbers <3400 cm^{-1} but must be considered when attempting quantitative analysis. Another important effect that occurs with different excitation wavelengths is that the absolute Raman shift obtained from a molecule scales with the laser wavelength. This is due to the fact that photons at the blue end of the optical spectrum carry a higher energy, while red photons carry a lower energy. Since the energy required to excite a molecular bond vibration is fixed, the ratio of energy transferred from the photon relative to the photon energy is smaller for blue photons than it is for red photons. Thus, if a Raman spectrum is plotted on a wavelength scale it will appear more compressed when obtained with a blue light source rather than a red light source. This is also schematically shown in Figure 8.5. Because of this effect, Raman spectra are typically shown plotted against a relative Raman shift, which removes the dependence on excitation laser wavelength. This is also the appropriate place to mention that the polarity of the sign in front of the relative Raman shift is rather randomly assigned. Different authors will plot the Stokes-shifted Raman line to the left or right of the laser line and also change polarity rather randomly. Strictly speaking, the sign of the Stokes-shifted Raman peaks should be negative since these represent the amount of energy transferred from the photon to the molecule. Most authors will, however, plot the Stokes-shifted Raman peaks with positive labels because the majority of spectroscopists only collect Stokes-shifted Raman lines, and showing negative Raman shifts might appear ambiguous and difficult to understand for the average reader.

Equation (8.7) can be simplified by equating the change in polarizability, $\partial\alpha/\partial r$, to the differential scattering cross-section, $d\sigma/d\Omega$:

$$\frac{\partial\sigma}{\partial\Omega} = \frac{\hbar}{2m\omega_\upsilon}\left(\frac{\partial\alpha}{\partial q}\right)^2 \frac{\omega^4}{c^4}. \tag{8.8}$$

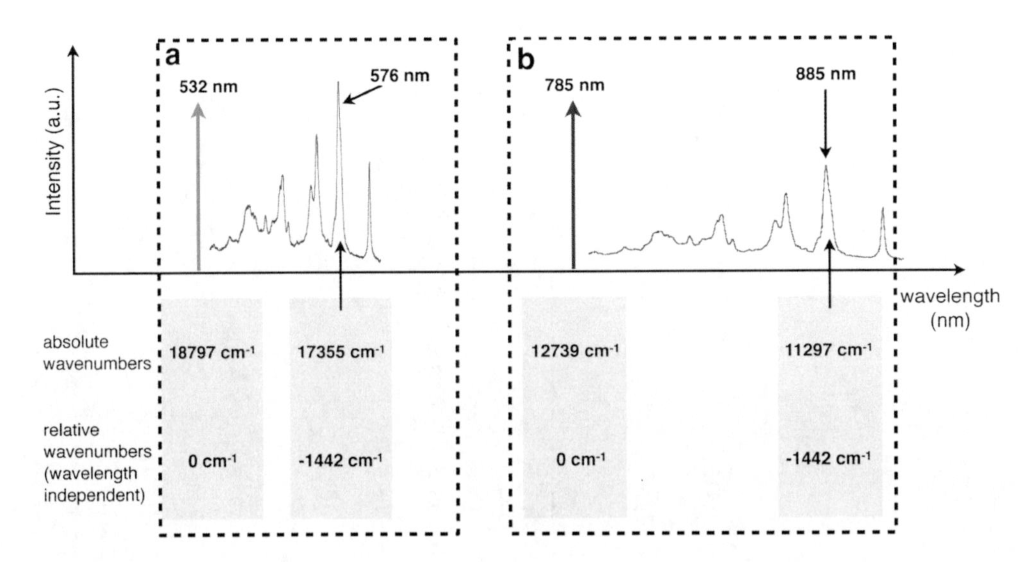

Figure 8.5. The strength of the Raman scattering process depends on the wavelength of the source. Since it is a scattering process, the intensity of the inelastically scattered light scales with λ^{-4}. This leads to Raman spectra obtained from the same compound to be more intense when probed with a green laser (box a) rather than a red laser (box b). Also, since green light has a higher photon energy than red light, Raman scattering off the same molecular bond can lead to more "condensed" or "stretched" spectra depending on the laser wavelength. Since Raman scattering by a specific bond always requires the same amount of energy, independent of the wavelength of the source, spectra obtained with a green laser are "denser" relative to those obtained with a red laser (when plotted against a wavelength axis). This is the reason why Raman spectra are typically plotted in terms of the energy difference between the source wavelength and the signal wavelength. These spectra are independent of the wavelength at which they were obtained. Please visit http://extras.springer.com/ to view a high-resolution full-color version of this illustration.

The intensity of the Raman line can then be expressed as a function of the scattering cross-section:

$$I_{\text{Raman}} = NL\Omega\left(\frac{\partial\sigma}{\partial\Omega}\right)I_l. \tag{8.9}$$

Given accurate estimates for the length of the focal volume, the collection angle, the laser intensity, and the bond density, the cross-section can be straightforward to determine. Alternatively, given an a priori knowledge of the cross-section (often found in look-up tables or published literature), the bond density can be determined with a high degree of accuracy. Although bond densities and cross-sections are useful to determine, they may not be sufficient for correctly assigning Raman lines or determining environmental effects on the Raman spectrum. For such determinations the symmetry of the vibration that results in the measured peak must often be determined.

If we return to Eq. (8.7) we can see the dependence of the scattering intensity on the tensor, $(\partial\alpha/\partial r)_0$. The elements of this tensor can be calculated for a molecule of interest using group theory, which can then be used to determine peak assignments. Usually these calculations are only useful when analyzing the spectrum of a neat compound or determining how that compound contributes to the spectrum of a noninteracting mixture. Often, unknown chemical products or interactions can contribute lines to a Raman spectrum, and in order to determine the origin of those lines the symmetry of the vibration must be determined without a priori knowledge.

If the excitation laser is polarized in the x direction and polarization-sensitive detection is used, then this symmetry can be determined. By measuring the intensity of a Raman peak, both parallel and perpendicular to the laser polarization, the depolarization ratio, ρ, can be determined:

$$\rho = \frac{I_{\text{perpendicular}}}{I_{\text{parallel}}}. \quad (8.10)$$

If ρ is found to be $0 \leq \rho < 3/4$ the vibration is said to be polarized or totally symmetric, and if $\rho = 3/4$ the vibration is said to be depolarized or non-totally symmetric. Classifying Raman peaks in this way provides an elegant method for determining the nature of the scattering signal. More complicated polarization geometries can be used to further characterize such parameters as reversal coefficients and degrees of circularity.

While fully characterized spectra are very useful and relatively straightforward to obtain in neat or bulk chemical samples, it is typically not possible to obtain them in complex chemical environments, such as the cytoplasm or nucleus of a living cell. In these cases, a useful alternative approach to take is difference spectroscopy, where spectra of a cell in an unaltered state and in an altered state are obtained and subtracted from each other to determine shifts in peak locations and changes in peak intensities due to controlled stimuli. The spectra from cells and tissue represent superimposed spectra from the many contributing constituents, the spectra of which can be obtained from neat compounds. In this case, cytometric measurements can also be made using spectral deconvolution algorithms, e.g., utilizing neural networks and multivariate statistical methods, such as principal component analysis.

An example of the spatial resolution afforded by confocal micro-Raman spectroscopy for the analysis of single cells is shown in Figure 8.6. Here the samples consist of frozen bull sperm cells that were thawed and diluted in salt solution, and then an aliquot was allowed to adhere to a calcium fluoride glass slide. Calcium fluoride and magnesium fluoride are excellent sample carriers for Raman analysis because they have essentially no Raman-active lines and thus add no background contributions throughout the entire fingerprint spectral range from of ~500–1800 cm^{-1}. These materials, are, however, relatively expensive to obtain as polished substrates in optical quality and they break rather easily. Quartz substrates work well, too, and can even be obtained in coverslip thickness (~170 μm), but add some background signals to the Raman spectrum that have to be carefully characterized and removed from the final sample spectra. After a few minutes the remaining solution of sperm cells that did not adhere to the substrate was washed off with deionized water. An autofluorescence image of the sperm cells was obtained by raster-scanning the CaF_2 substrate with a 488-nm laser beam focused to a diffraction-limited spot using a 100× long working distance air objective. The resulting image clearly displays the cells by their strong autofluorescence, as can be seen in Figure 8.6a. Sperm heads and tails are clearly visible by their different contrast and structure. The confocal Raman detection system is then used to probe different areas of the sperm cell (head and tail) to obtain Raman spectra from different parts of the cell. The spectra of the head reveals distinct peaks that can be assigned to

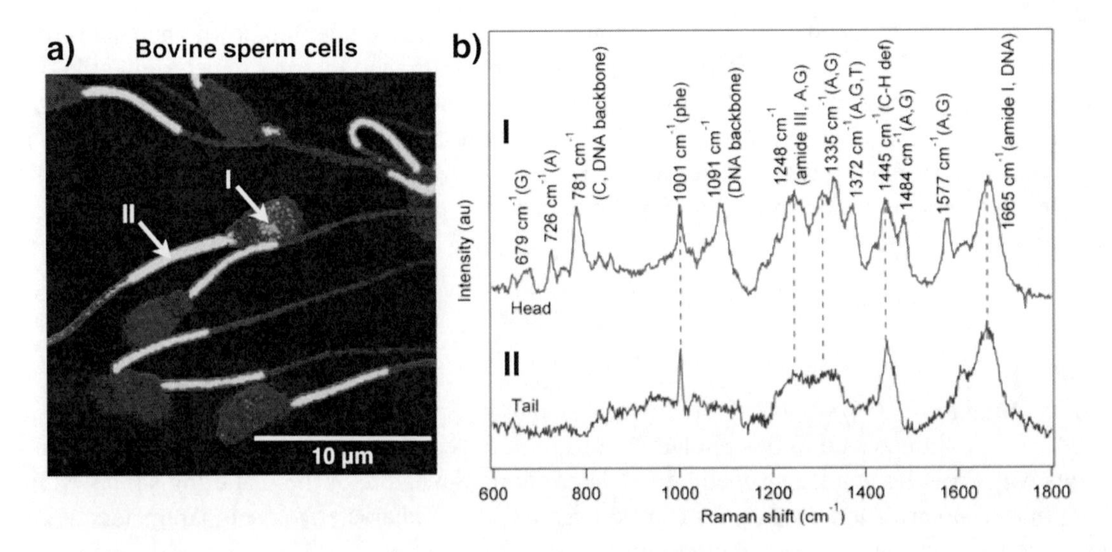

Figure 8.6. An example of Raman spectra observed from biological samples, in this case bovine sperm cells. (a) Intact, whole sperm cells are deposited onto a quartz cover slip and imaged in a confocal Raman microscope based on their autofluorescence. Specific areas, i.e., the sperm head (I) or sperm tail (II), can then be addressed by point spectroscopy. The resulting spectra and their assigned bond vibrations (see b) show that the sperm head is mostly composed of DNA and proteins (spectrum I), while the major constituent of the tail is just protein (spectrum II). Please visit http://extras.springer.com/ to view a high-resolution full-color version of this illustration.

DNA that is tightly packaged by protamine, an arginine-rich protein, whereas the tail only exhibits Raman peaks characteristic of lipids and protein and contains no evidence for DNA, as expected (see Fig. 8.6b). In this image, typical Raman acquisition times were approximately 5 minutes at 10-mW laser power.

Even though a blue laser was used to obtain these spectra, it should be noted that these data were first obtained about 10 years ago, when powerful near-infrared lasers were still scarce and expensive. Today these spectra would ideally be obtained with lasers in the near-infrared part of the optical spectrum, which reduces the risk of sample damage because biological samples have significantly less absorption in the near-IR. Contributions from autofluorescence excitation are also much lower or altogether absent in this wavelength range. This also results in much faster spectral acquisition times because the sample can tolerate higher laser powers.

8.2.4. Experimental Implementation of Micro-Raman Spectroscopy

One major advantage of Raman spectroscopy is the relative simplicity of its experimental implementation. All that is required to set up a basic Raman spectrometer is a single-line laser (e.g., an inexpensive helium–neon laser), a bandpass filter to remove residual laser lines, a steep-edge notch filter that blocks the elastically scattered and unscattered laser light, a dispersive element (grating or prism), and a sensitive detector or photographic film. The function of the filters, which are the most important elements of such a setup and require high-quality components, is briefly summarized schematically in Figure 8.7. In order to achieve the high spatial resolution desired for studying single cells or subcellular structures, this basic setup is, of course, complicated by the necessity of having to utilize confocal microscope platforms.

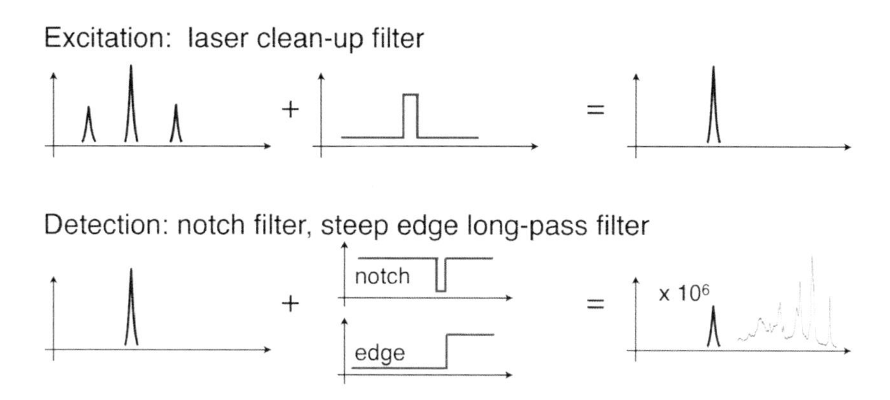

Figure 8.7. Modern Raman spectrometers require at least two sets of filters. A laser clean-up (narrow bandpass) filter on the excitation side removes plasma lines and other background contributions that can easily overwhelm the weak Raman signal. On the detection side a steep edge notch filter with high rejection for the laser line (optical density OD > 6) enables detection of both the Stokes and anti-Stokes-scattered light. Instead, lower-cost steep-edge longpass filters can also be used, but enable only detection of Stokes-shifted photons. Please visit http://extras.springer.com/ to view a high-resolution full-color version of this illustration.

As mentioned previously, Raman scattering is a rather inefficient process with typical Raman scattering cross-sections on the order of 10^{-30} cm^2 and with only 1 in 10^8 photons being inelastically scattered. It is this weak signal that has mainly limited its use as a tool for biological studies. Recent advances in instrumentation, however, have improved our ability to rapidly acquire high-quality reproducible data. Figure 8.8 depicts a typical micro-Raman setup that can be used for epi-detection of Raman signals using a CCD-based dispersive Raman spectrometer. This system has the sensitivity and spatial resolution to obtain Raman spectra even from single cells! In the following we will walk you through all the components, which should ideally enable you to build such a system yourself from scratch.

Starting with the laser source, any laser with reasonable long-term stability and relatively narrow spectral emission can be used. For reasons mentioned above, they should ideally be red or near-infrared lasers. The most common and convenient laser sources are helium–neon lasers (632.8 nm) or diode-pumped solid-state lasers. The beam profile of the laser beams should be Gaussian and might require shaping by spatial filters or fiber coupling. In our experience, it is also advantageous to add an optical isolator in the beam path behind the laser, which avoids potential backreflections from downstream optical elements being scattered back into the laser, which would lead to instability. The beam should then be expanded to a size that roughly fills or even slightly overfills the back aperture of the microscope objective used in your system. If you utilize oil-immersion microscope objective lenses with a numerical aperture of ~1.4, then this requires the beam diameter to be approximately 6 mm for most microscope lenses. Next, before the laser source can be used as a Raman excitation source, any residual background, such as laser plasma lines, need to be removed. This can be achieved by inserting narrow bandpass filters (3–10 nm width around the laser line) in the beam path. Several filters might have to be

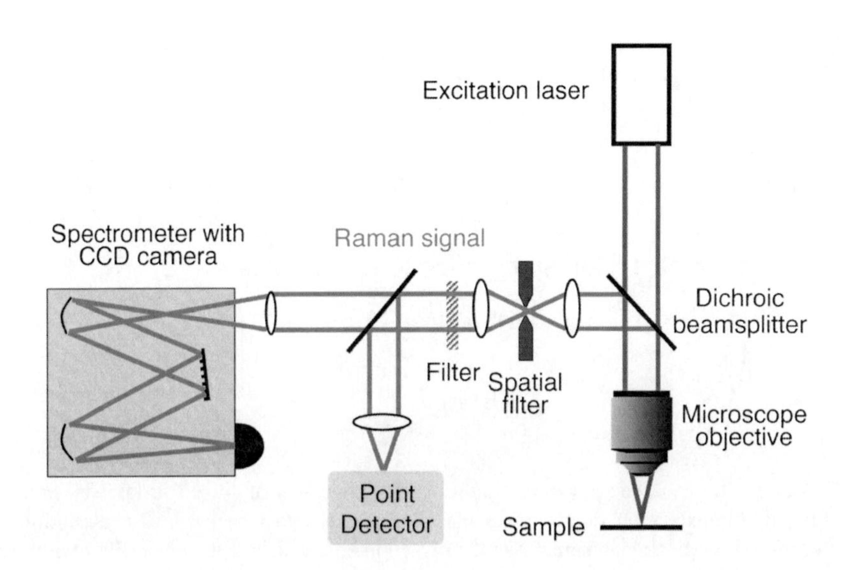

Figure 8.8. Diagram of a confocal micro-Raman spectroscopy setup. Light from a continuous wave laser passes a dichroic beamsplitter and is focused onto the sample through a microscope objective lens. The same lens collects the backscattered Raman photons, which are deflected to the left by the dichroic beamsplitter. A spatial filter assembly ensures that only Raman-scattered photons from the very focus of the laser beam at the sample are sent to the detectors, which could utilize a point detector for Raman imaging of a single vibrational mode, or a spectrometer for collecting the entire Raman spectrum. Please visit http://extras.springer.com/ to view a high-resolution full-color version of this illustration.

used in sequence in order to achieve the most efficient suppression of spectral side-wings. Now the light source is ready to be coupled into an optical microscope. Ideally, two mirrors in a periscope configuration should be used before coupling the laser beam into the microscope. These mirrors provide the flexibility required for steering the laser beams, which will be necessary to ensure that the beams are aligned along the optical axis of the microscope. Using an optical microscope as part of your setup is convenient because it allows for the use of other standard microscope components, such as phase contrast or differential interference contrast attachments to view samples, which greatly benefits biological applications. In confocal microscopy, the most widely used types of microscopes are inverted optical microscopes because they allow for the convenient placement of all optical components on an optical table below the sample level. Also, these types of microscopes are ideally suited for observing cells in culture dishes. Today, all microscopes have infinity-corrected optics, which allows for direct coupling of the excitation laser beam into the microscope. Infinity-corrected optics means that all the light beams inside the microscope are collimated, which allows microscope manufacturers to freely design their systems without having to pay particular attention to the distance between the microscope objective lenses and the eyepieces. This design also optimizes the performance of any filters inside the microscope because the beams traverse the filters at a perfect 90-degree angle. After guiding the laser beam into the microscope frame, it needs to be directed upwards toward the microscope objective. This is facilitated through the use of dichroic beamsplitters, which provide optimal response for each excitation wavelength. For spectroscopic applications, however, dichroic filters might not be the best choice because their spectral transmission profile is usually not flat, which can superimpose a structure on the sample spectra. In this case, if sufficient laser power is available, it is typically better to use a beamsplitter with a flat spectral profile. Such a beamsplitter reflects only a percentage of the laser beam up into the microscope objective, while the remaining light is transmitted and needs to be dumped and stopped in order to avoid reflections inside the microscope. Similarly, the Raman signal will be transmitted through the splitter with a loss. If a dichroic beamsplitter is chosen instead, its spectral profile has to be carefully considered—typically a steep edge as close as possible to the laser line and a spectral profile with the highest possible transmission and very few ripples are desired.

The laser beam is now entering the microscope objective lens. Objective lenses with a numerical aperture (NA) of 1.2 (water immersion lens), 1.3 (oil immersion lens), or better should be employed to obtain the smallest possible focus spot size and the highest possible collection efficiency. In order to obtain the tightest laser focus, the back aperture of the microscope needs to be efficiently filled by the laser beam. To avoid accidental eye exposure, the laser beam should only be viewed through CCD cameras. Sample scanning can be achieved by either scanning the sample itself utilizing piezoelectric translation stages, or galvanometric mirrors can be used to scan the beams, but then it is important to de-scan the beams (e.g., by tracing the signal back along the same path) before sending the signal into a spectrometer.

Once the laser beam is focused to the sample and Raman scattering is excited, the same microscope objective is typically used to collect the scattered light (epi-detection mode). This ensures that Raman-scattered light from the same sample volume is automatically collimated. This signal beam will then pass the dichroic mirror or beamsplitter and is then focused to an image spot by the tube lens built into the microscope. A longpass filter, bandpass filter, or notch filter can already be incorporated into the optical microscope to reject reflected laser light, or it can be added at any other point before the detectors. Retaining the microscope's tube lens in the beam path is a good idea because this allows for viewing the sample in regular microscopy mode. A confocal pinhole can then be added in the focus of the tube lens. Most commercial confocal microscopes use pinholes with diameters as small as 25 μm and less, which results in the highest possible spatial resolution. For Raman spectroscopy, however, somewhat larger pinhole diameters have to be used. Consider that the typical laser beam spot size at the sample is <0.5 μm. If a microscope objective with 100× magnification is used, the image size at the position of the pinhole will be ~50 μm. Thus, in order to ensure maximum signal-to-noise ratios, typical pinhole diameters should be >50 μm. This still provides efficient rejection of ambient light or spurious emission from other sample areas, while ensuring that most of the collected signal will reach the detectors. The pinhole has to be mounted on a stable *xyz* micromanipulation stage so that it can be properly aligned in the focus of the tube lens for maximum throughput. Typically, this pinhole needs to be checked regularly and possibly realigned. Alternatively, the signal can also be coupled into a multimode optical fiber with a core diameter of >50 μm. This typically allows for more compact and flexible setups. The Raman signal is then coupled into a monochromator with a cooled CCD camera attached by use of antireflection coated achromatic lenses. If necessary, the signal beam needs to be adjusted to the height of the monochromator input. Again, a periscope comes in handy for this purpose. Then an achromatic lens that ideally matches the F/# (ratio between the focal length and spatial dimension of the first spherical mirror in the monochromator) of the monochromator should be used to couple the signal beam into the spectrometer. Ideally, the mirrors and the grating in the spectrometer should be slightly underfilled, which sacrifices some spectral resolution for maximum throughput.

Most modern compact monochromators allow one to redirect the beam to another output port—typically by a built-in flip mirror. This enables you to mount a point-detector, such as a single-photon-counting avalanche photodiode (APD) or a photomultipler, to the second monochromator port, which can then be used in conjuction with the scanning stages to obtain confocal Raman images of the sample. If single-photon-counting APDs are used to detect the Raman signal, the signal has to be tightly focused onto the ~100 μm diameter detection area of the detector. Again, achromatic lenses deliver the best performance. The APD has to be mounted on another *xyz* micromanipulation stage in order to maximize the signal. Because APDs contain a Peltier cooler, the APD should first be mounted on a heat sink. A half-inch copper block has proven useful for this purpose. Inexpensive electronic counters with an adjustable time bin that continuously display the current detector count rate are ideal for optimizing spatial positioning. The photon counts can then be fed into a variety of commercial scan controllers or counter-timer boards for data acquisition.

Most of the initial alignment of such a system can be performed by using backreflections of the laser beam from the sample stage if sufficient laser power is available. Even the fine alignment of the detector position can be achieved with all filters in place if sufficient laser power is available to efficiently generate Raman-scattered light in the glass substrate.

8.2.5. Signal Calibration

Once able to obtain Raman spectra, as for example identified by their narrow lineshapes, we have to make sure that the spectra are carefully calibrated to verify their nature and enable quantitative analysis. Spectral calibration can most conveniently be achieved by utilizing spectral calibration lamps with a wide emission profile. Neon discharge lamps work well for the red part of the optical spectrum. The emission lines obtained from these lamps are well characterized and enable quick and easy wavelength calibration over a wide wavelength range. The conversion to the wavelength-independent wavenumber range used in Raman spectroscopy is then rather straightforward if the excitation laser wavelength is well known, as discussed earlier in this chapter. Spectral calibration lamps, however, cover a limited spectral range and typically do not extend their narrow emission lines to the near-infrared part of the spectrum. For this, and other reasons, it can be convenient to use a standard to calibrate the Raman spectrum. Ideally, such a standard should have a large number of distinct and intense Raman peaks that cover the entire spectral range of interest for biological materials, i.e., from ~200 to 3100 cm^{-1}. Organic solvents—such as toluene, ethanol, isopropylalcohol, etc.—or their mixtures have proven useful for this purpose. In our laboratory we typically use a neat toluene solution to quickly obtain reference spectra (to determine if the optical system is optimized, e.g., by establishing intensity values and by comparing Raman peaks to each other) and to calibrate the spectrometer. Figure 8.9 shows an example of the Raman spectrum of toluene obtained on a custom-build micro-Raman spectrometer. All the major Raman peaks are labeled with their wavenumber position, which should enable you to calibrate your own system with this standard. This particular spectrum of toluene was obtained by focusing ~300 μW of power from a red helium–neon laser (632.8 nm wavelength) through a 20× air objective lens into a solution of pure toluene. The solution was contained within a quartz cuvette with a teflon stopper. The Raman signal was collected in epi-direction, dispersed by a 600-line/mm grating blazed at 800 nm, and focused onto a CCD camera with 1340 pixels. The signal was integrated for 30 s.

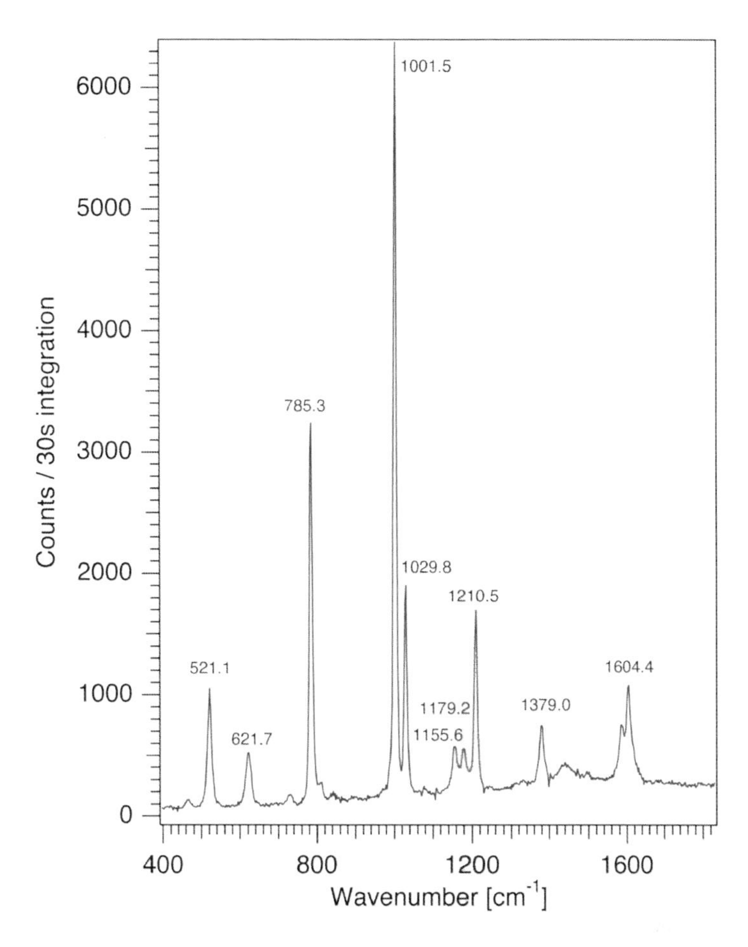

Figure 8.9. Calibration of Raman spectra typically utilize a well-characterized substance. Toluene and some alcohols, or mixtures thereof, of strong Raman-active modes spanning most of the fingerprint and high-wavenumber region. Their spectra are also relatively independent of temperature fluctuations, making them good calibration standards. The spectrum shown here was obtained from toluene by gently focusing ~300 μW of power from a red helium-neon laser (632.8 nm wavelength) into a solution of pure toluene. The spectrum was dispersed onto a CCD camera with 1340 pixels by a 600-line/mm grating and the signal was integrated for 30 s. Please visit http://extras.springer.com/ to view a high-resolution full-color version of this illustration.

8.2.6. Sample Preparation

As we learned above, pretty much any substance leads to Raman scattering. Very few materials have Raman peaks that are not relevant for the biological fingerprint spectrum and can thus be used as substrates (e.g., MgF_2 or CaF_2 as optically clear materials). Some materials have very broad and weak Raman spectra—e.g., quartz and water—and are still suitable as substrates. The majority, however, can lead to very significant background contributions, which is an issue in particular if very minute sample volumes are to be probed, as is the case in single-cell Raman spectroscopy. Here the use of confocal pinholes as detailed above helps reject background contributions; otherwise, substrates have to be very carefully chosen. In order to prepare samples for single-cell Raman experiments at room temperature, a few practical guidelines need to be followed. The substrates for your experiments have to be carefully selected. While glass substrates are acceptable when confocal detection techniques are employed, many glass substrates can contain impurities, leading to a distinct fluorescence background when excited with 785 nm. Typically, quartz glass is a good, albeit an expensive choice. Alternatively, one can try to obtain samples or to purchase small quantities of several different types of microscope coverslips from different vendors to determine which substrate has the lowest background. Also, before these substrates are used in an experiment they need to be carefully cleaned to remove organic residues and contaminants. In our day-to-day operations we have found a fairly effective non-hazardous cleaning protocol to be quite efficient. This protocol is based on a proprietary alkaline cleaning agent (Hellmanex, made by Hellma, a German company; distributed in the U.S. by Fisher Scientific). Ultrasonication of glass coverslips in a 2% solution of Hellmanex in DI water for approximately 20 minutes and subsequent rinsing and sonication in DI water works very well. Cleaned coverslips can be stored in DI water for up to 1 week. Sample aliquots of 10 μl or more are then dispensed with sterile plastic pipettes onto these substrates.

Alternatively, for single-cell Raman measurements, immobilization of single cells by optical tweezers has proven to be quite efficient. Optical trapping, or "laser tweezers," has become an established method in molecular and cell biology, involving the use of a laser beam to physically manipulate microparticles. In biology, it has been used for a wide array of applications, ranging from the micromanipulation of organelles within cells, measuring compliance of bacterial tails, stretching of DNA molecules, manipulation of sperm and application for in-vitro fertilization, and studying the forces of kinesin molecular motors. This technique is based on the principle that laser photons can impart their momentum onto the microparticle, and any refraction, scatter, or reflection of the photons results in a reactive force on the object. In its simplest arrangement, a single tightly focused laser beam with a Gaussian profile can impart both transverse and axial forces to a particle in the laser focus, resulting in a stable three-dimensional optical trap that immobilizes the particle within the focus and permits particle manipulation. The combination of both laser tweezers and Raman spectroscopy is straightforward since the same laser source and microscope objective used for confocal Raman spectroscopy can be used for optical trapping. The only requirement is that the laser beam used for optical trapping have sufficient power (typically >1 mW), and that it not harm the cells to be trapped (which requires careful control measurements). The most attractive features of this technique are minimization of sample preparation time and improved spectral acquisition time. It also lifts the cells of interest far from the substrate to further reduce background contributions. The only background sources that then remain and have to be dealt with are any potential autofluorescence from the cell and potential background signals from the aqueous suspension in which the cells are prepared. In our experience, phosphate buffered saline (PBS) makes an excellent medium in which cells can be prepared.

8.2.7. Advanced Forms of Raman Spectroscopy

As mentioned in the previous sections, a particular limitation of non-resonant Raman spectroscopy is the low Raman scattering cross-section of most molecular bonds, which, in order to acquire a sufficient number of photons for spectroscopic and imaging applications, requires long signal integration times. If the Raman resonance is excited within an electronic resonance (when the excitation wavelength or the energy of the Raman-scattered photon fall within an absorption band of the molecule), we speak of resonant Raman scattering. This combination can enhance Raman signals by up to 10^6, but typically requires visible or ultraviolet excitation that is resonant with an electronic transition of the analyte, which may potentially harm the biological sample. Also, even though the Raman signal is enhanced in this case, strong fluorescence background contributions excited by the same laser typically mask the Raman signals and have to be removed by elaborate techniques, such as optical time-gating. Fortunately, there are a few other options that enable the amplification of the very process of Raman scattering typically by enhancing the mechanism through which photons interact with the sample. Here, we will briefly summarize two of these that have significant applications for biological samples: surface-enhanced Raman scattering (SERS) and coherent anti-Stokes Raman scattering (CARS).

8.2.7.1. Surface-Enhanced Raman Scattering (SERS)

Surface-enhanced Raman scattering can enhance Raman signals by many orders of magnitude if the analyte is immobilized on noble metal surfaces (e.g., gold, silver, platinum) with a surface roughness on the nanometer scale or if the analyte is brought in contact with gold or silver nanoparticles. In this case, the strongly enhanced Raman scattering is attributed to specific interactions between the molecule and the noble metal surface enabled by optical excitations of the free electrons at the metal surface. This phenomenon was first discovered in 1977 by two groups working independently of each other [4]. By studying the adsorption of molecules to a roughened silver electrode, these two groups noticed an increase in Raman scattering by up to six orders of magnitude.

This large increase in the observed Raman scattering is the result of two enhancement mechanisms: an electromagnetic enhancement and a chemical enhancement [5,6]. The electromagnetic enhancement results from a localization of the local electric field at the particle surface by so-called surface plasmons. "Surface plasmons" is a term used to describe the excitation of collective oscillations of the conduction electrons near the metal surface. These electronic waves can be excited by coupling electromagnetic waves to the metal surface, which typically requires special optical configurations for coupling to extended surfaces, but which is particularly simple in the case of nanoparticles. Since surface plasmons are longitudinal waves based on coherent oscillations of the local surface electron density and optical waves are transverse waves, exciting surface plasmons in an extended surface requires coupling of electromagnetic waves to the surface typically by means of a glass prism. This allows the impulse of the electromagnetic wave to match the impulse of a surface plasmon, and thus its excitation. Precise matching of the impulse occurs at either a specific angle of incidence of the electromagnetic wave or a specific wavelength. The propagation of surface plasmons along the metal surface likewise results in an evanescent field near the surface. In the case of nanoparticles, these matching conditions are fulfilled much more easily because of the physical size and shape of such particles. The evanescent field at the metal–air interface, in addition to charge-transfer and resonant enhancements, then leads to a significantly enhanced Raman response if molecules are placed within the evanescent field.

The pure enhancement of the Raman scattering cross-section by surface plasmons has been shown to produce an enhancement of up to 10^{12}. The chemical enhancement arises from the formation of a charge-transfer complex that increases the polarizability of the molecule and produces an increase in the Raman scattering by another 1 to 2 orders of magnitude. Metal nanoparticles can be engineered, deposited, and processed on the nanometer scale in a fairly well-controlled manner, which provides a convenient means for efforts in trying to quantitatively control the SERS process. Through recent work with nanoparticles it has become increasingly clear that specific electromagnetic enhancing sites (e.g., resonant dipole structures) are needed to focus the electric field to small spatial scales [7,8]. In some special cases—e.g., for silica–gold nanoshells or hollow gold or silver nanospheres—the enhancement effect can also be directed to the entire surface of such an engineered nanoparticle [9,10]. In this case, the SERS signature is due to an average response of all molecules bound to the surface of the shelled particle rather than a single hotspot or junction, but can be further pronounced by the formation of dimerized particles [8].

The large increase in Raman scattering afforded by surface enhancement has significantly extended the range of applications for Raman spectroscopy. Of particular interest are applications in biosensing that directly detect molecular binding events, molecular reactions, or changes in molecular structure due to a change in their local biochemical environment [11,12]. Direct detection of many molecular species works well with SERS, but it is problematic because different chemicals compete for adsorption to the nanoparticle surface and the Raman signature obtained in this case might not always reflect the actual solution concentration of the molecules. Partitioning or conditioning of the surface can often provide higher specificity. An example for recent success in such an area has been the long-standing challenge of detecting glucose by SERS. Glucose by itself does not adsorb to noble metal surfaces. Recently, however, by adsorbing a partioning layer of long-chain alkanes onto the surface of a nanosphere substrate, a group at Northwestern University managed to selectively concentrate glucose within the range of the electromagnetic field enhancement and detect it by SERS [13–15]. This has now even led to the demonstration of in-vivo glucose sensing [16]. A slightly different approach to sensing with SERS was demonstrated even earlier and is now gaining significant popularity [12]. SERS particles have been turned into tags with highly multiplexed optical signatures by binding different molecules to the surface of solid nanoparticles and then embedding them in a glass matrix or some other biocompatible coating. The molecules—typically variations of fluorescent dyes—that provide additional resonant Raman enhancement produce a specific Raman spectrum when probed by laser Raman spectroscopy. Each molecule has a slightly different Raman spectrum and is unique to one type of nanoparticle, so that these tags can now be easily distinguished by their different spectra, opening up the possibility to probe hundreds or possibly thousands of particles simultaneously. The surrounding coating makes these particles soluble and inert and, after functionalizing the particles with antibodies or DNA targets, turns them into highly molecularly specific optical probes. This has been demonstrated by several groups over the last couple of years, who used this approach to, among other things, identify cancer genes, for immunoassays, or to identify viral DNA [17–19]. More recently, because of their brightness, robustness, and highly specific signature, SERS tags have also been used for in-vivo detection of cancer cells in live animals with extremely high sensitivity [20].

8.2.7.2. Coherent Anti-Stokes Raman Scattering (CARS) Microscopy

Another scheme to enhance Raman signals from biological samples that has gained much attention during the last decade and has made particularly strong inroads in biological imaging is coherent Raman scattering (CRS). Coherent Raman scattering is based on the broader concept of pump–probe spectroscopy. Here, a strong, typically short-pulsed, laser beam induces Raman scattered photons, while another, typically weaker, beam probes all molecular bonds that undergo the same type of Raman scattering simultaneously, leading to coherent emission of Raman-shifted photons. Similarly to spontaneous Raman scattering, this concept can be used to result in Stokes-shifted photons (stimulated Raman scattering (SRS)), as well as anti-Stokes-shifted photons (coherent anti-Stokes Raman scattering (CARS)). For the remainder of this chapter and for brevity, we will only discuss CARS microscopy, the currently more widely used CRS scheme, which has recently also become available in commercial implementations.

Even though spontaneous Raman scattering can be applied to image biological specimens such as cells and tissue sections, acquisition times can vary from several minutes up to several hours due to the rather weak signals [21–24]. Thus, the observation of fast dynamics within cells such as vesicular trafficking or diffusion is clearly not possible, and sample viability can also be compromised after such long data acquisition times. The use of CARS for chemically selective imaging of molecular constituents within a biological specimen circumvents these problems due to a significant increase in signal, which decreases the time necessary to acquire an image. This gain in signal can even enable CARS imaging at video rate speeds in vivo and ex vivo [25–39].

CARS is a nonlinear optical (NLO) phenomenon where the CARS signal arises from third-order induced polarization, $P^{(3)} = \chi^{(3)}E_pE_sE_p$, in the sample. As such, it occurs in any material exhibiting a nonzero third-order susceptibility $\chi^{(3)}$. Figure 8.10a illustrates the CARS process. When the frequency difference between pump (ω_p) and Stokes fields (ω_s) is tuned to (or is in resonance with) the frequency of a Raman-active vibration, $\omega_{vib} = \omega_p - \omega_s$, of a molecular bond, the anti-Stokes signal, ω_{as}, is significantly enhanced. When on resonance with ω_{vib}, all the molecular bonds represented by this vibrational frequency in a given excitation volume oscillate in phase, i.e., coherently. Scattering of a second photon at the pump laser frequency ω_p will then generate a CARS signal with frequency $\omega_{as} = 2\omega_p - \omega_s$. Since CARS is a four-wave mixing process, it is crucial that the phase-matching condition is met between all of the associated fields. Indeed, in the first demonstration of CARS microscopy a non-collinear beam geometry was employed [40]. This, however, severely limited the use of CARS in microscopic applications, as alignment was tedious and difficult to maintain. However, a renaissance in CARS was experienced when the relaxation of the phase matching condition was introduced under the conditions of highly focused laser beams [41]. This was accomplished through the use of collinear beams and high numerical aperture objective lenses, which improves the spatial resolution for imaging and decreases the average power necessary to achieve sufficient CARS signals. The very nature of the CARS signal generation process, however, does also lead to a non-resonant four-wave mixing signal due to the electronic properties of the material, which presents a significant source of background signal in any CARS application.

CARS spectroscopy and microscopy offers orders of magnitude ($\sim 10^4$) stronger vibrational signals due to the coherent nature of this process. The strong CARS signal enables rapid imaging of living biological samples without the use of exogeneous labeling. Living cells have, for example, been imaged using CARS microscopy based on their protein, DNA, and lipid distributions [41,42]. Because of its nonlinear optical nature, CARS signals are only generated within the small laser focus volume, allowing for high-spatial-resolution imaging and three-dimensional sectioning capability. An additional benefit is that, since the CARS signal is blue-shifted with respect to the excitation laser wavelengths, autofluorescence will not typically

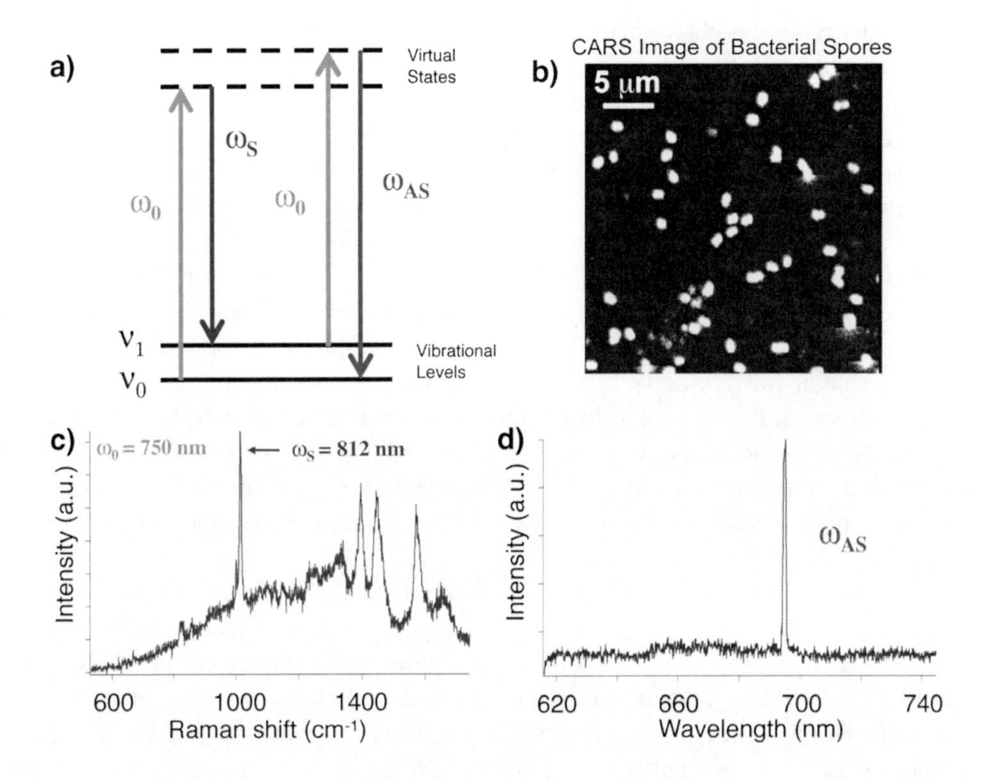

Figure 8.10. Rapid Raman imaging by coherent Raman scattering. (a) Energy diagram describing the nonlinear optical process of coherent anti-Stokes Raman scattering. To image bacterial spores by their intense 1013 cm^{-1} vibration from calcium dipicolinate, two short-pulsed lasers are synchronized and tuned to probe this molecular mode. (c) A pump laser at 750-nm wavelength (for instance) provides two photons that, when combined with a single photon from another laser at 812 nm, coherently probe the 1013 cm^{-1} vibration. (d) The resulting signal is an intense line at 697 nm composed of a Raman-resonant anti-Stokes Raman signal originating from the molecular vibration, as well as a non-resonant background signal due to four-wave mixing of the photons in the material. (b) The intense signal can then be utilized to obtain an image of bacterial spores prepared on a glass substrate. Please visit http://extras.springer.com/ to view a high-resolution full-color version of this illustration.

interfere with the detection of the CARS signal. These properties are summarized in the image of bacterial spores in Figure 8.10b. Here bacterial spores were deposited onto the surface of a glass coverslip and allowed to dry. The lasers of the CARS system were tuned to the 1013 cm^{-1} vibration (see Fig. 8.10c), which is attributed to calcium dipicolinate, a chemical present in significant amounts in the cortex of most *Bacillus* spores. Pump and laser wavelengths at 750 and 812 nm generated a CARS signal at 697 nm (see Fig. 8.10d). Laser powers were roughly 20 and 10 mW. The image (256 × 256 pixel) was acquired by scanning at a 2 ms/pixel rate.

Because CARS is a nonlinear optical effect, the use of pulsed laser sources with high peak power is necessary. Unlike two-photon fluorescence excitation, however, which typically requires the use of femtosecond pulsed laser sources to achieve the necessary power density for two-photon absorption, CARS only requires the use of picosecond laser pulses. In fact, it was found that the use of picosecond pulses is advantageous due to the narrow spectral width of the Raman transitions [43,44]. The broad spectral distribution of femtosecond laser pulses leads to an increase of the non-resonant signal, and conversely, the use of the spectrally narrower pulses decreases it. Even with the implementation of collinear beam excitation, CARS microscopy is somewhat challenging to realize in practice. Both the Stokes and pump–probe beams must be spatially and temporally overlapped at the sample. Furthermore, to cover the broad distribution of Raman vibrational transitions that are pertinent to biological systems, a laser source with a large range of wavelength tuning is required. This was first accomplished through the use of tunable dye lasers [40], and then with Ti–sapphire lasers [41], both of which have to be synchronized to maintain the temporal relationship between pulses at different wavelengths. In the past few years, with the development of new and more stable optical parametric oscillators (OPO), the CARS microscopy community has seen a transition to the use of OPO sources pumped, e.g., by a 1064-nm laser that can also serve as the Stokes beam [25,45–47]. Thus, a simple optical delay path inserted in the beam path of the Stokes beam is all that is necessary to account for the extra distance the pulses propagate inside the OPO cavity to achieve temporal overlap at the sample. CARS excitation is most commonly achieved with sources in the infrared wavelength range. This has several advantages for biological specimens. Since the light sources are at longer wavelengths, there is less absorption in the material and thus less potential for sample damage [48]. Furthermore, the longer wavelengths can penetrate deeper into thick tissue samples [49]. The background due to one-photon excited fluorescence is also substantially reduced. However, the use of longer wavelengths is not without its disadvantages. Specifically, it decreases the ultimate optical resolution that can be achieved. Due to the nonlinear optical nature of the CARS process, a slight improvement in the resolution is still achieved, especially in the axial directions, and thus CARS also has inherent optical sectioning capability. Finally, CARS can be combined with several other nonlinear optical imaging modalities, including two-photon excited fluorescence (TPEF) and second harmonic generation (SHG), to provide additional unique information [26,50].

8.3. SUMMARY

Raman spectroscopy was first discovered over 70 years ago by Sir Chandrasekhara Venkata Raman, but it has only found major biological applications over recent decades, in particular after the invention of the laser. Since biological processes are accompanied by changes in intracellular biochemistry, resulting both in compositional and conformational changes, vibrational spectroscopy is ideally suited for monitoring such processes. Spontaneous Raman spectroscopy has been used to obtain a variety of intrinsic disease markers, for example, for cancer and cardiovascular disease. Its sensitivity is high enough to enable cancer detection at the single-cell level and even allows for the chemical analysis of submicrometer-sized particles, such as triglyceride-rich lipoproteins. The only caveat to Raman spectroscopy is the weak nature of the process, which requires long signal acquisition times. This can be overcome by an enhancement mechanism, such as surface-enhanced Raman scattering, or, almost more importantly because of its significant implications for biochemical imaging, coherent Raman scattering.

CARS microscopy enables rapid spectral data acquisition with high temporal resolution with the potential for monitoring fast (millisecond) dynamics of individual biological particles. Raman-based biophysical methods are currently experiencing a very high level of activity, resulting in continued improvements in instrumentation, techniques, and applications. There is much in store for this technique in basic biophysical research.

PROBLEMS

8.1. By making a number of generous assumptions, estimate the detection limit (concentration limit) of spontaneous micro-Raman spectroscopy and discuss factors that might improve this limit (e.g., increased signal accumulation time, higher laser power). Remember, the Raman scattering cross-section of a typical Raman-active bond is (on average) $\sigma \sim 10^{-30}$ cm^2. Assume that a reasonable (safe) laser power for most substances is 1 mW at 632.8 nm focused to a diffraction-limited spot.

8.2. Assuming that the Raman peak for the CH stretching vibration appears at approximately 2900 cm^{-1}, estimate the new location of the peak if the hydrogen (H) is replaced with deuterium (D). Use the fact that deuterium has approximately twice the mass of hydrogen.

8.3. SERS results in stronger signals by essentially increasing the scattering cross-section for inelastic light scattering (the probability of photons to become Raman-scattered). Assume you have 100,000 molecules illuminated in a tightly focused laser spot (approximated by a cube with a volume of 1 μm^3), that the Raman scattering cross-section of the molecules is 10^{-30} cm^2, and that the laser produces 4×10^{17} photons per second. You can also assume for this problem that you collect all of the scattered photons.

a. Use the following variation of Eq. (8.9) to estimate the efficiency of Raman scattering:

$$\Phi_{\text{Raman}} = \Phi_{\text{Laser}} N_{\text{M}} L \Omega \frac{\partial \sigma}{\partial \Omega},$$

where the photon flux, Φ, in photons/sec/cm^2 is used instead of intensity.

b. If surface enhancement increases the scattering efficiency by 10^{10}, estimate by how much the scattering cross-section is increased by SERS for every molecule. To solve this problem assume that the 100,000 molecules are now adsorbed onto a single gold sphere (diameter of 50 nm) in your focused laser spot.

FURTHER READING

For further reading we suggest the reader consult some of the latest review papers on Raman spectroscopy. These can frequently be found in the *Journal of Raman Spectroscopy* or by searching for appropriate phrases on *Web of Science*. Unfortunately, the number of textbooks related to biological Raman spectroscopy is fairly small. A very good and more recent introductory books is [2], and a specific chapter on micro-Raman imaging can be found in the frequently updated [3].

REFERENCES

1. Raman CV, Krishnan KS. 1928. A new type of secondary radiation. *Nature* **121**:501–502.
2. Smith E, Dent G. 2005. *Modern raman spectroscopy: a practical approach*. New York: Wiley.
3. Pawley JB, ed. 2006. *Handbook of biological confocal microscopy*. 3rd ed. New York: Springer.

4. Jeanmaire DL, Vanduyne RP. 1977. Surface Raman spectroelectrochemistry, 1: heterocyclic, aromatic, and aliphatic-amines adsorbed on anodized silver electrode. *J Electroanal Chem* **84**:1–20.
5. Otto A, Mrozek I, Grabhorn H, Akemann W. 1992. Surface-enhanced Raman-scattering. *J Phys Condens Matter* **4**:1143–1212.
6. Moskovits M. 1985. Surface-enhanced spectroscopy. *Rev Mod Phys* **57**:783–828.
7. Michaels AM, Jiang J, Brus LE. 2000. Ag nanocrystal junctions as the site for surface-enhanced Raman scattering of single rhodamine 6G molecules. *J Phys Chem B* **104**:11965–11971.
8. Talley CE, Jackson JB, Oubre C, Grady NK, Hollars CW, Lane SM, Huser TR, Nordlander P, Halas NJ. 2005. Surface-enhanced Raman scattering from individual Au nanoparticles and nanoparticle dimer substrates. *Nano Lett* **5**:1569–1574.
9. Jackson J, Halas N. 2004. Surface-enhanced Raman scattering on tunable plasmonic nanoparticle substrates. *PNAS* **101**:17930–17935.
10. Schwartzberg AM, Oshiro TY, Zhang JZ, Huser T, Talley CE. 2006. Improving nanoprobes using surface-enhanced Raman scattering from 30-nm hollow gold particles. *Anal Chem* **78**:4732–4736.
11. Haynes CL, McFarland AD, Van Duyne RP. 2005. Surface-enhanced Raman spectroscopy. *Anal Chem* **77**:338A–346A.
12. Vo-Dinh T, Yan F, Wabuyele MB. 2005. Surface-enhanced Raman scattering for medical diagnostics and biological imaging. *J Raman Spectrosc* **36**:640–647.
13. Lyandres O, Shah NC, Yonzon CR, Walsh Jr JT, Glucksberg MR, Van Duyne RP. 2005. Real-time glucose sensing by surface-enhanced Raman spectroscopy in bovine plasma facilitated by a mixed decanethiol/mercaptohexanol partition layer. *Anal Chem* **77**:6134–6139.
14. Stuart DA, Yonzon CR, Zhang X, Lyandres O, Shah NC, Glucksberg MR, Walsh JT, Van Duyne RP. 2005. Glucose sensing using near-infrared surface-enhanced Raman spectroscopy: gold surfaces, 10-day stability, and improved accuracy. *Anal Chem* **77**:4013–4019.
15. Yonzon CR, Haynes CL, Zhang X, Walsh Jr JT, Van Duyne RP. 2004. A glucose biosensor based on surface-enhanced Raman scattering: improved partition layer, temporal stability, reversibility, and resistance to serum protein interference. *Anal Chem* **76**:78–85.
16. Stuart DA, Yuen JM, Shah N, Lyandres O, Yonzon CR, Glucksberg MR, Walsh JT, Van Duyne RP. 2006. In vivo glucose measurement by surface-enhanced Raman spectroscopy. *Anal Chem* **78**:7211–7215.
17. Vo-Dinh T, Allain LR, Stokes DL. 2002. Cancer gene detection using surface-enhanced Raman scattering (SERS). *J Raman Spectrosc* **33**:511–516.
18. Allain LR, Vo-Dinh T. 2002. Surface-enhanced Raman scattering detection of the breast cancer susceptibility gene BRCA1 using a silver-coated microarray platform. *Anal Chim Acta* **469**:149–154.
19. Cao YC, Jin R, Mirkin CA. 2002. Nanoparticles with Raman spectroscopic fingerprints for DNA and RNA detection. *Science* **297**:1536–1540.
20. Qian X, Peng XH, Ansari DO, Yin-Goen Q, Chen GZ, Shin DM, Yang L, Young AN, Wang MD, Nie S. 2008. In vivo tumor targeting and spectroscopic detection with surface-enhanced Raman nanoparticle tags. *N Biotechnol* **26**:83–90.
21. Crane NJ, Morris MD, Ignelzi MA, Yu GG. 2005. Raman imaging demonstrates FGF2-induced craniosynostosis in mouse calvaria. *J Biomed Opt* **10**:8.
22. Kazanci M, Wagner HD, Manjubala NI, Gupta HS, Paschalis E, Roschger P, Fratzl P. 2007. Raman imaging of two orthogonal planes within cortical bone. *Bone* **41**:456–461.
23. Zhang GJ, Moore DJ, Flach CR, Mendelsohn R. 2007. Vibrational microscopy and imaging of skin: from single cells to intact tissue. *Anal Bioanal Chem* **387**:1591–1599.
24. Zhang GJ, Moore DJ, Sloan KB, Flach CR, Mendelsohn R. 2007. Imaging the prodrug-to-drug transformation of a 5-fluorouracil derivative in skin by confocal Raman microscopy. *J Invest Dermatol* **127**:1205–1209.
25. Burkacky O, Zumbusch A, Brackmann C, Enejder A. 2006. Dual-pump coherent anti-Stokes Raman scattering microscopy. *Opt Lett* **31**:3656–3658.
26. Fu Y, Wang HF, Shi RY, Cheng JX. 2007. Second harmonic and sum frequency generation imaging of fibrous astroglial filaments in ex vivo spinal tissues. *Biophys J* **92**:3251–3259.
27. Cheng JX, Jia YK, Zheng GF, Xie XS. 2002. Laser-scanning coherent anti-Stokes Raman scattering microscopy and applications to cell biology. *Biophys J* **83**:502–509.
28. Cheng JX, Volkmer A, Xie XS. 2002. Theoretical and experimental characterization of coherent anti-Stokes Raman scattering microscopy. *J Opt Soc Am B* **19**:1363–1375.

29. Cheng JX, Xie XS. 2004. Coherent anti-Stokes Raman scattering microscopy: instrumentation, theory, and applications. *J Phys Chem B* **108**:827–840.
30. Djaker N, Lenne PF, Marguet D, Colonna A, Hadjur C, Rigneault H. 2007. Coherent anti-Stokes Raman scattering microscopy (CARS): instrumentation and applications. *Nucl Instrum Methods Phys Res A* **571**:177–181.
31. Evans CL, Potma EO, Puoris'haag M, Cote D, Lin CP, Xie XS. 2005. Chemical imaging of tissue in vivo with video-rate coherent anti-Stokes Raman scattering microscopy. *Proc Natl Acad Sci USA* **102**:16807–16812.
32. Huff TB, Cheng JX. 2007. In vivo coherent anti-Stokes Raman scattering imaging of sciatic nerve tissue. *J Microsc Oxford* **225**:175–182.
33. Kano H, Hamaguchi H. 2005. Vibrationally resonant imaging of a single living cell by supercontinuum-based multiplex coherent anti-Stokes Raman scattering microspectroscopy. *Opt Express* **13**:1322–1327.
34. Muller M, Zumbusch A. 2007. Coherent anti-Stokes Raman scattering microscopy. *Chemphyschem* **8**:2157–2170.
35. Nan XL, Potma EO, Xie XS. 2006. Nonperturbative chemical imaging of organelle transport in living cells with coherent anti-Stokes Raman scattering microscopy. *Biophys J* **91**:728–735.
36. Rodriguez LG, Lockett SJ, Holtom GR. 2006. Coherent anti-Stokes Raman scattering microscopy: a biological review. *Cytometry A* **69A**:779–791.
37. Tong L, Lu Y, Lee RJ, Cheng JX. 2007. Imaging receptor-mediated endocytosis with a polymeric nanoparticle-based coherent anti-Stokes raman scattering probe. *J Phys Chem B* **111**:9980–9985.
38. Volkmer A, Cheng JX, Xie XS. 2001. Vibrational imaging with high sensitivity via epidetected coherent anti-Stokes Raman scattering microscopy. *Phys Rev Lett* **8702**:4.
39. Wang HF, Fu Y, Zickmund P, Shi RY, Cheng JX. 2005. Coherent anti-Stokes Raman scattering imaging of axonal myelin in live spinal tissues. *Biophys J* **89**:581–591.
40. Duncan MD, Reintjes J, Manuccia TJ. 1982. Scanning Coherent anti-Stokes Raman Microscope. *Opt Lett* **7**:350–352.
41. Zumbusch A, Holtom GR, Xie XS. 1999. Three-dimensional vibrational imaging by coherent anti-Stokes Raman scattering. *Phys Rev Lett* **82**:4142–4145.
42. Cheng JX, Xie XS. 2004. Coherent anti-Stokes Raman scattering microscopy: Instrumentation, theory, and applications [Review]. *J Phys Chem B* **108**:827–840.
43. Cheng JX, Volkmer A, Book LD, Xie XS. 2001. An epi-detected coherent anti-Stokes raman scattering (E-CARS) microscope with high spectral resolution and high sensitivity. *J Phys Chem B* **105**:1277–1280.
44. Hashimoto M, Araki T. 2000. Molecular vibration imaging in the fingerprint region by use of coherent anti-Stokes Raman scattering microscopy with a collinear configuration. *Opt Lett* **25**:1768–1770.
45. Lee ES, Lee JY, Yoo YS. 2007. Nonlinear optical interference of two successive coherent anti-Stokes Raman scattering signals for biological imaging applications. *J Biomed Opt* **12**:5.
46. Potma EO, Evans CL, Xie XS. 2006. Heterodyne coherent anti-Stokes Raman scattering (CARS) imaging. *Opt Lett* **31**:241–243.
47. Toytman I, Cohn K, Smith T, Simanovskii D, Palanker D. 2007. Wide-field coherent anti-Stokes Raman scattering microscopy with non-phase-matching illumination. *Opt Lett* **32**:1941–1943.
48. Fu Y, Wang HF, Shi RY, Cheng JX. 2006. Characterization of photodamage in coherent anti-Stokes Raman scattering microscopy. *Opt Express* **14**:3942–3951.
49. Ganikhanov F, Carrasco S, Xie XS, Katz M, Seitz W, Kopf D. 2006. Broadly tunable dual-wavelength light source for coherent anti-Stokes Raman scattering microscopy. *Opt Lett* **31**:1292–1294.
50. Le TT, Langohr IM, Locker MJ, Sturek M, Cheng J-X. 2007. Label-free molecular imaging of atherosclerotic lesions using multimodal nonlinear optical microscopy. *J Biomed Opt* **12**:10.

PROBLEM SOLUTIONS

CHAPTER 1

1.1. a. What are usually the strongest interactions (in amplitude) in a molecule in its native conformation?
   1. bonded interactions
   2. VdW interactions
   3. electrostatics interactions
   4, interaction with the solvent

   b. When would you use homology modeling to predict the structure of a protein?
   1. when the best homologous protein with known structure found by BLAST has a *P*-value of 8 or above
   2. when the best homologous protein with known structure found by BLAST has a *P*-value of 10^{-4} or below
   3. when BLAST cannot identify a homologous protein
   4. always, as the other techniques are too complicated

   c. To build a model for the structure of a target protein using homology modeling, you need:
   1. the structure of a homologous protein (the template), and the alignment between the sequence of the target and template proteins
   2. the sequence of the target protein only
   3. a lot of imagination
   4. the structures of proteins that interact with the target protein

   d. CASP is
   1. a series of meetings set to assess programs that predict protein–protein interactions
   2, the best method currently available for homology modeling
   3. an experiment set to assess protein structure prediction techniques
   4. an experiment set to assess RNA structure prediction techniques

Answers

a. The correct answer is 1; chemical bonds are very strong and cannot be broken easily; all the other interactions are much weaker. The repulsive term of the vdW interactions can be very large, but only if the molecule is deformed and there are atomic overlaps.
b. The correct answer is 2. BLAST is used to find a possible template structure for the target protein; only significant hits are retained, and these hits have very low *E*-values (the *E*-value is a measure of the probability that a random sequence would yield a similar alignment score as the identified template).
c. The correct answer is 1. Homology modeling uses the structure of a homologue protein to build a model for the structure of the target protein.
d. The correct answer is 3. CASP stands for Critical Assessment of Structure Prediction and is geared toward protein structure prediction only.

1.2. Using the Chou and Fasman propensity values given in Table 1.2 and the prediction rules defined in Section 1.5.3.1, predict the secondary structures of the following peptide sequences:

a. WHGCITVYWMTV
b. CAENKLDHVRGP

Answers

a. WHGCITVYWMTV

To predict the secondary structure of this peptide, we use the Chou and Fasman propensities:

	W	H	G	C	I	T	V	Y	W	M	T	V
$P(\alpha)$	0.99	1.22	0.56	1.11	0.97	0.82	0.91	0.72	0.99	1.47	0.82	0.91
$P(\beta)$	1.14	1.08	0.92	0.74	1.45	1.21	1.49	1.25	1.14	0.97	1.21	1.49

We test both for the presence of helices and strands in the sequence:

Predicting helices:

Step 1: we search for a "seed": 6 consecutive residues, 4 of them with $P(\alpha) > 1$: no seed in the sequence. This peptide is not predicted to contain any helices

Predicting strands:

Step 1: we search for a "seed": 5 consecutive residues, 3 of them with $P(\beta) > 1$: there are several options, we choose the first five residues, i.e., WHGCI.

Step 2: extension: we can safely add TVYW to the seed WHGCI. To test the following residues:

M: the segment of four residues VYWM has an average propensity to form a strand of 1.21; we can add M.

T: the segment of four residues YWMT has an average propensity to form a strand of 1.14; we can add T.

V: the segment of four residues WMTV has an average propensity to form a strand of 1.21; we can add V.

The longest region is therefore: WHGCITVYWMTV.

Step 3: final check: The average strand propensity of the sequence WHGCITVYWMTV is 1.17, larger than 1. This segment is therefore predicted to be a strand.

Conclusion: The predicted secondary structure of the sequence WHGCITVYWMTV is EEEEEEEEEEEE, where E stands for Extended or Strand.

b. CAENKLDHVRGP

The procedure is the same as for part a) above:

	C	A	E	N	K	L	D	H	M	R	G	C
$P(\alpha)$	1.11	1.29	1.44	0.90	1.23	1.30	1.04	1.22	1.47	0.96	0.56	1.11
$P(\beta)$	0.74	0.90	0.75	0.76	0.77	1.02	0.72	1.08	0.97	0.99	0.92	0.74

Briefly:

Predicting helices:

The whole peptide CAENKLDHVRGC is predicted to be helical, with an average propensity of 1.14

Predicting strands:

There are no seeds for strands, hence the peptide is not predicted to contain any strands.

Conclusion: the predicted secondary structure of the sequence CAENKLDHVRGC is HHHHHHHHHHHH, where H stands for Helix.

1.3. The following eukaryotic DNA sequence is given to you:

5′-CCCTTAATGCGTATCGCTCACGAGATGTTGGGCGGCTAA-3′

a. You are told that this sequence, or its complementary, codes for one gene. Find the longest "gene," or open reading frame (ORF), corresponding to this DNA sequence; remember that there are 6 possibilities, i.e., 3 possible reading frames for one strand and 3 possible reading frames for its complementary. Transcribe this ORF into an RNA sequence.
b. As this is a eukaryotic sequence, it may contain an intron. For simplicity, we will assume that introns always start with GU and end with AG. Identify all possible introns, and explain why their removal would result in loss of the gene.
c. Based on question (b) just above, we know that the RNA is not spliced. Find the sequence of the "protein" it encodes.
d. Predict the secondary structure of this "protein" using the Chou and Fasman method, based on the propensities given in Table 1.2.
e. Can you find a single mutation at the DNA level of that gene that will modify the corresponding "protein" such that it is predicted to be fully extended (i.e., predicted to be a strand by Chou and Fasman)? (**Hint**: there are several possible answers.)

Answers

The following eukaryotic DNA sequence was given to you:

5′-CCCTTAATGCGTATCGCTCACGAGATGTTGGGCGGCTAA-3′

a. Finding the gene:

We don't know if the sequence given corresponds to the coding strand, so we need to check both this sequence S, and its complementary C:

5′-TTAGCCGCCCAACATCTCGTGAGCGATACGCATTAAGGG-3′

The complementary strand C does not contain any ATG (Start codon).

The initial sequence S contains two ATG in phase, and one TAA (stop codon), in phase with both ATG:

5′-CCCTTA**ATG**CGTATCGCTCACGAG**ATG**TTGGGCGGC**TAA**-3′

Consequently, the longest ORF goes from the first ATG to the stop codon:

5′ **ATG** CGT ATC GCT CAC GAG ATG TTG GGC GGC **TAA**-3′

The corresponding RNA sequence is

5′ AUG CGU AUC GCU CAC GAG AUG UUG GGC GGC UAA-3′

b. Checking for introns:

There are two GU and one AG in the RNA sequence:

5′ AUG C**GU** AUC GCU CAC G**AG** AU**G** **U**UG GGC GGC UAA-3′

Based on the positions of these markers, there is only one putative intron:

GU AUC GCU CAC G**AG**

If we remove this putative intron from the RNA sequence:

5′ AUG CAU GUU GGG CGG CUA A-3′

The start and stop codons are no more in phase, and we have lost the gene. Therefore this is not an intron, and the RNA sequence remains intact.

c. Translation to protein:

The protein sequence is obtained directly using the genetic code:

Nter – Met Arg Ile Ala His Glu Met Leu Gly Gly – Cter

d. Secondary structure prediction:

To predict the secondary structure of this peptide, we use the Chou and Fasman propensities:

	M	R	I	A	H	E	M	L	G	G
$P(\alpha)$	1.47	0.96	0.97	1.29	1.22	1.44	1.47	1.30	0.56	0.56
$P(\beta)$	0.97	0.99	1.45	0.90	1.08	0.75	0.97	1.02	0.92	0.92

We test both for the presence of helices and strands in the sequence:

Predicting helices:

nucleation sequence: IAHEML

extension: add R and M on Nter side, and first G on Cter side

computed average over 9 first residues: 1.11 > 1.0

The 9 first residues are predicted to be part of a helix.

Predicting strands:

no nucleation site

The prediction is therefore: HHHHHHHHHO.

Note: if you use the nucleation sequence HEMLGG to predict the helical content, you find that the whole peptide is helical! This is a problem of the Chou and Fasman scheme. Both answers are correct.

e. Testing mutations:

Both Glu and Met are strong helix stabilizers according to the Chou and Fasman propensities.

If we replace Glu with a strong strand stabilizer, we create a strand nucleation site that can be extended up to Met at the N-terminal end of the protein and to the first Gly on the C-terminal end of the sequence. Glu can be mutated to Val with a single mutation (GAG $\rightarrow$ GUG), and Val has a small $P(\alpha)$ (0.91) and a strong $P(\beta)$ (1.47).

With the new sequence MRIAHVMLGG, the propensities are

	M	R	I	A	H	V	M	L	G	G
$P(\alpha)$	1.47	0.96	0.97	1.29	1.22	0.91	1.47	1.30	0.56	0.56
$P(\beta)$	0.97	0.99	1.45	0.90	1.08	1.49	0.97	1.02	0.92	0.92

and the peptide MRIAHVMLG has a propensity to be a strand of 1.08. The same sequence, however, has a propensity to be a helix on 1.18, and the Chou and Fasman method would still predict the conformation of the peptide to be helical.

If instead we replace Met with Val:

	M	R	I	A	H	E	V	L	G	G
$P(\alpha)$	1.47	0.96	0.97	1.29	1.22	1.44	0.91	1.30	0.56	0.56
$P(\beta)$	0.97	0.99	1.45	0.90	1.08	0.75	1.44	1.02	0.92	0.92

This time, the whole peptide (MRIAHEVLGG) is predicted to be a strand, with an average propensity of 1.04. The first 9 residues of the peptide would still be predicted to form a helix, but the average propensity to form a helix over the whole peptide is 0.97, lower than the average strand propensity. The Chou and Fasman technique would therefore predict the conformation of this peptide to be a strand.

Met can be replaced with Val with a single mutation: AUG $\rightarrow$ GUG. Consequently, one possible mutation is

5′ AUG CGU AUC GCU CAC GAG **G**UG UUG GGC GGC UAA-3′

We obtain the same result if we replace Met with Ile, in which case three different single mutations are possible:

5′ AUG CGU AUC GCU CAC GAG AU**U** UUG GGC GGC UAA-3′

or

5′ AUG CGU AUC GCU CAC GAG AU**C** UUG GGC GGC UAA-3′

or

5′ AUG CGU AUC GCU CAC GAG AU**A** UUG GGC GGC UAA-3′

CHAPTER 3

3.1. Draw an energy level scheme similar to that in Fig. 3.2 (disregard the nuclear Zeeman interaction) for a system with one unpaired electron, two equivalent nuclei of spin $I_1 = ½$, and one nucleus of spin $I_2 = 1$ with a hyperfine coupling four times greater for I_1 than for I_2. How many EPR lines should you obtain?

Answer

We should obtain six lines. The simplified energy diagram is drawn as follows:

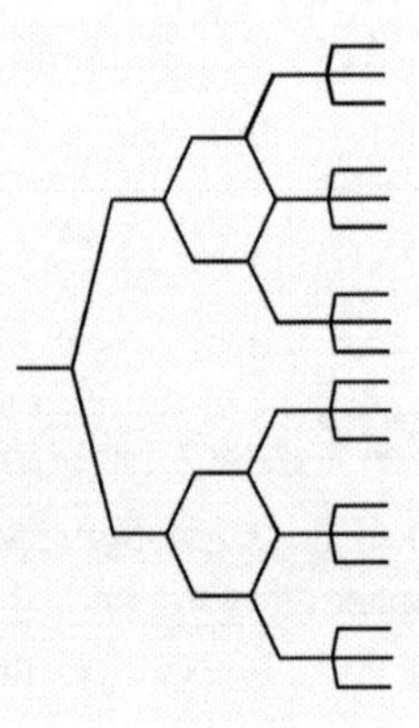

3.2. The spectral line corresponding to $g = 2.001$ is observed in an EPR spectrum at a field of 3467 G recorded in X-band (Microwave Freq. = 9.986 GHz). At what resonant field should this line be seen if the spectrum is recorded at high frequency on a W-band instrument (Microwave Freq. = 95.210 GHz)?

Answer

We know that

$$h\upsilon = g_e \beta_e B_0 .$$

Since g_e is a constant, we can write

$$\frac{h\nu_1}{\beta_e B_0^1} = \frac{h\nu_2}{\beta_e B_0^2} .$$

After simplification and substituting numerical values into the equation, we obtain

$$B_0^2 = \frac{B_0^1 \nu_2}{\nu_1} = \frac{3467 \cdot 95.210}{9.986} = 33055.6 .$$

The line will appear at a field of 33056 [G] or ~3.3 [T].

3.3. A phosphinyl radical (one unpaired electron on phosphorus atom R–P•H) gives a doublet of doublets ($I_P = 1/2$, $I_H = 1/2$, four lines) with a hyperfine coupling constant of 300 MHz for phosphorus P and another coupling constant of 30 MHz for proton H. If this compound is deuterated (H is replaced by a deuteron D), a new pattern is observed. The coupling constant remains the same for phosphorus but changes for the deuteron. Can you calculate a priori the coupling constant for D? Draw the corresponding spectra with H and with D.

Answer

From Eq. (18) we know that the isotropic part of the hyperfine coupling is proportional to $g_N\beta_N$:

$$A_{\text{iso}} \div g_N \beta_N \,.$$

For proton (nuclear spin of 1/2) and deuteron (nuclear spin of 1) we have the magnetic moments, which are, respectively,

$$\mu_H = g_H \beta_H \cdot \frac{1}{2}, \quad \mu_D = g_D \beta_D \cdot 1 \,.$$

Therefore, the ratio of the two A_{iso} will be provided by the ratio of the magnetic moments of the two isotopes multiplied by two:

$$\frac{A_{\text{iso}}^{H}}{A_{\text{iso}}^{D}} = \frac{g_D \beta_H}{g_D \beta_D} = \frac{\mu_H \cdot 2}{\mu_D} = 6.513 \,.$$

To get the corresponding hyperfine coupling constant for D, the hyperfine constant for H has to be divided by the factor 6.513.

The spectra for RPH and RPD are drawn below:

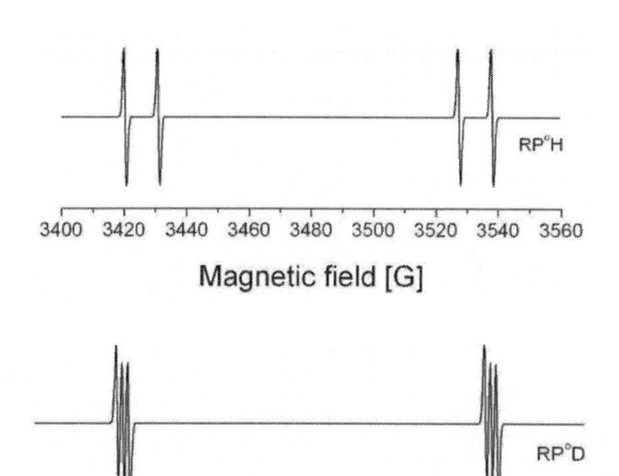

3.4. The spectrum below was simulated for a mixture of CHD_2 and CH_2D radicals. Identify the lines belonging to each species and give the values for each of the hyperfine splittings. Compute the ratio a_H / a_D from the spectrum using a ruler. Compare it with the expected numerical value.

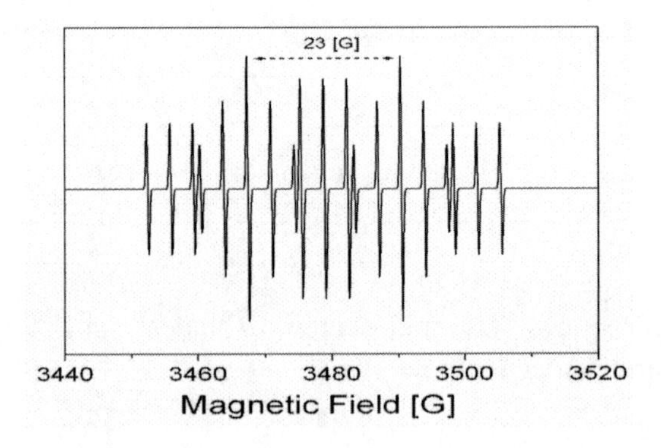

Answer

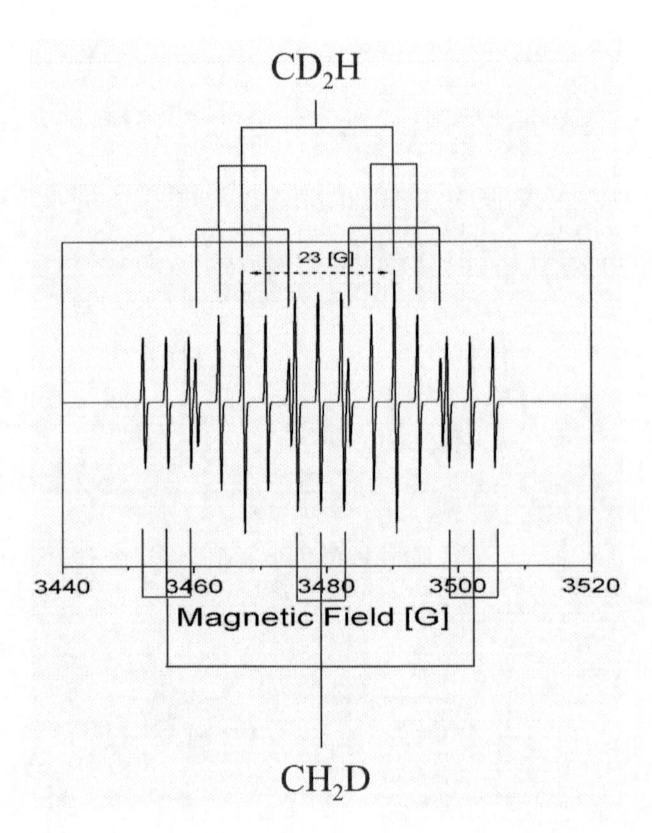

3.5. The spectrum below was simulated for a 1,3-butadiene anion radical. Draw schematically the molecule and identify the lines belonging to the equivalent protons. Draw a "stick diagram" for this spectrum.

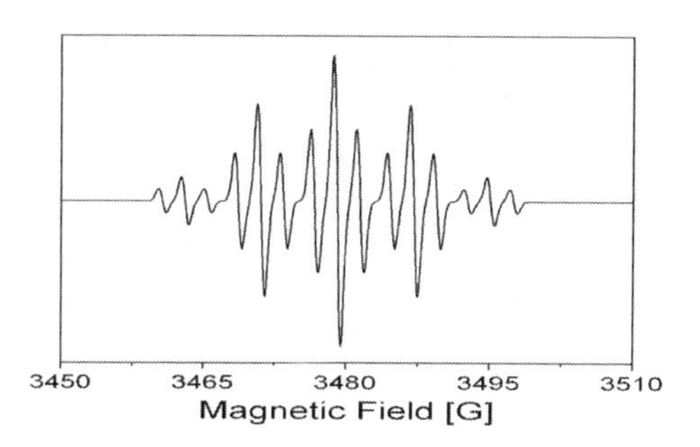

Answer

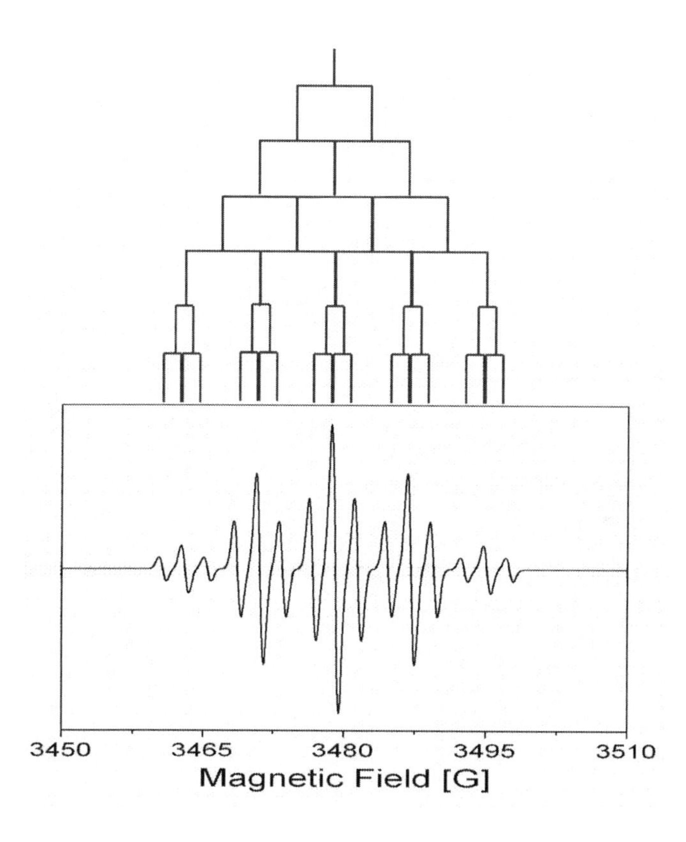

3.6. Calculate the spacing (in gauss) between two EPR transitions (lines at g = 1.999 and 2.001), measured at the X-, W-, and D-bands. Which one gives a better resolution, and thus higher accuracy?

Answer

The X-, W-, and D-band correspond to spectrometer frequencies of approximately 9.5, 95, and 130 GHz. The difference of the resonant fields for the transitions observed at two different g values is proportional to the ratio of the frequencies of the spectrometers:

$$h\upsilon = g_e \beta_e B_0 .$$

For a given frequency of the spectrometer *f*,

$$h\upsilon_f = g_1\beta_e B_0^1, \quad h\upsilon_f = g_2\beta_e B_0^2.$$

For the same operating frequency of the spectrometer we have a conversion factor of

$$\frac{h}{\beta_e} = \frac{2.0023\cdot 3390}{9.5} = 714.505.$$

We can at this point calculate the magnetic fields corresponding to the two transitions for X-, W-, and D-band:

$$B_0^1 = \frac{9.5h}{1.999\beta_e} = 3395.6, \quad B_0^2 = \frac{9.5h}{2.001\beta_e} = 3392.2.$$

The difference between two lines is 3.4 [G].

We can now calculate the same for W-band:

$$B_0^1 = \frac{95h}{1.999\beta_e} = 33956, \quad B_0^2 = \frac{9.5h}{2.001\beta_e} = 33922.$$

The difference is 34 [G]

And, finally, for D-band we have

$$B_0^1 = \frac{130h}{1.999\beta_e} = 46466.1, \quad B_0^2 = \frac{130h}{2.001\beta_e} = 46419.6.$$

The difference is 46.5 [G]

We can see that the best resolution in terms of the *g*-factor is obtained at higher spectrometer frequencies, in this case at D-band (130 GHz).

CHAPTER 4

4.1. Using Boltzmann's equation, calculate the equilibrium ratios of populations in the two spin states of ^{1}H, ^{15}N, and ^{29}S in a 5.87-T field (250 MHz 1H) at 298 K. How does the sign of the gyromagnetic ratio (positive for ^{1}H and negative for ^{15}N and ^{29}Si) affect your calculations?

Answer

$$\frac{P_\beta}{P_\alpha} = \exp\frac{-\Delta E_{\alpha\beta}}{kT}, \quad k = 6.626\times 10^{-34}\,J\cdot s, \quad \Delta E_{\alpha\beta} = h\gamma B_0,$$

$$\gamma_{1H} = 26.7522\times 10^7\,\text{rad}\cdot\text{s}^{-1}, \quad \Delta E_{\alpha\beta}(^1\text{H}) = 6.626\times 10^{-34}\gamma(5.87\ \text{Tesla})/2\pi = 1.656\times 10^{-25}\,\text{J},$$

$$\gamma_{15N} = -2.7126\times10^{7}\,\text{rad}\cdot\text{s}^{-1},\ \ \Delta E_{\alpha\beta}(^{15}\text{N}) = 1.68\times10^{-26}\,\text{J},$$

$$\gamma_{29Si} = -5.319\times10^{7}\,\text{rad}\cdot\text{s}^{-1},\ \ \Delta E_{\alpha\beta}(^{29}\text{Si}) = 3.29\times10^{-26}\,\text{J},$$

$$\frac{P_\beta}{P_\alpha}(^{1}\text{H}) = \exp(-4.025\times10^{-5}) = 0.9999597,$$

$$\frac{P_\alpha}{P_\beta}(^{15}\text{N}) = \exp(-4.08\times10^{-6}) = 0.9999959,$$

$$\frac{P_\alpha}{P_\beta}(^{29}\text{Si}) = \exp(-7.99\times10^{-6}) = 0.9999920.$$

The change in sign for the gyromagnetic ratio going from ^{1}H to ^{15}N or ^{29}Si does not have any affect on the relative populations of the two states, only which orientation with respect to the magnetic field has the higher energy. In the classical model of Larmor precession, one can think of spins with opposite signs of gyromagnetic ratio precessing in opposite directions in the applied field.

4.2. Consider the following system: proton A is coupled to proton B, which exchanges with site C to which A is not coupled. All chemical shifts are distinct, with an equilibrium constant K_{BC} = 1 for site exchange. Sketch the spectrum you expect for each of the following situations:

a. Slow exchange on the chemical shift and J-coupling timescale.
b. Fast exchange on the J-coupling scale, but slow on the chemical shift timescale.
c. Fast exchange on both timescales.

Answers

a. Slow exchange on the chemical shift and J-coupling timescale:

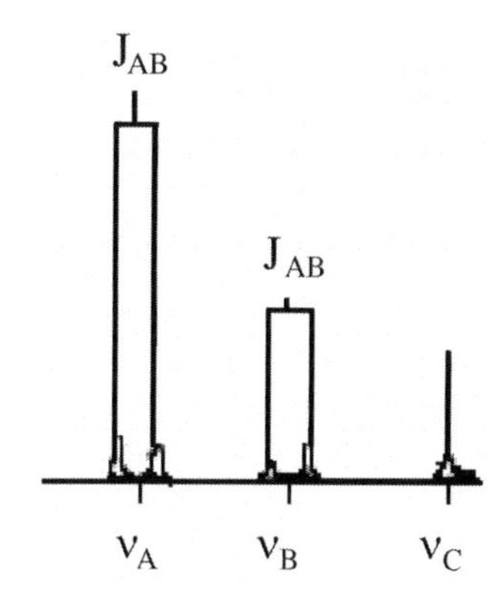

b. Fast exchange on the coupling scale, but slow on the chemical shift timescale:

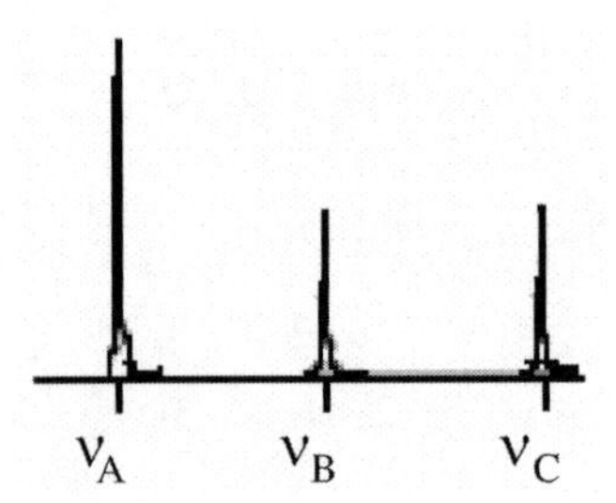

c. Fast exchange on both timescales:

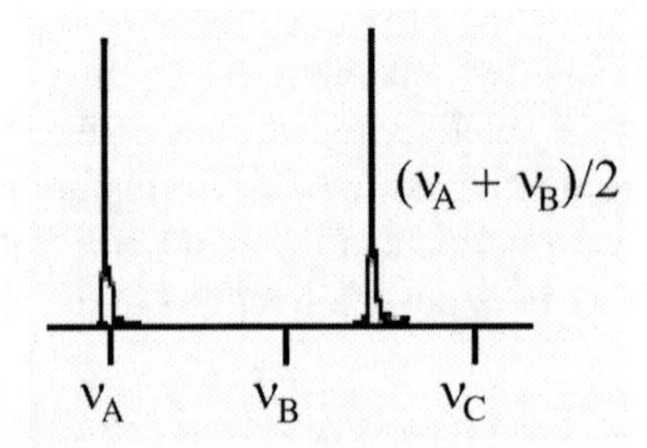

4.3. Spin populations can be thought of in a very chemical way in terms of "concentrations" of spins (e.g., [α] or [β]). The usual diagram for an $I = ½$ system using a chemist formalism is

β
k1 k2
α

$$[\alpha] \underset{k2}{\overset{k1}{\rightleftharpoons}} [\beta]$$

where $k1$ and $k2$ are rate constants for interconversion of spin populations with concentrations given by [α] and [β]. Calculate ΔG^0 for this reaction for ^{1}H in an 11.74-T magnetic field. Also, calculate the equilibrium constant for this reaction ($K_{eq} = [\alpha]_{eq}/[\beta]_{eq}$). Note that $\Delta S = 0$ since there is no disorder change associated with a single spin flip.

Answer

$$K_{eq} = \frac{[\beta_0]}{[\alpha_0]} = \exp\left(\frac{-\Delta E_{\alpha\beta}}{k_B T}\right) = \exp\left(\frac{-\gamma h B_0}{k_B T}\right),$$

$$k_B = 1.38\times10^{-23} \text{ J/K},$$

$$h = 1.05\times10^{-34} \text{ J}\cdot\text{s},$$

$$\gamma_{1H} = 26.7522\times10^{7} \text{ rad/s}\cdot\text{Tesla},$$

$$K_{eq} = \exp\left(\frac{-\gamma h B_0}{k_B T}\right) = \exp(-8.01\times10^{-5}) = 0.9999198,$$

$$\Delta G^0 = -RT \ln K_{eq} = -RT(-8.01\times10^{-5}) = 0.1987 \text{ J/mol}.$$

CHAPTER 5

The figure below (please visit http://extras.springer.com/ to view a high-resolution full-color version) illustrates structural changes associated with the binding of cGMP to the ligand-binding domain of a cyclic nucleotide-gated (CNG) channel: the α–helix shown in red moves to the left, becoming closer to the rest of the ligand-binding domain. The distance between the base of the ligand-binding domain and three positions (1, 2, and 3) along the α–helix in the unliganded state (left) are 35, 25, and 16 Å, respectively; they become 22, 19, and 14 Å, respectively, when cGMP is bound (right). A FRET pair with R_0 = 40 Å is chosen to monitor the structural change with FRET. If the donor (D) is attached to the base of the ligand-binding domain as shown, where should the acceptor be attached to? How much FRET change do you expect to see?

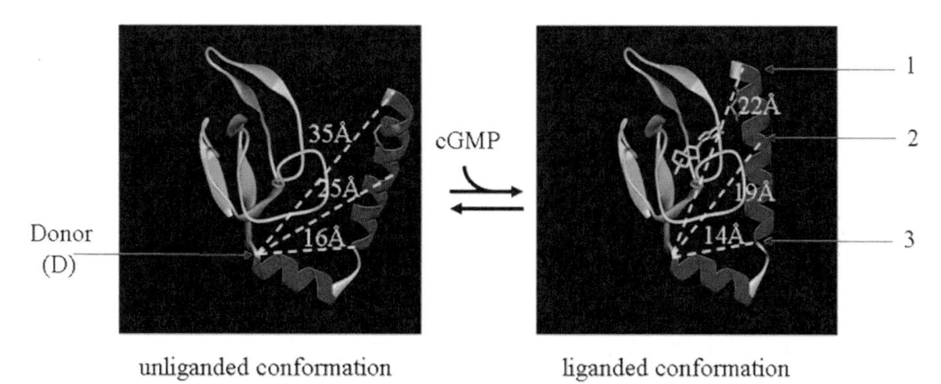

Please visit http://extras.springer.com/ to view a high-resolution full-color version of this illustration.

Answer

Recall that $E = \frac{R_0^6}{R^6 + R_0^6}$. Here R_0 = 40 Å. We can calculate the FRET efficiency values as follows:

FRET efficiency	Position 1	Position 2	Position 3
Unliganded state	0.690	0.944	0.996
Liganded state	0.973	0.989	0.998

From this table it can be seen that for position 1 the change in FRET efficiency is 0.973 – 0.690 = 0.283. For positions 2 and 3 the values are 0.989 – 0.944 = 0.045

and 0.998 – 0.996 = 0.002, respectively. Clearly, the acceptor should be attached to position 1, as it gives the largest detectable FRET change. In general, donor–acceptor distances around R_0 are preferable, as this is the most sensitive range for detecting movements.

CHAPTER 6

6.1. Find the radius of gyration of a low-energy ion of *m/z* 1000 in a 9.4 T magnetic field. Let $v - 100$ m/s and $\omega = 6.7542 \times 10^5\ \text{s}^{-1}$.

Answer

Rearrange the cyclotron equation:

$$\omega = \frac{v}{r} = \frac{qB}{m},$$

$$r = \frac{v}{\omega} = v\left(\frac{m}{qB}\right),$$

$$r = \left(\frac{100\ \text{m/s}}{6.7542 \times 10^5\, s^{-1}}\right) = 1.5 \times 10^{-4}\,\text{m},$$

$$r = 0.15\ \text{mm}.$$

6.2. Find the translational energy for an ion of *m/z* 1257 when it has been accelerated to a radius of gyration of 1 cm.

Answer

Substitute cyclotron frequency, ω, into the equation for translational energy:

$$T = \frac{mv^2}{2} = \frac{m(\omega r)^2}{2},$$

$$m = 1257 \times\ 1.6605 \times\ 10^{-27}\ \text{kg},$$

$$\omega\ =\ 6.7542 \times 10^5\ \text{s}^{-1},$$

$$r = 0.02m,$$

$$T = \frac{\left(1257 \times 1.6605 \times 10^{-27}\,\text{kg}\right)\left(\left(6.7542 \times 10^5\ \text{s}^{-1}\right)\left(0.01\ \text{m}\right)\right)^2}{2},$$

$$T = 4.761 \times 10^{-17}\,\text{Joules}.$$

Convert Joules to eV:

$$T = \left(4.761 \times 10^{-17}\,\text{J}\right)\left(\frac{1\ \text{eV}}{1.602 \times 10^{-19}\,\text{J}}\right) = 297\ \text{eV}\ .$$

CHAPTER 7

7.1. The energy of an electron is related to its mass and velocity by

$$\text{eV} = \frac{m_0 v^2}{2}\ .$$

Substituting momentum, p, with $m_0 v$ gives

$$P = m_0 v = (2m_0\ \text{eV})^{1/2}\ .$$

Substituting h/λ for p gives

$$\lambda = \frac{h}{(2m_0\ \text{eV})^{1/2}}\ .$$

Considering relativistic effects, the above equation becomes

$$\lambda = \frac{h}{\left[2m_0\ \text{eV}\left(1 + \frac{\text{eV}}{2\text{m}_0 c^2}\right)\right]^{1/2}}\ .$$

Substituting the appropriate accelerating voltages into these equations, the fraction of the speed of light of an electron accelerated to 200 kV is

$$\left(\frac{2.086 \times 10^8\ \frac{\text{m}}{\text{sec}}}{2.998 \times 10^8\ \frac{\text{m}}{\text{sec}}}\right) = 0.696\ ,$$

while the relativistic mass of an electron accelerated to 1000 kV is 2.957 × m_0. Clearly, electrons within an electron microscope are traveling at significant fractions of the speed of light and experience large increases in relativistic mass, which must be taken into consideration when modeling their interactions with samples.

7.2. Given

$$\gamma(s) = -\sin\left[\frac{\pi}{2} C_S \lambda^3 s^4 + \pi \Delta z \lambda s^2\right],$$

a $-\pi/2$ phase shift is modeled by

$$\gamma(s) = -\sin\left[\left(\frac{\pi}{2}C_s\lambda^3 s^4 + \pi\Delta z\lambda s^2\right) - \frac{\pi}{2}\right],$$

which is equivalent to

$$\gamma(s) = \cos\left[\frac{\pi}{2}C_s\lambda^3 s^4 + \pi\Delta z\lambda s^2\right].$$

7.3. Resolution is limited by the condition $2m \geq 2N + 1$ if only m views are available. Since the Bessel function $J_n(x)$ is approximately 0 for $|n| - 2 > x > 0$, the maximum value for n (in which the G_n contribution is nonzero for an annulus of radius R) can be approximated by

$$2\pi Rr \doteqdot n - 2.$$

Substituting for N with $2m \geq 2N + 1$, the largest reciprocal spacing given m views is

$$R_{\max} \doteqdot \frac{2m-5}{4\pi r};$$

with large m,

$$R_{\max} \simeq \frac{2m-5}{4\pi r}$$

giving

$$m \simeq \frac{\pi D}{d}.$$

CHAPTER 8

8.1. By making a number of generous assumptions, estimate the detection limit (concentration limit) of spontaneous micro-Raman spectroscopy and discuss factors that might improve this limit (e.g., increased signal accumulation time, higher laser power). Remember, the Raman scattering cross-section of a typical Raman-active bond is (on average) $\sigma \sim 10^{-30}$ cm^2. Assume that a reasonable (safe) laser power for most substances is 1 mW at 632.8 nm focused to a diffraction-limited spot.

Answer

To keep things simple we assume that that laser spot is a square with a 1-μm edge length (a "realistic" diffraction-limited laser spot would be only about 1/4 of this dimension when we use microscope objectives with the highest possible NA (~1.4)—so, we are underestimating the numbers, which keeps us on the safe side). The power density then is

$$P = 1\frac{\text{mW}}{(\mu\text{m})^2} = 1\frac{10^{-3}\ \text{W}}{10^{-12}\ \text{m}^2} = 1\frac{10^{-3}\ \text{W}}{10^{-8}\ \text{cm}^2} = 1\times10^5\ \text{Wcm}^{-2} = 1\times10^5\ \text{Js}^{-1}\text{cm}^{-2}.$$

A photon at wavelength $\lambda = 632.8$ nm has an energy of

$$E_p = h\nu = h\frac{c}{\lambda} = 6.626\times10^{-34}\ \text{Js}\ \frac{2.9979\cdot10^8\ \text{ms}^{-1}}{632.8\cdot10^{-9}\ \text{m}} = 3.14\times10^{-19} J \approx 2\ \text{eV}.$$

The photon flux through the diffraction-limited spot is

$$\phi = \frac{1\ \text{mW}}{E_p} = \frac{1\times10^{-3}\ \text{Js}^{-1}}{E_p} = 3.2\times10^{15}\ \text{s}^{-1}.$$

This is the flux through a two-dimensional spot, but we can assume that this is also the flux through the full confocal detection volume, i.e., $V = 1\ \mu\text{m}^3$.

A concentration of 1 nM is ~1 Molecule/μm^3.

If that molecule has 1 Raman-active bond, then the number of Raman-scattered photons produced per second is

$$N_R = \frac{P}{E_p}\sigma = \frac{1\times10^5\ \text{Js}^{-1}\text{cm}^{-2}}{3.14\times10^{-19}\ \text{J}}\times10^{-30}\ \text{cm}^2 = 3.2\times10^{-7}\ \text{s}^{-1}.$$

That means that at a wavelength of 632.8 nm the number of Raman-scattered photons produced per second from a sample at 1-nM concentration (with only 1 Raman-active bond per molecule) is 3.2×10^{-7}. Fortunately, most biological samples (i.e., proteins) have thousands of CH bonds per molecule, which brings these numbers up a bit.

The typical collection efficiency of a confocal microscope with the highest possible NA (~1.4, immersion oil objective) and detectors with the highest possible quantum efficiency is ~5%. This is valid only if the Raman scattering is isotropic, i.e., is generally only true for unordered samples. So as to detect a useful Raman signal, we have to stay above the signal-to-noise ratio for a typical liquid nitrogen–cooled detector, which in the case of all modern detectors (CCDs) is essentially only shot-noise limited. This means we have to produce a minimum of 3 photon counts per pixel in order to qualify as signal. Better yet, 10 (the S/N is then ~3). These 10 counts can be generated over an arbitrarily long timescale, since the detector is shot-noise limited. If we wanted to generate these 10 photon counts with a 1-mW probe laser power within a 1-s integration time, the concentration would have to be at least $10\ \text{s}^{-1} = 3.2\times10^{-7}\ \text{s}^{-1}\times0.05\times c$, where c is the concentration in nM. This means that the minimum detectable concentration under these conditions is $c = 625$ mM.

Let's assume we had 10 mW of laser power and we wanted to detect a signal from a 1-μM solution. How long would we have to accumulate the signal? ~62500 s or ~17.4 h! With 100 mW of laser power we could cut the integration time down to ~1.7 h, but still, this clearly tells us that (unenhanced) Raman spectroscopy is only useful for concentrations of mM and up, or for molecules with many tens of the same Raman-active bonds. Another possibility for detecting lower concentrations is to increase the volume while keeping the power density the same; that's why most gas-phase spectroscopy is done in multipass cells.

8.2. Assuming that the Raman peak for the CH stretching vibration appears at approximately 2900 cm^{-1}, estimate the new location of the peak if the hydrogen (H) is replaced with deuterium (D). Use the fact that deuterium has approximately twice the mass of hydrogen.

Answer

Utilizing the equation shown in Figure 8.1b, the vibrational frequency of a bond vibration can be approximated. If we assume that the spring constant, k (the strength of the bond) is the same for both CH and CD and we approximate the mass of C, H, and D in atomic mass units to find the reduced mass for each situation, we obtain

$$\frac{1}{\lambda_{\mathrm{CH}}} = \frac{1}{2\pi c}\sqrt{\frac{k}{\mu_{\mathrm{CH}}}}, \quad \frac{1}{\lambda_{\mathrm{CD}}} = \frac{1}{2\pi c}\sqrt{\frac{k}{\mu_{\mathrm{CD}}}},$$

$$m_{\mathrm{C}} \approx 12, \quad m_{\mathrm{H}} \approx 1, \quad m_{\mathrm{D}} \approx 2,$$

$$\mu_{\mathrm{CH}} = \frac{m_{\mathrm{C}} m_{\mathrm{H}}}{m_{\mathrm{C}} + m_{\mathrm{H}}} \approx \frac{12}{13}, \quad \mu_{\mathrm{CD}} = \frac{m_{\mathrm{C}} m_{\mathrm{H}}}{m_{\mathrm{C}} + m_{\mathrm{D}}} \approx \frac{24}{15}.$$

We can the solve both equations for k and set them equal to each other:

$$\mu_{\mathrm{CH}}\sqrt{2\pi c\frac{1}{\lambda_{\mathrm{CH}}}} = \mu_{\mathrm{CD}}\sqrt{2\pi c\frac{1}{\lambda_{\mathrm{CD}}}}.$$

By solving for $1/\lambda_{\mathrm{CD}}$ we get

$$\frac{1}{\lambda_{\mathrm{CD}}} = \frac{1}{\lambda_{\mathrm{CH}}}\sqrt{\frac{\mu_{\mathrm{CH}}}{\mu_{\mathrm{CD}}}}.$$

Plugging in the values for the reduced masses as calculated above, as well as the vibrational frequency of the CH bond, we find

$$\frac{1}{\lambda_{\mathrm{CD}}} \approx 2200 \text{ cm}^{-1}.$$

8.3. SERS results in stronger signals by essentially increasing the scattering cross-section for inelastic light scattering (the probability of photons to become Raman-scattered). Assume you have 100,000 molecules illuminated in a tightly focused laser spot (approximated by a cube with a volume of 1 μm^3), that the Raman scattering cross-section of the molecules is 10^{-30} cm^2, and that the laser produces 4×10^{17} photons per second. You can also assume for this problem that you collect all of the scattered photons.

a. Use the following variation of Eq. (8.9) to estimate the efficiency of Raman scattering:

$$\Phi_{\text{Raman}} = \Phi_{\text{Laser}} N_{\text{M}} L \Omega \frac{\partial \sigma}{\partial \Omega},$$

where the photon flux, Φ, in photons/sec/cm^2 is used instead of intensity.

b. If surface enhancement increases the scattering efficiency by 10^{10}, estimate by how much the scattering cross-section is increased by SERS for every molecule. To solve this problem assume that the 100,000 molecules are now adsorbed onto a single gold sphere (diameter of 50 nm) in your focused laser spot.

Answer

Using the values

$$L = 10^{-4}\,\text{cm}, \quad \Phi_{\text{Laser}} = 4 \times 10^{25} \frac{\text{photons}}{\text{cm}^2\text{s}},$$

$$N_{\text{M}} = \frac{10^5 \text{ molecules}}{10^{-12} \text{ cm}^3}, \quad \Omega = 4\pi, \quad \frac{\partial \sigma}{\partial \Omega} \approx 10^{-30} \text{ cm}^2$$

By plugging into the equation we get

$$\Phi_{\text{Raman}} \approx 5 \times 10^9 \frac{\text{photons}}{\text{cm}^2\text{s}}, \quad \frac{\Phi_{\text{Raman}}}{\Phi_{\text{Laser}}} \approx 1.3 \times 10^{-16}.$$

Here, the ratio $\Phi_{\text{Raman}}/\Phi_{\text{Laser}}$ is the efficiency with which laser photons are being converted into Raman-scattered photons. For the relatively low concentration of 100,000 molecules contained in a volume of 1 μm^3 this means that only 1 in 10^{16} laser photons is Raman-scattered.

Using the equation we can solve for scattering cross-section in terms of efficiency:

$$\frac{\partial \sigma}{\partial \Omega} = \frac{1}{N_{\text{M}} L \Omega} \frac{\Phi_{\text{SERS}}}{\Phi_{\text{Laser}}}.$$

Plug in the following values where the number density and interaction length have been adjusted to reflect the fact that the molecules are adsorbed onto the gold surface:

$$N_{\mathrm{M}} = \frac{10^5 \text{ molecules}}{\frac{4}{3}\pi(25\times 10^{-7}\text{cm})^3}, \quad L = 50\times 10^{-7} \text{ cm}, \quad \Omega = 4\pi,$$

$$\frac{\Phi_{\mathrm{SERS}}}{\Phi_{\mathrm{Laser}}} = \frac{\Phi_{\mathrm{Raman}}}{\Phi_{\mathrm{Laser}}} 10^{10} \approx 1.3\times 10^{-6}, \quad \frac{\partial\sigma}{\partial\Omega} \approx 1.4\times 10^{-23}.$$

These approximation shows that an increase of the scattering cross-section by 7 orders of magnitude due to SERS will result in a signal that is roughly 10 orders of magnitude more intense.

INDEX